PERGAMON INTERNATIONAL LIBRARY
of Science, Technology, Engineering and Social Studies
The 1000-volume original paperback library in aid of education, industrial training and the enjoyment of leisure
Publisher: Robert Maxwell, M.C.

COMBUSTION AND MASS TRANSFER

A Textbook with Multiple-Choice Exercises for Engineering Students

Other Pergamon Titles of Interest

CHIGIER	Pollution Formation and Destruction in Flames
DUNN & REAY	Heat Pipes, 2nd Edition
EDE	An Introduction to Heat Transfer Principles and Calculations
GILCHRIST	Fuels, Furnaces and Refractories
GRAY & MULLER	Engineering Calculations in Radiative Heat Transfer
KUTATELADZE & BORISHANSKII	Concise Encyclopaedia of Heat Transfer
SPALDING	GENMIX - A General Computer Program for Two-Dimensional Parabolic Phenomena
WHITAKER	Elementary Heat Transfer Analysis
WHITAKER	Fundamental Principles of Heat Transfer
ZARIC	Heat and Mass Transfer in Flows with Separated Regions

Pergamon Related Research and Review Journals

International Journal of Heat and Mass Transfer

Letters in Heat and Mass Transfer

Progress in Energy and Combustion Science

COMBUSTION AND MASS TRANSFER

A Textbook with Multiple-Choice Exercises for Engineering Students

BY

D BRIAN SPALDING

Reilly Professor of Combustion, Purdue University,
Thermal Sciences and Propulsion Center, Chaffee Hall, West Lafayette, Indiana 47907, USA

and

Professor of Heat Transfer, Imperial College of Science & Technology
Exhibition Road, London SW7, England

PERGAMON PRESS

OXFORD · NEW YORK · TORONTO · SYDNEY · PARIS · FRANKFURT

U.K. Pergamon Press Ltd., Headington Hill Hall, Oxford OX3 0BW, England

U.S.A. Pergamon Press Inc., Maxwell House, Fairview Park, Elmsford, New York 10523, U.S.A.

CANADA Pergamon of Canada, Suite 104, 150 Consumers Road, Willowdale, Ontario M2 J1P9, Canada

AUSTRALIA Pergamon Press (Aust.) Pty. Ltd., P.O. Box 544, Potts Point, N.S.W. 2011, Australia

FRANCE Pergamon Press SARL, 24 rue des Ecoles, 75240 Paris, Cedex 05, France

FEDERAL REPUBLIC OF GERMANY Pergamon Press GmbH, 6242 Kronberg-Taunus, Pferdstrasse 1, Federal Republic of Germany

---

First edition 1979

Library of Congress Cataloging in Publication Data

Spalding, Dudley Brian.
Lectures on combustion (1-10)
1. Combustion. I. Title.
QD516.S73 1977 541'.361 77-8110

ISBN 0-08-022105-X (Hardcover)
ISBN 0-08-022106-8 (Flexicover)

In order to make this volume available as economically and as rapidly as possible the author's typescript has been reproduced in its original form. This method unfortunately has its typographical limitations but it is hoped that they in no way distract the reader.

Reproduced, printed and bound in Great Britain by Cox & Wyman Ltd, London, Fakenham and Reading

CONTENTS

Each chapter is accompanied by a set of tutorial exercises, partly analytical, and partly multiple-choice.

PREFACE

The following series of twenty lectures represents the outcome of several years of teaching the subjects of combustion and mass transfer to final-year undergraduate students of mechanical engineering at Imperial College.

The subject of combustion is a large and important one. Because of its size, a twenty-lecture course can include only a small fraction of the available useful information; and it is hard to make the selection. I have here selected the topics which I have personally found interesting and helpful, in the course of an academic and consulting career in which combustion has been an abiding theme. Naturally, the knowledge possessed by the students entering the course, and their need to integrate combustion studies with other parts of the curriculum, have exerted limiting influences.

Although the advent of the digital computer has recently transformed the extent to which combustion theory can be applied to engineering, I have confined attention in the present lectures to analyses which can be made without the computer's aid. The reason for this restriction is that I regard it as important for students of combustion to gain the clear understanding of phenomena which formulae and graphs can provide, before they become immersed in the welter of information (and misinformation) which the computer can generate.

Each of the twenty chapters is provided with a set of exercises, designed to assist the student to digest the material of the lectures. Most of these are of the multiple-choice variety; and many are of the "P is Q because R is S" kind, in which the student has to ask himself: "Is P Q?"; "Is R S?"; and "If so, is R S because P is Q?". I have found these to be valuable in stimulating and refining thought about "why" as well as "what".

It is a pleasure to thank Colleen King and Christine MacKenzie for the assistance which they have given me with the preparation of the book. That the numerous and overlapping hand-written versions have finally taken a legible and orderly form is largely due to their efforts.

I must also thank Drs F C Lockwood, A S C Ma and W M Pun, who have assisted me, at various times, with the delivery of the course; and the students whose critical comments have also had a constructive influence.

D BRIAN SPALDING
London
September 1978

CHAPTER 1

INTRODUCTION TO COMBUSTION

1.1 THE IMPORTANCE OF THE SUBJECT

A) THE PLACE OF COMBUSTION IN ENGINEERING

(i) Power production

There are only a few means by which man produces the power with which he dominates his environment; and almost all of them involve the combustion of a fuel, either solid, liquid or gas.

Coal is burned in central power stations to raise the steam for the turbines.

Oil is used for the same purpose, and also as the source of energy for vehicles of all sorts - automobiles, aircraft and ships.

Natural gas may be used as the fuel for gas turbines or for reciprocating engines, as well as for steam-raising.

Although nuclear power will certainly become increasingly significant as an energy source in industrial societies, and although solar energy, and wind and wave power, are being actively developed, combustion will remain the predominant source of power for many generations.

(ii) Process industry

Much fuel is also burned as an essential ingredient of the production of engineering materials, for example:-

- iron, steel, and many non-ferrous metals;

- glass and ceramics;

- cement;

- refined fuels, carbon black, and other hydrocarbon derivatives.

Sometimes the fuel must be specially processed before it is employed, as when coke is produced from coal, by a carefully-controlled heating process, in order that it can be later burned in an iron-ore-reducing "blast furnace".

(iii) Domestic and industrial heating

Residences, factories, offices, hospitals and other buildings require to be heated, in many parts of the world. In the majority of cases, the preferred source of heat is a fuel. The camp fire radiates heat within the tent; while the basement furnace distributes its heat through the agency of ducted air or piped water; but combustion is the foundation of both means of heat transfer.

(iv) The "energy crisis"

The "energy crisis" is first and foremost a "fuel crisis", and still more a "combustion crisis". The world's supplies of fuel are being rapidly diminished; and those fuels which are easiest to burn are being used up most rapidly.

If engineers were able to burn heavy fuel oil in domestic heating appliances, or to use coal dust as the fuel for automobile engines, mankind's difficulties would be greatly alleviated. Thus it is in part a deficiency in combustion technology which gives the "energy crisis" its urgency.

(v) Unwanted combustion

Forests, buildings, and even clothing, can act as fuels; and fire-prevention engineers have to ensure that they do not, or that means are to hand for extinguishing such conflagrations as do inadvertently arise. The damage caused by fires can be so great that much attention must be given by specialist engineers to the techniques of prevention and extinguishment.

B) SOME FACTS ABOUT EXPECTED ENERGY USAGE

The following table gives an indication of the magnitudes of the annual energy usages expected in Europe and the U.S.A. Evidently, combustion-derived energy is likely to dominate.

	OECD EUROPE		U S A	
YEAR	1980	1985	1980	1985
Coal	94.0	90.0	178.1	237.3
Oil	477.8	603.5	470.1	547.0
Gas	92.6	127.1	265.7	282.0
Nuclear	41.5	114.6	58.5	123.4
Hydro & Geothermal	118.2	20.5	14.7	16.5
Units are 10^{18} joules				

C) REASONS FOR THE ENGINEER'S CONCERN WITH COMBUSTION

(i) Procurement of combustion efficiency

Combustion appliances do not necessarily burn all the fuel which is supplied to them; and that which is not burned completely is often discharged into the atmosphere.

Fuel is too expensive to waste in this way; and the products of incomplete combustion are often noxious. Appliance designers therefore have a double reason for ensuring that combustion efficiency is very close to 100%.

(ii) The reduction of costs

Fuel is expensive; and so is the equipment for preparing it for combustion and for burning it. This equipment must therefore be designed from the point of view of cost reduction as well as from that of efficiency.

Costs are of two kinds: capital, and running. To reduce the former, the combustion engineer may well try to bring the chemical reaction to completion in a small space; to reduce the former, he may try to reduce the pressure drop experienced by the gases in passing through the device. Often the two aims are incompatible; an enlightened compromise must be sought.

(iii) Temperature and composition control

Sometimes it is not sufficient that the fuel should burn: there may be an additional requirement that the products of combustion should have a particular temperature or a specified composition. This can occur when the combustion products are to engender some physical or chemical transformation in another material, as when metals are being "heat-treated"; or it may be that the combustion products must not exceed a certain temperature, lest damage ensue to nearby structural elements.

The engineer who designs combustion equipment must therefore be able to predict what the temperature and composition of the combustion products will be, and to bring them under his control.

D) SOME SPECIAL FEATURES OF COMBUSTION AS AN ENGINEERING SPECIALISATION

(i) Multiplicity of constituent disciplines

The engineer who specialises in compressors, pumps and turbines needs to understand the laws of thermodynamics and fluid mechanics, but little more.

The heat-transfer specialist must understand thermodynamics and fluid mechanics; but, in addition, he must be fully conversant with the laws of conduction and radiation, with the thermal properties of materials, and with mass transfer.

The combustion engineer is concerned with fluid flow, because his fuel may be a fluid, as is certainly the air in which it burns. He must understand heat transfer because combustion both produces and is influenced by variations in temperature. In addition, however, he must understand the laws governing chemical transformations, both in respect of their speeds (chemical kinetics) and their effects (chemical thermodynamics).

It follows that, if he is to perform his function efficiently, a specialist in combustion must have a rather extensive grounding in science. Understandably, the proportion of engineers who can afford the time for these studies is rather small; so well-trained combustion engineers are in much demand. At a time when many branches of the engineering profession are over-staffed, this consideration is worthy of attention.

(ii) The role of the computer

The combustion engineer needs to make quantitative predictions of what will occur in the equipment which he is designing; and this, because of the multiplicity of processes which he must master, is a formidable task.

Until recently, it has been an impossible one. However, the development of the digital computer, and of methods of using it to solve complex mathematical problems, has transformed the situation. Now it is possible, to an increasing extent, to make realistic and useful predictions of combustion phenomena, with the aid of computers of modest size.

(iii) The role of the "mathematical model"

Whether the prediction is made by way of a computer or of more primitive devices, the focus of study is always a "mathematical model" of the process, i.e. of an idealisation possessing some of the features of reality, but not all.

Such "mathematical models" are, of course, an essential feature of all engineering analyses; but the concept of "modelling" must be given special attention when there are numerous constituent processes, as is true of combustion; for care is needed to select for scrutiny only those which play the major roles.

The publications cited at the end of this chapter (see Section 4) contain many references to mathematical models of combustion; and the present book is, indeed, an account of those models which are necessary to an understanding of the majority of combustion processes.

1.2 THE STRUCTURE OF THE SUBJECT OF COMBUSTION

A) CLASSIFICATION INTO LEVELS

(i) Engineering practice

The subject matter of concern in the present book appears at three levels, those of:-

- engineering practice;
- mathematical models; and
- fundamental science.

The first level involves the description of items of engineering equipment, discussion of their modes of operation, and delineation of their desired and undesired features of performance.

This might be regarded as the user's and designer's level. The user of the equipment is concerned with how well it performs; the designer's attention is concentrated on how the shape, size, configuration, materials, etc., can influence its performance.

(ii) Mathematical models

The human mind, being limited, must select for attention only those parts of reality which are immediately relevant to its purpose; the rest must be excluded, or at least subordinated. So it comes about that all engineering analyses are concerned with models of reality, rather than reality itself; and, when quantitative predictions are in question, the models are necessarily <u>mathematical</u> in character.

These mathematical models take the form of sets of differential and algebraic equations, the solutions of which agree, in important respects, with the behaviour of the pieces of equipment, or the processes, which are being modelled. The models can be regarded as "idealisations", or "essences", of the interactions which actually exist between the design and operating characteristics of the equipment on the one hand, and the fundamental laws of physics and chemistry on the other.

The mathematical-model level of study is therefore intermediate between the level of engineering practice and the level of fundamental science.

Mathematical models often have names, e.g. "the burning droplet", "the well-stirred reactor", "the laminar-propagating flame". Many will be encountered in the present book.

(iii) Fundamental science

As already mentioned, combustion processes are affected by the laws of nature which appear, in the usual classification of scientific knowledge, under the headings of:-

- thermodynamics;
- fluid mechanics;
- heat and mass transfer;
- and chemical kinetics.

Differently classified, the knowledge may be organised under the headings of:-

- conservation laws (of mass, energy, etc.);
- transport laws (of momentum, mass, chemical

species, energy, etc.); and

source laws (of the same entities).

(iv) Linkages

The following table contains entries at each of the above three levels; and two of the mathematical models appearing at level 2 are linked, by way of illustration, with the relevant fundamental disciplines at the lower level and the fields of engineering application at the upper one.

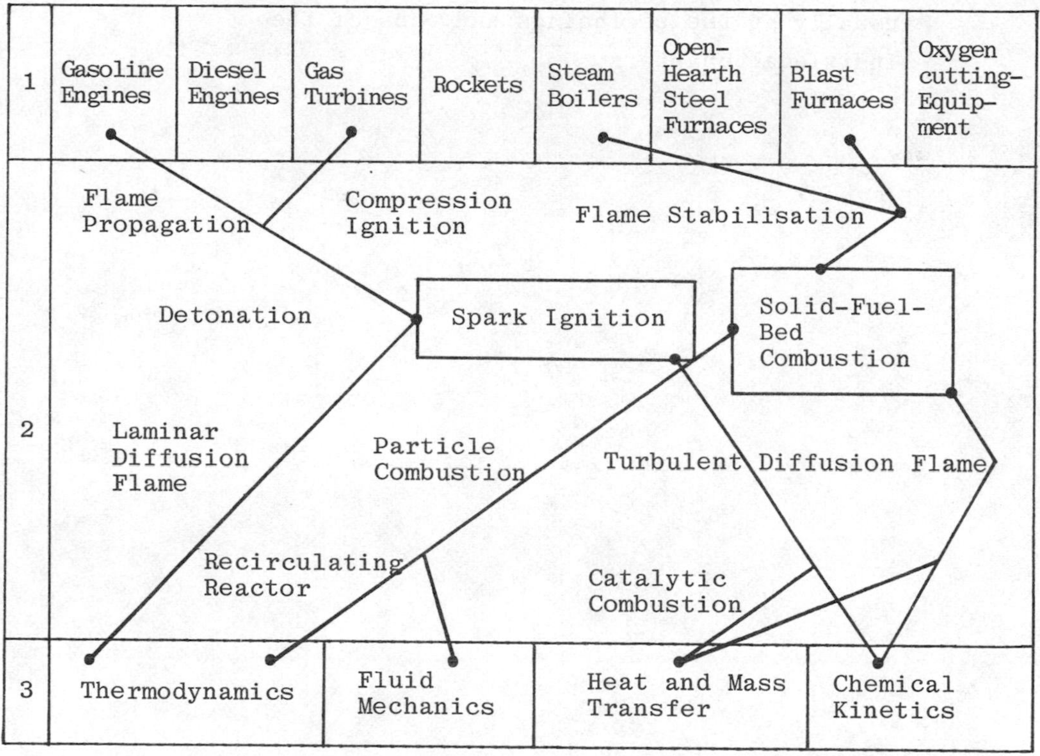

It may prove to be interesting to the reader to return to this table, after he has concluded his study of the book, so as to draw on the table the link lines for all the other models.

B) SOME FEATURES OF CURRENT ENGINEERING PRACTICE

(LEVEL 1)

There is no space to provide here more than a few notes on what are the common components of engineering combustion practice. The following tables provide these notes, first for steady-flow equipment and secondly for that in which the processes vary significantly with time.

The connexions of combustion theory with items of engineering equipment are discussed in the remainder of this book as opportunity offers, usually at the beginnings and ends of the individual chapters.

(i) Steady-flow equipment

FUEL	FORM & MEANS OF SUPPLY TO EQUIPMENT	APPLICATION
Coal	Lumps on grate	Domestic Small industrial boilers Cupola and blast furnaces
	Powder, suspended in air stream	Large industrial boilers Cement kilns
Kerosine	Vapour, from heated tube or pot	"Primus" stove Some aircraft gas turbines Small domestic boilers
	Spray of droplets from rotating cup	Domestic boilers
	Spray from swirl atomiser	Aircraft gas turbines
Gas Oil	Spray from swirl atomiser	Industrial gas turbines Larger domestic boilers
Residual Oil	Jet of droplets impelled by steam	Open-hearth furnaces
	Swirl atomisers	Large industrial boilers
Liquid Rocket Propellants	Spray formed by impingement of liquid jets	Large rocket engines, especially for re-use
Solid Rocket Propellants	As perforated blocks, the burning occurring at the free surface	Small rocket engines, and those (e.g. Polaris) requiring instant readiness

(ii) Unsteady-flow equipment

FUEL	FORM & MEANS OF SUPPLY TO EQUIPMENT	APPLICATION
Gasoline	Vaporised by contact with air and hot surfaces	Spark-ignition engines
Gas Oil	Spray formed by injection through small orifice into compressed air	Compression-ignition (Diesel) engines

c) SOME MATHEMATICAL MODELS (LEVEL 2)

The following notes are provided as a further introduction to some of the mathematical models which will be described elsewhere in this book.

(i) The single carbon particle, burning in still air

This concerns the interactions of diffusion of oxygen, of heat conduction and radiation, and of chemical kinetics, so as to determine the rate and time of burning.

Ordinary differential equations are involved.

The model is discussed in Chapter 20.

(ii) The single-liquid fuel droplet, burning in still air

This mathematical model is discussed in Chapter 7 because, although it might appear more complex than that for carbon-particle combustion, it is actually simpler; for chemical-kinetic aspects are of little significance.

A model of a droplet which vaporises without burning is presented in Chapter 3.

(iii) The one-dimensional liquid-propellant rocket

Droplets interacting with a gas stream comprising their combustion products feature in the mathematical model of a liquid-propellant rocket which is discussed in Chapter 8.

This model, provided that certain simplifying assumptions are made, is described by ordinary differential equations which are capable of analytical solution. Algebraic formulae are therefore derivable for the length of rocket which is needed for complete combustion.

The processes concerned are those of droplet combustion, together with that of droplet drag.

The model is a composite one; it embodies that of Section (ii), and combines it with one for one-dimensional gas flow in ducts.

(iv) The turbulent diffusion flame

A mathematical model of relevance to many industrial processes is that of the turbulent diffusion flame, or "gas torch". It can be represented mathematically by partial differential equations, expressing the interplay of the processes of transport of mass, momentum and energy; and these equations may be solved analytically if simplifying assumptions are made.

The model is presented in Chapter 12, after the simpler cases of laminar and non-burning jets have been discussed in Chapters 9, 10 and 11.

(v) The well-stirred reactor

Chemical kinetics is subordinate, in diffusion flames, to heat, mass and momentum transfer. The reverse is true of the mathematical model known as the "well-stirred reactor": here conditions are governed by the interaction of the supply rate of gas with the rate of the chemical reaction; the laws of thermodynamics also play a part.

This model is described at length in Chapter 16, after preliminary studies of chemical kinetics have been presented in Chapters 13, 14 and 15.

Because of the simplicity of the flow system, from which all non-uniformities are supposed absent, the model involves only algebraic equations. However, they require graphical or numerical solution.

(vi) The baffle-stabilised flame

Although well-stirred reactors never occur in practice, the flame which exists behind a "baffle" in a high-velocity stream of pre-mixed gas and air has some similarities of behaviour. "Baffles", also called "flame-holders" or "flame-stabilisers", are widely used in combustion equipment: they are obstacles which, because of their "bluff" shape, create regions of recirculating gas in their immediate wake.

The relevant mathematical model is described in Chapter 17.

(vii) Spark ignition

Flames require to be ignited; and frequently this ignition is effected by way of a spark, i.e. of a brief and localised discharge of electrical energy.

In order that this important process can be properly understood, a mathematical model of spark ignition is presented in Chapter 19. The differential equations are partial ones, requiring a computer for their exact solution. However, approximate solution procedures have been employed which permit the most important features to be revealed by simple hand calculations.

The model involves the full interaction of physical and chemical processes, albeit in laminar flow only. It is perhaps the most advanced model discussed in the present book.

D) THE FUNDAMENTAL SCIENCES (LEVEL 3)

In this section, some introductory notes will be provided to the fundamental scientific disciplines which will be touched on in the remainder of the book.

(i) Thermodynamics

Only few concepts from thermodynamics will be employed; and they will appear in their simplest forms. Thus, specific heats will be taken as independent of temperature, and it will be presumed that equilibrium exists only when one of the two participants in a reaction has been completely consumed.

This practice is adopted so as to permit the student to concentrate his mind on aspects of combustion which are (for present purposes) more interesting than the fact that specific heats actually vary with temperature.

The main use of thermodynamics in the present book is to link the rise of temperature resulting from combustion quantitatively with the amount of fuel which is consumed. Knowledge of the First Law of Thermodynamics, usually in its steady-flow form, is all that is demanded.

(ii) Fluid mechanics

A few of the mathematical models draw upon knowledge of the laws of fluid mechanics, particularly as expressed by the so-called "boundary-layer" equations. In these, the effects of inertia and of viscosity are of similar orders of magnitude; and their interplay governs the flow pattern. These equations first make their appearance, in the present book, in Chapter 9.

Despite the fact that the viscosities of gases are strongly temperature-dependent, the analyses presented in the present book neglect the variation. Once again, the motive is to procure clarity at the expense of quantitative exactitude; for the latter can always be obtained from computer studies, once the sought-for understanding has been achieved.

Some elementary ideas about turbulence will be presented in Chapter 11. They suffice for present purposes.

(iii) Heat transfer

Chemical reactions proceed so slowly at low temperatures, that, if it were not for the transfer of heat from hot reacted to cold unreacted gases, flames would seldom exist. This is sufficient reason for combustion theory to make extensive use of the science of heat transfer

All three main modes are important, namely conduction, convection and radiation; but the first has especial significance because its rate is proportional to the gradient of temperature, which is often large in a flame. The relevant law, that of Fourier, will be introduced in Chapter 2; but it will be frequently returned to thereafter.

In some processes, conduction heat transfer is the dominant mechanism. Droplet vaporisation is one of these.

(iv) Mass Transfer

Whereas heat transfer is a subject which is studied by engineers of nearly all disciplines, mass transfer features regularly in the undergraduate courses only of chemical engineers. It is however of great importance to combustion specialists; so it is allotted a rather extended treatment in the present book. The main concepts are introduced in Chapters 2 and 4; and they are used in many places.

Mass transfer is similar in many ways to heat transfer: the place of conduction in the latter is taken by diffusion in the former; convection takes similar forms in both processes; but mass transfer exhibits no counterpart to radiation.

Diffusion is the process by which differences in composition even themselves out. Both laminar and turbulent diffusion can occur; and both make their appearance in the present book.

(v) Chemical kinetics

Chemical reactions involve the re-arrangement of atoms between molecules: molecules of one set (the reactants) must be broken; and the fragments recombine in a different pattern to form molecules of another kind (the products).

The subject of chemical kinetics describes how the rates of these rupture and re-combination processes depend upon local conditions such as: pressure, temperature, gas composition, etc.

This subject is introduced, in an elementary way, in Chapters 13 and 14 of the present book. Just sufficient is provided to permit the ways in which chemical-kinetic constants influence flame behaviour to be quantitatively understood.

1.3 SOME TRENDS

A) TRENDS IN ENGINEERING PRACTICE

(i) Pollution control

Public concern with the influence of combustion processes on the quality of the environment has become so intense in most industrial countries that fuel usage has been subjected to restrictive legislation.

The automobile may pollute the atmosphere with unburned hydrocarbons. In order to meet new or impending regulations, several pollutant-reducing measures are being introduced, including:-

- the provision of "catalytic after-burners" which mix the exhaust gases with further air, and oxidise them in a chamber inserted in the exhaust pipe;
- the use of gasoline which is free from lead tetra-ethyl (otherwise added to diminish "knock"), because the lead oxide in the exhaust gases would prevent the catalyst from working;
- the development of "stratified-charge" combustion chambers, the main feature of which is their ability to work with a larger ratio of air to fuel.

Large power stations are significant producers of oxides of nitrogen, a noxious gas, the emission of which is now limited by law. The oxide is produced mainly in the highest-temperature regions of the furnace; so designs and operating procedures are being changed so that the maximum temperatures are reduced. One means of doing this is the re-introduction into the furnace of exhaust gases which have already once traversed the heat-transfer sections.

"Smokeless zones" have been imposed in many cities, in which the emission of smoke from chimneys is expressly forbidden. Such regulations have caused many users to change completely their fuels, their combustion appliances, or both.

(ii) Efficiency of combustion

In most cases, what reduces pollution also improves combustion efficiency; for pollutants comprise, in part, fuel which is being wasted.

However, the saving of fuel remains as a distinct motive for the improvement of combustion equipment; and its importance can only increase in the course of time.

(iii) Change of fuel type

The same desire to reduce the cost of fuel is also causing large users to shift, when they can, to cheaper fuels. This means, in many countries, a preference for coal rather than oil or natural gas.

The shift may be accompanied by the introduction of new kinds of combustion equipment. For example, large research and development efforts are being devoted, at the present time, to "fluidised-bed" combustion of coal. In this novel process, the coal particles "float" in an upward-rising stream of air; and heat-transfer surfaces are embedded with the "floating" cloud. Temperatures are held low, so that the nitrogen-oxide production rate is small.

B) TRENDS IN MATHEMATICAL MODELS

(i) Computer methods

Since the late 1960's, computer programs have been increasingly used for the prediction of combustion processes. At first, attention was given to handling the complexities of real chemical reactions, which involve scores of individual species rather than the two or three which feature in the models described in the present elementary book. Typical examples concerned the accurate calculation of the equilibrium composition of the gases in a rocket nozzle, followed by the kinetically-influenced variation of the composition of the gases, as they flow out through the nozzle.

Later, two- and three-dimensional variations in space, and variations in time, were taken into account; and, the invention of "turbulence models" permitted realistic representation of the interactions of flow pattern and chemical reaction.

The contributions of computer methods to combustion modelling are far from complete; for the demands of the designer for increases in predictive accuracy and reduced cost are constantly increasing. Thus it will probably be many years before a computer model exists for the whole injection and combustion process in a diesel engine; for this exhibits the following complications:-

- three space dimensions, with domain boundaries which vary cyclically with time;
- unsteadiness;
- the presence of two phases, represented by the liquid fuel and the gaseous air and combustion products;
- turbulence;
- multiple chemical reactions;
- heat-transfer to walls by convection and radiation.

All of these features do exist in current computer models; but they have never all been put together in a single practical computer program.

(ii) Validation

It is one thing to produce a predictive method, and quite another to ensure that its predictions are correct, i.e. in agreement with experimental observations.

Consequently, a considerable effort is being made by research workers concerned with industrial applications to "validate" the computer models. The method is to make detailed comparisons between predictions and experiments; to interpret whatever discrepancies are discovered in terms of computational inaccuracies, inadequacies of the assumptions, and imprecisions of measurement; and then to invent and implement improvements which result finally in the reduction of the discrepancies to acceptably-small values.

Validation work is time-consuming, and therefore expensive; and it can be successfully performed only by scientists who are both industrious and of high integrity. Among every hundred who are capable of having "bright ideas", only a handful are persistent enough to check their ideas against reality, and then courageous enough to discard those ideas when they are proved to be unfounded.

(iii) Utilisation

Some computer models have been sufficiently validated for designers and operators of equipment to be using them successfully. These models could, therefore, already be widely employed, with great advantage.

However, no models are yet in widespread use; and the reasons are two-fold:

- there are too few persons, adequately trained in mathematical modelling generally, and computer modelling of combustion in particular, to supply the needs of the potential users;
- the users and designers of combustion equipment have tasks to perform of such seriousness that failure is not to be contemplated; they are therefore unwilling to employ predictive means

in which they do not have complete faith.

Both these obstacles to the utilisation of computer models of combustion will, of course, be eroded in the course of time by improvements in education and in the dissemination of information. The present book is a modest contribution to these ends.

c) TRENDS IN THE FUNDAMENTAL SCIENCES

(i) Thermodynamics

Thermodynamics has been a well-developed subject for many decades. Apart from the increase in knowledge about the thermodynamic properties of relevant materials, there is little development in the subject for the combustion engineer to take note of.

(ii) Fluid mechanics

The most striking and relevant development in fluid mechanics, in recent years, is the bringing of turbulent flows within the range of quantitative predictive methods. This trend will no doubt continue, albeit not necessarily smoothly.

At present, computational procedures for turbulent flows often involve sophisticated methods for computing a rather *un*sophisticated quantity; the effective viscosity. These procedures suffice for many purposes; yet the implied likening of turbulent flows to laminar ones certainly leaves some important features out of account.

The interactions of turbulence and chemical reaction display features of especial interest; and, indeed, the current approach to "turbulence modelling" is far less successful when combustion is present than when it is not. There are two questions:

- How, both qualitatively and quantitatively, does the presence of the chemical reaction affect the turbulent motion?

- How does the turbulent motion affect the rate of chemical reaction?

Because the second influence is stronger than the first, current attention is concentrated upon it. No complete answer yet exists.

Whereas most studies of fluid-mechanical phenomena have, in the past, been confined to single-phase flows, methods are now becoming available for analysing multi-phase ones. This is especially fortunate for combustion engineers; for their flow phenomena often are multi-phase, as has been explained above.

(iii) Heat Transfer

The developments in heat-transfer theory which are most needed by those concerned with combustion relate primarily to the radiative model of heat transfer. Both physical and mathematical developments are needed.

What makes radiation hard to handle mathematically is that one needs to know at every point in a flame not merely the radiation intensity, but also its distribution with respect to angle and wave-length; and the physical complexity resides in the fact that materials interact differently with radiation of different wave lengths.

There is research to perform, of the kind which accumulates more reliable data for the radiative properties of materials; but still more to be desired is the "conceptual break-through" which will enable predictions to be made which are both realistic and economical. In short, a better mathematical model of radiation is needed.

(iv) Mass transfer

There appears to be little need for major developments under this heading; existing knowledge suffices, for the most part.

(v) Chemical kinetics

The work of the world's physical chemists has resulted in enormous acquisitions of knowledge about the chemical-kinetic constants of the reactions which are of importance to combustion scientists. Their task will perhaps never be finished; for the number of interacting chemical species is vast. However, whereas it was customary a few years ago to deplore the paucity of reliable chemical-kinetic data, the situation nowadays is that more data exist than the computational procedures can easily handle.

(vi) Numerical mathematics

Finally, it should be mentioned that the computer can be used for predictions only after reliable and economical methods have been developed for solving the equations by which the mathematical models are defined.

The current position is that methods exist for solving equations of all the kinds which are known to arise in combustion predictions. However they sometimes need a large amount of "hand-tuning" to be successful; and they are nearly always more expensive than their users find congenial.

Further developments in this area are needed; and, since only application and ingenuity are required, they will surely be forthcoming.

4 REFERENCES

4.1 TEXTBOOKS

BEER J M and CHIGIER N A
"Combustion and Aerodynamics"
Applied Science Publishers, London, 1972

FRISTROM R M and WESTENBERG A A
"Flame Structure"
McGraw Hill, New York, 1965

GLASSMAN I
"Combustion"
Academic Press, New York, 1977

GÜNTHER R
"Verbrennung und Feuerungen"
(In German) Springer-Verlag, Berlin, 1974

KENNEDY L A (Editor)
"Turbulent Combustion"
AIAA, New York, 1978

LEWIS B and von ELBE G
"Combustion, Flames and Explosions in Gases"
Academic Press, New York, 1961

LEWIS B, PEASE R N and TAYLOR H S
"Combustion Processes. Vol II of High Speed Aerodynamics and Jet Propulsion"
Oxford University Press, 1956

MURTY KANURY A
"Introduction to Combustion in Phenomena"
Gordon and Breach, New York, 1975

SPALDING D B
"Some Fundamentals of Combustion"
Butterworth's, London, 1955

SPALDING D B
"Convective Mass Transfer"
Edward Arnold, London, 1963

SPALDING D B
"Combustion Theory Applied to Engineering"
Imperial College, London, Heat Transfer Section Report No HTS/77/1, 1977

THRING M W, DUCARME J and FABRI J (Editors)
"Selected Combustion Problems II"
Butterworth's, London, 1956

WILLIAMS F A
"Combustion Theory"
Addison-Wesley, 1965

4.2 GENERAL SOURCES

THE PROCEEDINGS of the International Symposia on Combustion, published every two years since about 1949

WEINBERG F (Editor)
"Combustion Institute European Symposium 1973"
Academic Press, London, 1973

COMBUSTION AND FLAME
Journal published by Elsevier

COMBUSTION SCIENCE AND TECHNOLOGY
Journal published by Gordon and Breach, New York.

PALMER H B and BEER J M (Editors)
"Combustion Technology: Some Modern Developments"
Academic Press, New York, 1974

4.3 PUBLICATIONS CONCERNED WITH THE MODELLING OF COMBUSTION PHENOMENA

A) PHYSICAL MODELS

CHESTERS J H, HOWES R S, HALLIDAY I M D and PHILIP A R
J Iron & Steel Institute vol 162 p 385 1949

CLARKE A E, GERRARD A J and HOLLIDAY L A
"Some Experiences in Gas Turbine Combustion Practice Using Water Flow Visualization Techniques"
Ninth Symposium (International) on Combustion
Academic Press, New York, p 878 1963

DAMKOHLER G
Der Chemie-Ingenieur (A EUCKEN and M JAKOB (Editors)
Bank III Teil 1 Akademie Verlag 1937
Reprinted by VDI-Fachgruppe Verfahrenstechnik, Leverkusen 1957

GROUME-GRJIMAILO W E
"The Flow of Gases in Furnaces"
Wiley, New York, 1923

HOTTEL H C, WILLIAMS G C, HENSEN W P, TOBEY A C and BURRAGE P M R
"Modelling STudies of Baffle-Type Combustors"
Ninth Symposium (International) on Combustion
Academic Press, p 923 1963

LeFEBVRE A H and HALLS G A
"Simulation of Low Combustion Pressures by Water Injection"
Seventh Symposium (International) on Combustion
Butterworths, London, p 654 1959

PUTNAM A A and JENSEN R A
"Application of Dimensionless Numbers to Flash-back and Other Combustion Phenomena"
Third Symposium (International) on Combustion
Williams and Wilkins, Baltimore, pp 89-98 1949

SPALDING D B
"Analogue for High-Intensity Steady-Flow Combustion Phenomena"
Proceedings of Institution of Mechanical Engineers
vol 171 no 10 pp 383-411 1957

SPALDING D B
"The Art of Partial Modelling"
Ninth (International) Symposium on Combustion
Academic Press, New York, pp 833-843 1963

SPALDING D B
"Some Thoughts on Flame Theory"
Combustion and Propulsion,
Third AGARD Colloquium, Pergamon Press, London,
pp 369-406 1959

STEWART D G
"Scaling of Gas Turbine Combustion Systems"
Selected Combustion Problems Volume 2
Butterworths, London, pp 384-414 1956

TRAUSTEL S
"Modellgesetze der Vergasung und Verhüttung"
Akademie-Verlag 1949

B) ZERO-DIMENSIONAL MATHEMATICAL MODELS

BRAGG S L and HOLLIDAY J B
"The Influence of Altitude Operating Conditions on Combustion Chamber Design"
Selected Combustion Problems Volume 2
Butterworths, London, pp 270-295 1956

HOTTEL H C and STEWART I McC
"Space Requirement for the Combustion of Pulverized Coal"
Ind Eng Chem, vol 32, pp 719-730, 1940

LONGWELL J P, FROST E E and WEISS M A
"Flame Stability in Bluff-Body Recirculation Zones"
Ind Eng Chem, Vol 45, p 1629, 1953

SPALDING D B
"The STability of Steady Exothermic Chemical Reactions in Simple Non-Adiabatic Systems"
Chemical Engineering Science, vol 11, pp 53-60, 1958

VAN HEERDEN C
"Autothermic Processes"
Ind Eng Chem, vol 45, p 1242, 1953

C) ONE-DIMENSIONAL MATHEMATICAL MODELS

CHESTERS J H, HOYLE K H, PEARSON S W and THRING M W
"Comparison of Calculated and Actual Heat Transfer"
Iron and Steel Institute Special Report No 59 p 65 1965

CSABA J and LEGGETT A D
"Prediction of the Temperature Distribution along a Pulverized-Coal Flame"
J Institute of Fuel, vol 37, pp 440-448, 1964

FIELD M A and GILL D W
"A Mathematical Model of the Combustion of Pulverized Coal in a Cylindrical Combustion Chamber"
BCURA Members Information Circular Number 318 1967

SPALDING D B
"The One-Dimensional Theory of Furnace Heat Transfer"
University of Sheffield Fuel Society Journal vol 10, pp 8 - 17, 1959

THRING M W
"The Effect of Emissivity and Flame Length on Heat Transfer in the Open-Hearth Furnace"
J Iron and Steel Institute, vol 171, pp 381-392, 1952

THRING M W and SMITH D
"An Improved Model for the Calculation of Heat Transfer in the Open-Hearth Furnace"
J Iron and Steel Institute, vol 179, p 227, 1955

D) TWO-DIMENSIONAL MATHEMATICAL MODELS

GIBSON M M and MORGAN B B
"Mathematical Model of Combustion of Solid Particles in a Turbulent Stream with Recirculation"
J Institute of Fuel, vol 43, pp 517-523, 1970

KENT J H and BILGER R W
"Turbulent Diffusion Flame"
Fourteenth Symposium (International) on Combustion
Combustion Institute, Pittsburgh, pp 615-625, 1973

LOWES T M, HEAP M P, MICHELFELDER S and PAI B R
"Mathematical Modelling of Combustion Chamber Performance"
Proc. Fourth Flames & Industry Symposium, London, 1972

PUN W M and SPALDING D B
"A Procedure for Predicting the Velocity and Temperature Distributions in a Confined, Steady, Turbulent, Gaseous, Diffusion Flame"
XVIII International Astronautical Congress, Belgrade, 1967, Pergamon Press, London, pp 3-21, 1968

RICHTER W and QUACK R
"A Mathematical Model of a Low-Volatile Pulverized-Fuel Flame"
Heat Transfer in Flames, edited by Afgan N H & Beer J M, pp 95-110, Scripta Book, Co., Washington, 1974

SALA R and SPALDING D B
"A Mathematical Model for an Axi-Symmetrical Diffusion Flame in a Furnace"
La Rivista dei Combustibili, vol XXVII, pp 180-196, May 1973

SCHORR J, BERMAN K and WORNER G
"Modelling of High-Energy Gaseous Combustors for Performance Prediction"
Bell Aerospace Company, 1971

SCHORR C J, WARNER G A and SCHIMKE J
"Analytical Modelling of a Spherical Combustor including Recirculation"
Fourteenth Symposium (International) on Combustion, Combustion Institute, Pittsburgh, pp 567-574, 1973

SPALDING D B
"Mixing and Chemical Reaction in Steady Confined Turbulent Flames"
Thirteenth Symposium (International) on Combustion, Combustion Institute, Pittsburgh, p 649, 1971

E) THREE-DIMENSIONAL MATHEMATICAL MODELS

PATANKAR S V and SPALDING D B
"A Computer Model for Three-Dimensional Flow in Furnaces"
Fourteenth Symposium (International) on Combustion, Combustion Institute, Pittsburgh, pp 605-614, 1973

PATANKAR S V and SPALDING D B
"Simultaneous Predictions of Flow Pattern and Radiation for Three-Dimensional Flames"
Heat Transfer in Flames, edited by Afgan N H & Beer J M, pp 73-94, Scripta Book Co., Washington, 1974

ZUBER I
"Ein Mathematisches Modell des Brennraums"
Monographs and Memoranda No 12 Staatliche Forschungs Institute für Maschinenbau, Bechovice, Czechoslovakia, 1972

PROF. D. B. SPALDING, M.A., Ph.D., Sc.D., F.I.Mech.E., F.Inst.F.*

Combustion as applied to engineering†

The paper reviews the needs of the combustion engineer, the offerings of the combustion scientist and the discrepancy between the two. It is argued that, though the scientist has in the past been able to provide only qualitative understanding, he is now making available quantitative means of prediction that the design engineer can use. The main reason for the advance is that computers are available, and programs are being developed for their use, which allow the mathematical obstacles to be removed. The physical obstacle, ignorance of the physics of turbulence, is crumbling under the impact of research. And, though chemical kinetics still lags, the use of chemical-kinetic knowledge in prediction methods is likely to stimulate new advances.

This reprint is supplied as an Appendix to Chapter 1. It describes the applicability of combustion theory to engineering in 1970. By comparing it with the notes at the end of the chapter, the rate of advance of the subject can be observed.

1. Some combustion problems of engineering

My theme is combustion and its applications to engineering. Fire first raised man from the animals. With earth, air and water, it was one of the four elements of the Greeks. Through fire man made metals for weapons and tools; and the engines of our power-based civilization would be impossible without practical mastery of the laws of combustion.

So the subject is ancient, and of central importance to technology. I intend to review our scientific knowledge of it, but from the engineer's point of view. What use, I want to ask, can the engineer make of our current understanding of combustion? And in what directions should research proceed in order that this use can be extended?

The gasoline engine

flame

The combustion system must:

- be free from "knock",
- give certain ignition by spark,
- reduce heat transfer,
- burn fuel completely under all conditions.

FIG. 1 The gasoline engine.

Let us first examine some of the problems to which engineering practice gives rise. Fig. 1 reminds us of the gasoline engine, the most universally used of all prime movers. It exists; it works; what more need we know about the combustion process? In the early days of the gasoline engine, the main need was to understand and control the process of 'knock'; for efficiency demanded a high compression ratio; and it was soon found that this could lead to an uncontrolled ignition in the cylinder, to excessive heat transfer, to noise, and to damage to the engine.

The diesel engine

fuel jet

The combustion system must:

- provide smooth ignition,
- use all available oxygen,
- reduce heat transfer,
- reduce pumping power,
- emit no smoke.

FIG. 2 The diesel engine.

Engineers introduced changes of design which minimized the tendency to knock; and chemists introduced lead-based additives which effected further drastic reduction. So the gasoline engine prospered, and multiplied mightily. But machinery is subject to the same ecological laws as are animal species. The increased population of gasoline engines has begun to affect its environment; the atmosphere is becoming polluted, and the human species, on which the gasoline engine lives in symbiosis, like a parasite on a host, is beginning to suffer and resist.

The combustion engineer therefore has now especially to concentrate on *completeness* of combustion. There must be no unburnt fuel, or poisonous side-products, from the engine exhaust. This is a severe problem, difficult to solve.

The designer of diesel engines is not in exactly the same position, as Fig. 2 makes clear. He *wants* his fuels to ignite spontaneously; for there is no spark in his engine, and combustion is controlled by the rate of injection of fuel. Whatever additives he introduces will therefore promote ignitability, not resist it.

However the diesel engine pollutes the atmosphere as severely as, and more obviously than, its gasoline rival. The smoke which pours from the exhausts of heavy trucks is a result, it may be said, of the engineer's desire to maximize the power of the engine, by using up all the oxygen it induces. But this smoke exacts a heavy price, paid by the public at large; it is a matter of time only before legislation forbids smoke production.

The engineer therefore has to find means of increasing the power output without generating smoke. This will necessitate a better understanding and control of the combustion process. For smoke is a regular by-product of hydrocarbon combustion. It is nearly always produced somewhere in the flame; the task is to ensure that it is consumed again, before the exhaust gases escape.

The gas-turbine engine (see Fig. 3) also produces smoke; but, since most of it is ejected into the atmosphere at a high altitude, this pollutant is more tolerable to mankind (those of you who live near to airport runways might think differently about this).

The combustion chamber of a gas turbine is an important part of the machine. Not only does it perform an

* Professor of Heat Transfer, Department of Mechanical Engineering, Imperial College of Science and Technology, London.

† Lecture presented at the Conference on Mechanical Engineering and Mechanics, July 1970. Technion, Israel Institute of Technology, Haifa.

The gas turbine

The combustion system must:

- burn fuel completely,
- cause little pressure drop
- produce gases of nearly uniform temperature,
- occupy small volume,
- maintain stable combustion over wide range of pressures.

FIG. 3 The gas turbine.

essential function; its size and pressure drop are both regrettably high. If the combustion chamber could be shorter, the problems of mechanical design would be less severe; if the pressure drop could be eliminated, the efficiency of the engine would be very significantly improved.

What is found in practice is that, if the combustion chamber is made too short, the flame can be too easily extinguished. Further, measures which lower the pressure drop usually worsen the degree of mixing of the gases which enter the turbine.

As usual, therefore, the engineer has to seek an optimum path between irreconcilable desiderata. He often finds this path by trial and error. The task of research is to provide him with a map which leads him to his objective less expensively.

Power-plant engineers are a rather special group; and perhaps their needs have attracted excessive attention from those of us who are concerned with combustion. Certainly the furnace designer is as much in need of assistance from combustion science. His plant is no less expensive; and the sheer number of particular designs is very great indeed. Fig. 4 must therefore serve to represent innumerable particular configurations.

The industrial furnace

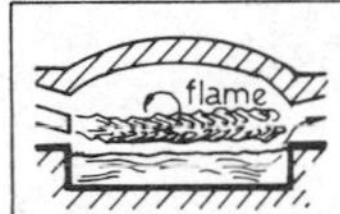

The flame must provide:

- specified distributions of radiant and convective heat transfer,
- complete combustion,
- freedom from noise and oscillation,
- insensitivity to fuel changes.

FIG. 4 The industrial furnace.

Whereas the engine designer would like to be entirely rid of smoke, the furnace designer needs it. For, without smoke, the radiant heat fluxes would be much smaller; and often the designer wants them to be as large as possible. Of course, the smoke must be consumed again before it leaves the furnace.

It is not usually enough that the total heat-transfer rate should be high. The *distribution* of the flux over the surface of the processed material is also important. Neither the material nor the furnace roof must be overheated; and often a particular distribution of temperature has to be procured, for optimum quality of output. The well-equipped designer must know how to shape his furnace, and position the burners, so as to achieve these effects.

2. What combustion science can contribute

Let us now consider what the combustion scientist can provide. It is not negligible; but you will see that a gap remains. In the end I will discuss how it can be filled.

Scientists have to simplify, often to over-simplify, in order to reduce their observations to quantitative order. Therefore, what I shall at first show you will be very much less complex than what has appeared in the previous diagrams.

Fig. 5 shows the first example. It is a single droplet of fuel, burning under quasi-steady conditions in an oxidizing atmosphere. The flame forms an envelope around the droplet; fuel vapour diffuses towards the flame from inside; oxygen diffuses from outside; the heat which is liberated by their reaction is conducted in both directions, and keeps the process going. The theory of this phenomenon is complete,[1,2] and explains well what happens to isolated droplets. Of course, *sprays* of droplets are a bit more complicated.

Burning-rate formula for an isolated droplet

$$\text{burning time} \approx K \times (\text{initial diameter})^2$$

where K depends in known ways on:

- oxygen content of air,
- oxygen requirement of fuel,
- heat of combustion of fuel
- latent heat of vaporisation of fuel,
- transport properties of gas.

FIG. 5 Burning-rate formula for an isolated droplet.

No-one has yet worked out a complete calculation procedure for sprays of droplets, though it could now easily be done. (That word 'easily', by the way, is one that my research students have learned to treat with suspicion. It means that success is certain, because no difficulties of principle remain; but there may be many practical obstacles to overcome; and a steady optimism and resourcefulness will be needed to surmount them.)

However, large flames of liquid-fuel sprays often behave as though the fuel were injected as a gas. The reason is that the time of vaporization of a droplet is small compared with its residence time in the flame. In these circumstances one can employ the theory of turbulent jets for the prediction of the length of the flame, of the temperature distribution, and so on (see Fig. 6).

In a crude way, but with fair accuracy, we have been able to make this calculation for many years. Very recently, much-refined methods have come into existence. I shall be mentioning these in some detail below.

Both the single droplet and the turbulent diffusion flame fall into that important group of combustion phenomena which we call 'diffusion-controlled'. This label means that the rate of combustion is almost uninfluenced by the chemical–kinetic properties of the materials; additives, and temperature-level changes, are

Turbulent diffusion flame

Flame height can be calculated from:

- oxygen content of air,
- oxygen requirement of fuel,
- nozzle diameter,
- theory of jets.

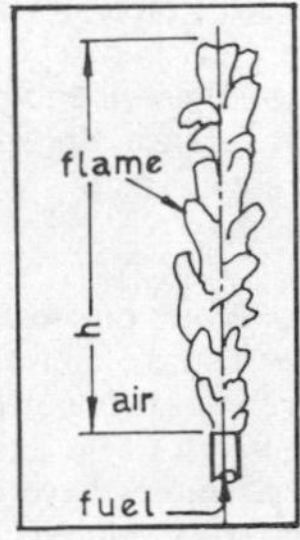

FIG. 6 Turbulent diffusion flame.

without effect. Fig. 7 shows a phenomenon which is definitely *not* in that class: spontaneous ignition. It represents a laboratory apparatus for measuring the time of reaction of a fuel with a hot-air stream, after it has been rapidly and uniformly mixed[3]. One simply measures the distance between the mixing region and the plane at which the flame disappears, and divides by the flow velocity.

Spontaneous-ignition delay

Ignition distance can be calculated from:

- fuel-air ratio, • initial temperature,
- thermodynamic properties,
- chemical-kinetic properties.

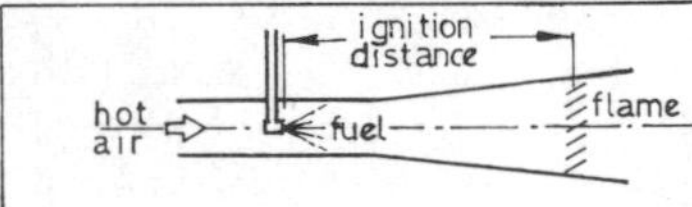

FIG. 7 Spontaneous-ignition delay.

Because the mixture is uniform, and the velocity of flow is too high for heat-transfer effects to play any part, the theory is especially simple. Knowledge of the chemical–kinetic and thermodynamic properties leads, through the solution of ordinary differential equations, directly to the ignition delay. Knowledge of thermodynamic properties we possess. But chemical–kinetic properties, too? Alas no. Fig. 8 will enlarge on this.

It is a great pity that Nature, which is elegant and straightforward in its conduct of physical processes, should become so circuitous and elaborate in its chemical ones. One might think that, to form a simple chemical

Kinetics of hydrogen-oxygen reaction

Values of rate constants are approximately known for the reactions:

(1) $H_2 + O_2 \rightleftharpoons 2OH$ (6) $2H + M \rightleftharpoons H_2 + M$

(2) $H_2 + OH \rightleftharpoons H_2O + H$ (7) $2O + M \rightleftharpoons O_2 + M$

(3) $O_2 + H \rightleftharpoons O + OH$ (8) $H + O + M = OH + M$

(4) $H_2 + O \rightleftharpoons H + OH$ (9) $H + OH + M = H_2O + M$

(5) $H_2O + O \rightleftharpoons 2OH$

FIG. 8 Kinetics of hydrogen-oxygen reaction.

compound like water from its constituents hydrogen and oxygen, the latter molecules would merely have to meet and combine, in simple straightforward encounters. This does not happen at all.

Instead, a whole sequence of reactions occurs, between various fragments and ephemeral compounds of H_2 and O_2, of which the substance H_2O is a remote end-product. Fig. 8 shows nine of the more important participating reactions. You see that the monatomic H and O, the radicle OH, and the anonymous stranger M, all make their appearance.

In order to calculate the rate of burning of hydrogen and oxygen, therefore, it is necessary to know the rate constants of each of the individual reactions; and usually it is impossible to isolate each of these, for separate study. The result is that, even for these reactions, the constants are not all known to within a factor of five. For reactions involving compounds of carbon, the ignorance is very much greater.

If one cannot study the reactions separately, at least one can measure their rates in combination; and perhaps, if the conditions of the experiment are varied sufficiently, so that different reactions occupy in turn the dominant position, a careful analysis will allow the individual rate constants to be determined. If not, at least one will possess a large amount of useful quantitative information.

One of the pieces of apparatus that provide data of this kind is the 'Longwell bomb',[4] which is shown in Fig. 9. It is a steady-flow apparatus, in which the flow pattern is arranged so that the temperature and gas composition in the reactor are almost uniform. Because of this, the equations governing the process are algebraic, not differential. This greatly eases the task of analysis.

The stirred reactor (Longwell bomb)

The extinction condition can be predicted from:

- fuel-air ratio,
- initial temperature,
- thermo-dynamic properties,
- chemical-kinetic properties.

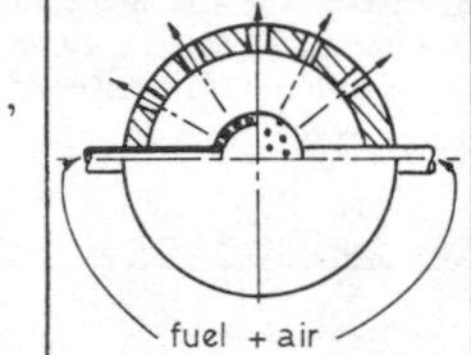

FIG. 9 The stirred reactor (Longwell bomb)

The Longwell bomb has been more productive of data for 'global' reactions than for individual ones. However, because the conditions in the bomb are quite similar to those in practical steady-flow equipment, these unanalysed 'global' data can be used directly without introduction of serious error.

Fig. 10 shows my last example of what I will call 'classical' combustion research. The system illustrated is the well-known Bunsen burner; and the question which is to be answered with its aid is: what is the speed of flame propagation in the gas under laminar conditions?

This flame speed is easily related to the angle α and the flow velocity in the pipe. But how is it connected with the chemical–kinetic, thermodynamic and transport properties of the mixture? A large part of the older literature on combustion theory has concerned itself with this question. I will just say that mathematical methods now exist that are capable of predicting laminar flame speeds, no matter

Speed of laminar flame propagation

Angle α can be calculated fr m:

- approach velocity,
- fuel-air ratio,
- initial temperature,
- thermodynamic properties,
- transport pr perties.
- chemical-kinetic properties.

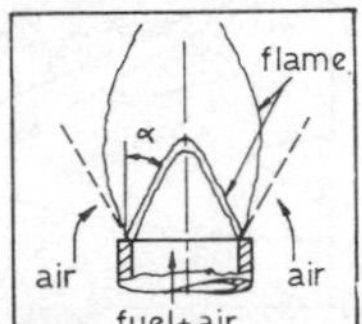

FIG. 10 Speed of laminar flame propagation.

how complex the reaction mechanism, provided of course that the relevant data are provided.[5] Until these mathematical methods were available, all that one could do was to accumulate experimental data, and interpret them quantitatively. Unfortunately, this habit has persisted. If chemists could be persuaded to use the mathematical methods that now exist, I believe that they would soon clear broad pathways through the too-little-explored jungle of their subject.

3. The calculation of turbulent flows

3.1. Introduction

Now let us mentally compare the examples of practical problems with those from the research scientist's repertoire. If you reflect upon the earlier diagrams, you will be struck by the complexities of the practical problems, as compared with what the theoretician can master. There is a substantial gap; and it must be filled before the scientist can truly be said to be helping the designer. What prevents this from being done?

Fig. 11 indicates an answer. There are three obstacles: one chemical, one mathematical, and one physical. Of the first, the lack of chemical–kinetic data, I shall say little more than that we should not be dismayed by it. We can do quite well with 'global' kinetic expressions for the time being; and the chemists will catch up eventually.

The mathematical obstacle is crumbling fast. Already we can handle steady two-dimensional phenomena, and are moving on to unsteady ones. The majority of practical flames are, it is true, three-dimensional in character; and for these we shall need larger computers than are widely available as yet. Nevertheless, they will soon be with us; so we must already prepare ourselves for using them.

You will be seeing examples of what our present computers can do in the remainder of the lecture. Therefore, in the present section, I shall concentrate attention on the third obstacle: that of predicting the behaviour of turbulent flows.

Chemical-kinetic

Very few of the reaction-rate constants are known for hydrocarbon systems.

Turbulent transport

Laws are only just appearing.
Effects on reactions are uncertain.

Computational

2D problems can now be solved.
3D problems require bigger computers.

FIG. 11 The main obstacles to further progress.

Variables:

k ≡ velocity fluctuations (energy)
W ≡ vorticity fluctuations
g ≡ concentration fluctuations

Processes affecting values of k, W and g:

- Convection
- Diffusion
- Generation
- Dissipation

These processes are linked by differential equations.

FIG. 12 The *kWg* turbulence model; nature.

3.2. Predicting the behaviour of turbulent flows

Fig. 12 describes what I will call the *kWg* model of turbulence. The name indicates the three variables which play prominent parts in it: the turbulence energy *k*, the vorticity fluctuations *W*, and the concentration fluctuations *g*. Of course, one cannot truly describe so complex a phenomenon as turbulence in terms of just three quantities; yet that is what we hopefully assume.

In the *kWg* model, as in other mathematical models of this kind, one supposes that the three entities *k*, *W*, and *g* are affected by the four processes of convection, diffusion, generation, and dissipation. Since these processes can be expressed mathematically by differential expressions, the distributions of *k*, *W* and *g* throughout a turbulent fluid can be determined by solving differential equations. I shall show you solutions of this kind below.

One might express the approach as treating the turbulent fluid like a non-Newtonian laminar one, of which the properties have to be determined, at each point, by solving special additional differential equations.

In a short time, I can show you only the main features of the model. Fig. 13 picks out some especially important ones, e.g. the generation and dissipation terms of the three variables.

Variable	Generation	Dissipation
k	$\propto \mu_t \left(\frac{\partial u}{\partial y}\right)^2$	$\propto - W^{\frac{1}{2}} k$
W	$\propto \mu_t \left(\frac{\partial}{\partial y}\,\frac{\partial u}{\partial y}\right)^2$	$\propto - W^{\frac{1}{2}} W$
g	$\propto \mu_t \left(\frac{\partial f}{\partial y}\right)^2$	$\propto - W^{\frac{1}{2}} g$

y ≡ distance, u ≡ velocity, f ≡ concentration
$\mu_t \propto$ turbulent viscosity $\propto k W^{-\frac{1}{2}}$

FIG. 13 The *kWg* turbulence model; some details.

I hope that it emphasizes sufficiently the similarities between the terms for the three variables. Let us look first at the dissipation term, which reveals the parallelism most clearly. *W* has the dimensions of frequency squared, so that its square root has those of the reciprocal of time. We note therefore that the rates of dissipation of *k*, *g*, and *W* can each be regarded as proportional to the local magnitude of the relevant entity, times the characteristic rate of the local turbulence, namely $W^{\frac{1}{2}}$. The generation terms also possess uniformity of structure. Each is the product of the local effective viscosity of the fluid times the square of a gradient; and the gradient is of time–mean velocity, time–mean vorticity, and time–mean concentration, respectively.

The effective viscosity, μ_t, is itself proportional to density times the product of k and $W^{\frac{1}{2}}$; this combination, you can easily work out, has the right dimensions. The various expressions are arrived at by a judicious mixture of analysis, intuition, and empiricism. However we may try to disguise the fact, there is, in turbulence research, no other way forward.

Now for some results. Fig. 14 concerns a simple turbulent flow, the axi-symmetrical jet, injected into stagnant surroundings.[8] There is no chemical reaction in this case; but we have already seen that jets are highly relevant to some combustion processes.

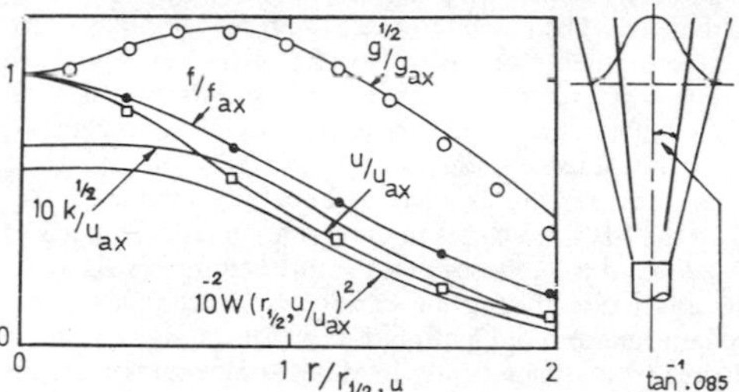

FIG. 14 The kWg turbulence model; results for a jet.

What happens when a stream of fluid is injected into a reservoir of similar fluid is that a conical jet is formed. Some tens of initial diameters from the orifice, the profiles of velocity, concentration, turbulence energy and concentration fluctuations take up fixed shapes, though the breadth and height of each profile varies with longitudinal position.

Fig. 14 shows calculated and experimental shapes of these 'self-similar' profiles. The former are the curves which result from solving the differential equations for k, W and g, together with the corresponding ones for the longitudinal velocity u, and for the time-average concentration f. We have fast standard procedures for doing this now. The experimental data are represented by crosses. You will agree, I think, that the agreement between theory and experiment is fairly satisfactory. It is typical of what we can currently achieve.

4. Some computations of turbulent combustion phenomena

4.1. Some simple flames

Is there any reason for restricting our calculations to non-reacting fluids? None at all. Once we have adopted numerical methods of computation, we might just as well introduce all the factors which promote realism. I showed the non-reacting case first solely because that is the one for which we possess experimental data. Fig. 15 shows what happens when we make the injected fluid a fuel, and suppose that it reacts instantaneously with the first oxygen it encounters. The left-hand diagram shows profiles for the section indicated by the right-hand sketch.[9]

If you think about these profiles, you may at first be puzzled. For, did I not say that fuel and oxygen were

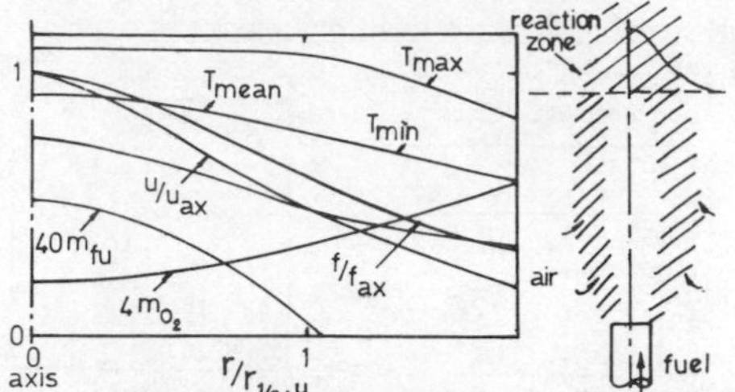

FIG. 15 The kWg turbulence model; the diffusion flame.

presumed to react simultaneously? Yet the profiles show them both to have finite concentrations at the same place. How can this be? The answer is that, in the calculations as in practice, the fuel and oxidant are present at the locations in question *at different times*; this is a consequence of the *fluctuations*, which, of course, we are calculating. Another consequence is that we can draw *three* temperature profiles: one for the time-mean temperature, one for the maximum, and one for the minimum. Fig. 15 shows this feature too.

You may now be asking: can we also calculate flames in which the chemical reaction rate is *not* infinitely fast? The answer is yes; but I must make it clear that, in all these matters, we are at the frontiers of the subject, liable at any time to be confronted with new knowledge that will restore to chaos what we had reduced to order. Let us suppose that chemical kinetics tells us the reaction-rate function, R, of a mixture. This connects the instantaneous rate with the instantaneous temperature and composition. However, it does *not* connect the *time–mean* rate with the *time–mean* temperature and composition.

The problem about reaction rate

Instantaneous rate = $R\{m_{fu}, m_{ox}, T, p\}$,

But time-mean rate $\neq R\{\overline{m}_{fu}, \overline{m}_{ox}, \overline{T}, \overline{p}\}$.

A hypothesis

In fully turbulent flow,
reaction rate cannot exceed

$$\frac{m_{fu}}{k} \times \text{rate of energy (k) dissipation.}$$

FIG. 16 Turbulence-controlled pre-mixed combustion.

How should we calculate this time–mean rate? The lower part of Fig. 16 shows a hypothesis that I am working on. It implies that, in many circumstances, the reaction rate is controlled by the rate at which large lumps of fluid are broken down until they are small enough for the molecular processes to play their part. Why does the turbulence-energy dissipation rate come into this? Because that rate is *also* controlled by the rate of reduction of the fluid lumps to dimensions at which viscous processes can prevail.[10]

What are the implications of this hypothesis? Here are some. Consider, if you will, the process of turbulent flame spread from behind a baffle, when the flow is confined in

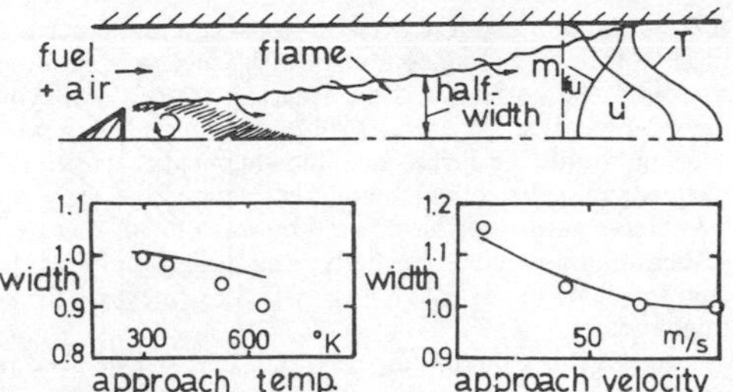

FIG. 17 Flame spread in a duct.

a duct (see Fig. 17). The experimental data show a remarkable feature: the angle of spread of the flame is almost entirely uninfluenced by the temperature and pressure levels, by the fuel–air ratio, by the fuel type, and by the approach velocity. Can our hypothesis explain this? The sketches in the lower half of the diagram show that it can. They reveal that it predicts correctly how the flame width decreases very slightly with increase in both unburnt-gas temperature and approach velocity. The neglect of the 'breakdown-control' of chemical reaction would have led us to predict that the width *rose* with increase in T_u, and decreased much more sharply with increase in V.

We are still just at the beginning of our studies of this topic. I dare not therefore positively assert that the key to turbulent combustion kinetics has been found. However, I am strongly hopeful. I think it likely that this line will throw especially strong light on the pollution problem; for the hypothesis shows how fuel gets through the flame without burning.

4.2. More complex flames

The flames I have just been describing are, from the computational point of view, exercises in the solution of coupled, non-linear, parabolic differential equations. These are easy to solve by a single sweep, in a downstream direction across the field of flow. This 'marching integration' is permissible because no flow reversal is present. Such flames correspond to the first entry in Table 1.

Table 1. Computational features of practical problems

Type	Example	Possible?
2D, steady, no reversal	Axi-symmetrical flame in converging duct	Yes
2D, steady, with reversal	'Gutter' flame holder	Yes
3D, steady, with reversal	Power-station boiler	No
Axi-symmetrical, unsteady	Idealized diesel engine	Yes
3D, unsteady	Gasoline engine	No

We can, however, also make predictions for flows *with* flow reversal. These give rise to *elliptic* differential equations. I will show some examples shortly. The distinctions between problems which affect their solubility from the computational point of view are: steadiness or unsteadiness; and two- or three-dimensionality. Generally speaking, we are better at steady than at unsteady calculations; but it is just a matter of time before this distinction will cease to be significant. Axi-symmetrical processes count as two-dimensional, I should say. It is the three-dimensional ones with flow reversal that are most difficult. No research worker should fear that progress with them is so far advanced that there will be no contribution for him to make.

Fig. 18 shows an example of an axi-symmetrical turbulent combustion-chamber flow which exhibits a flow reversal. You see two examples: in the top one, the fuel and the air enter with only axial components to their velocity; in the lower one, the air possesses also a strong component of swirl about the symmetry axis. The diagram shows both streamlines and isotherms. The especially thickened isotherm corresponds to the maximum temperature in the field, and so indicates the location of the flame. The calculations, I should mention, were carried out some years ago by my colleague, Dr. Pun.[11] The turbulence model which we then used was rather simple.

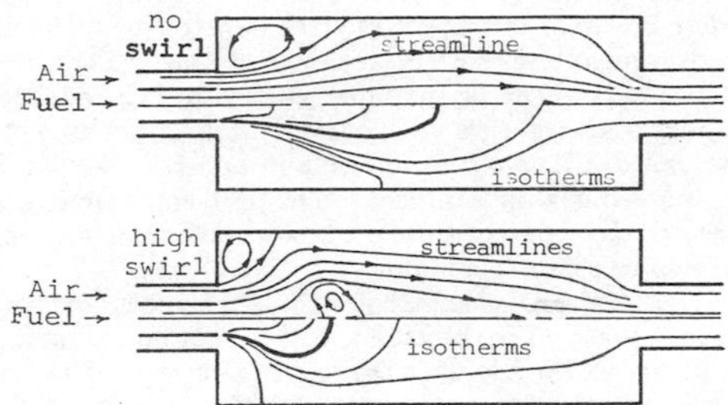

FIG. 18 A diffusion flame with flow reversal.

I think that the most interesting feature of Fig. 18 is that it shows how strong the effect of swirl can be. The calculations predict, in agreement with experiment, that a pronounced swirl provokes an additional recirculation region, near the axis of the chamber. This modifies the flow pattern profoundly; among other things, it shortens the flame.

Fig. 19 shows another example[12] in which, however, the process is chemically rather than physically controlled, because the fuel and air are premixed before entry. The flow is again turbulent, but steady apart from random fluctuations. Those of you who are familiar with flame-stabilization phenomena will recognize the predictions as being highly plausible. There is a region of recirculation just behind the baffle; this acts as a flame-stabilizer, from which the hot gas spreads into the centre of the duct.

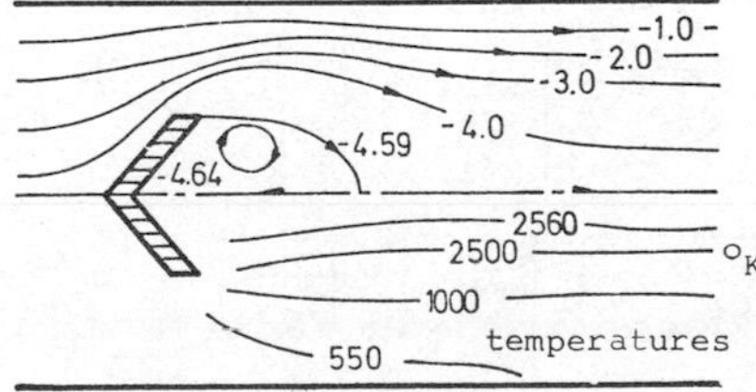

FIG. 19 Flame stabilization by a baffle.

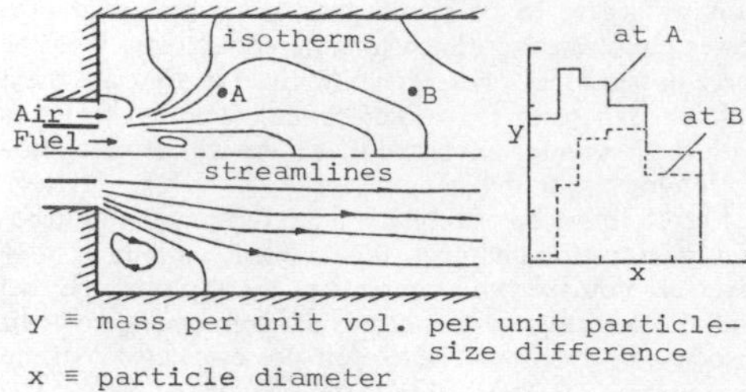

FIG. 20 Pulverized-coal combustion.

So far, we have not had sufficient resources to enable us to refine the turbulence model or to make detailed comparisons of our predictions with the experimental data. No doubt, when we come to do so, we shall have to modify somewhat our present ideas. That, however, is what research is for. We *want* to have them modified, and brought more into conformity with fact.

Even two-phase flows can be handled by existing techniques when we use a little ingenuity. Fig. 20 shows what we can do. It relates to the combustion of a stream of pulverized coal in a furnace.[13] The problem is similar to the one that you saw earlier; but now we have something new to calculate in addition.

The left-hand sketch shows the streamlines and isotherms that you are becoming used to; but on the right is a sketch which relates to the 'something new'. This has coal-particle size as abscissa; and the ordinate is the amount of coal that lies, at a particular point in the furnace, within the various particle-size intervals. What we are calculating, in addition to the usual hydrodynamic and thermal properties, is the particle-size distribution at every point.

This example represents just another beginning, which has not yet been exploited. There is much work still to do in this field, and many applications in engineering. Anyone looking for a rich vein of research to mine could well find what he wants here.

I sometimes think, as I look over the whole field of research in heat transfer and combustion, that we research workers are too timid. We fly too close to the ground, along familiar hedgerows, trying to snap up the same insects that all the other birds are after too. Sometimes we should use our wings to climb a little higher; we may well then discern nearby fields which we can enjoy alone.

Fig. 21 is the result of one such flight: the conception that we might immediately begin to make a mathematical model of diesel-engine combustion.[14] True, we shall have to round the corners a little, treating as truly axi-symmetrical what is, in reality, only partly so. But why not? We can handle the turbulence, the chemical reaction, the unsteadiness, the changing chamber shape, and even the two-phase nature of the flow. Why not put all these together and construct a predictive and design tool that the engineer can really use? Fig. 21 illustrates what we might get.

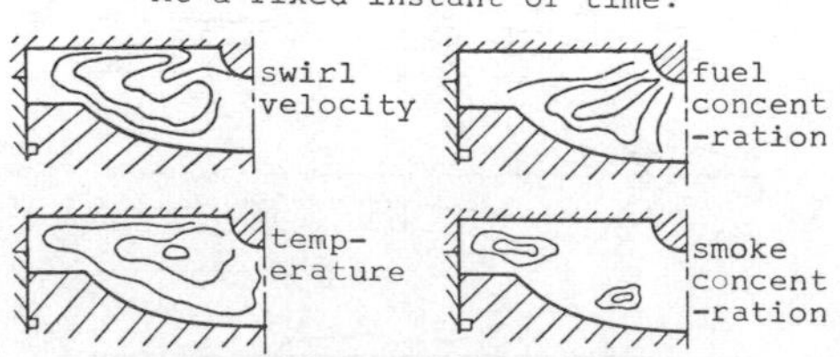

FIG. 21 For the near future, a diesel-engine computation.

I do not think there is any real obstacle. All we need to do is to find some means of gathering the necessary resources and of organizing the co-operative work. When this is done, we may begin to make a really important contribution to the reduction of air pollution.

5. Conclusions

Fig. 22 gives us a summary of what we can currently achieve. Our question, you will recall, concerns the extent to which combustion research is really applicable to the problems of real engineering. I am hopeful that you will now agree with me that, even if there has been comparatively little application so far, we are entering a period in which the engineer will be able to make real use of the results of research.

Computational

- Steady 2D problems are now soluble.
- Unsteady 2D problems soon will be.

Turbulence models

- Models with 2 or 3 differential equations for turbulence quantities permit fairly good predictions to be made.

Chemical kinetics

- Turbulence-limited reaction can be calculated.

FIG. 22 Summary of current capabilities.

It is of course the coming of the computer, and of the numerical methods that exploit it, that have made the difference. They allow us to regard as soluble all those problems which possess only two independent variables.[15,16]

The progress with turbulence models is also, in my view, very encouraging. We can now link together, through a few general hypotheses, a very large number of previously disconnected experimental observations. This progress actually disturbs some purists, who seem to believe that what is admittedly an over-simplified model simply *should* not work; and if it does, that it is ungentlemanly to exploit the fact.

Well, if we took that view with the chemical kinetics as well, we should not predict *any* combustion phenomena. Fortunately, our simplified models are fairly realistic in their implications, whatever they are in their assumptions.

In Fig. 23 I have provided a list of some of the things that remain to be done, still under the three main headings of computation, turbulence, and chemical kinetics.

It is three-dimensional problems, typified by the rectangular power-station boiler or the side-valve gasoline engine, which present the major challenge to the numerical analyst. They are *just* soluble, on the largest of today's

Computational
- Methods and computers are needed for 3D problems (steady and unsteady).

Turbulence
- Existing models need refinement;
- Reliable measurements are needed.

Chemical kinetics
- A simple model is needed for soot formation and consumption.

FIG. 23 Tasks for further research.

computers; but great ingenuity will be needed if the expense of making a prediction is not to be excessive. It is not too early to begin work in this field.

In the turbulence field, we should be most benefited by new and reliable experimental data. Especially needed are data of a new type, which allow a direct assessment of the quantitative behaviour of individual terms in the turbulence equations.

Finally, under chemical kinetics I will refrain from stating the obvious, that we need more accurate data of all kinds. Instead, I draw special attention to the kinetics of smoke formation. Here we want a simple model, good enough to guide designers reliably. If we are soon to master air pollution, this is very urgent indeed.

6. References

1. GODSAVE, G. A. E. Studies of the combustion of drops in a fuel spray—the burning of single drops of fuel. Fourth Symposium on Combustion, p. 818. Williams and Wilkins, Baltimore, 1953.
2. SPALDING, D. B. The combustion of liquid fuels. Fourth Symposium on Combustion, p. 847. Williams and Wilkins, Baltimore, 1953.
3. MULLINS, B. P. Studies on the spontaneous ignition of fuels injected into a hot air stream. *Fuel*, 1953, **32**, 214.
4. LONGWELL, J. P., and WEISS, M. A. Heat release rates in hydrocarbon combustion. I. Mech. E./ASME Joint Conference on Combustion, 1955, 334 to 340.
5. SPALDING, D. B., STEPHENSON, P. L., and TAYLOR, R. G. A calculation procedure for the prediction of laminar flame speeds. Imp. Coll. Mech. Eng. Dept. Rep., 1970.
6. SPALDING, D. B. The prediction of two-dimensional steady, turbulent, elliptic flows. Imp. Coll. Mech. Eng. Dept. Rep. EF/TN/A/16., 1969. (Paper delivered at International Seminar on Heat and Mass Transfer in Flows with Separated Regions, Herceg-Novi, Yugoslavia, Sept. 1969.)
7. SPALDING, D. B. Turbulence models and the theory of the boundary layer. Imp. Coll. Mech. Eng. Dept. Rep. BL/TN/A/24, 1969.
8. SPALDING, D. B. Concentration fluctuations in a round turbulent free jet. Imp. Coll. Mech. Eng. Dept. Rep. BL/TN/A/30, 1970. (To be published in *Chem. Eng. Sci.*)
9. SPALDING, D. B. Mathematische Modelle turbulenter Flammen. V.D.I. Tagung Verbrennung und Feuerungen, Karlsruhe, October 1969. (To be published in V.D.I.-Berichtsheft, 1970.)
10. SPALDING, D. B. Mixing and chemical reaction in steady confined turbulent flames. Thirteenth Symposium on Combustion, Salt Lake City, August 1970.
11. PUN, W. M., and SPALDING, D. B. A procedure for predicting the velocity and temperature distributions in a confined, steady, turbulent, gaseous, diffusion flame. Imp. Coll. Mech. Eng. Dept. Rep. SF/TN/11. Also presented at the International Astronautical Federation Meeting at Belgrade, 1967 (September).
12. ODLOZINSKI, G. A procedure for predicting the distributions of velocity and temperature in a flame stabilized behind a bluff body. Imp. Coll. Mech. Eng. Dept. Rep. EF/R/G/2, October 1968.
13. GIBSON, M. M., and MORGAN, B. B. Mathematical model of combustion of solid particles in a turbulent stream with recirculation. BCURA Research Report No. 361, October 1969. BCURA Industrial Laboratories, Leatherhead, Surrey.
14. SPALDING, D. B. Predicting the performance of diesel-engine combustion chambers. Closing lecture at Institution of Mechanical Engineers Symposium on Diesel-Engine Combustion, London, 1970.
15. GOSMAN, A. D., PUN, W. M., RUNCHAL, A. K., SPALDING, D. B., and WOLFSHTEIN, M. Heat and mass transfer in recirculating flows. Academic Press, London, 1969.
16. PATANKAR, S. V., and SPALDING, D. B. Heat and mass transfer in boundary layers. Second edition. Intertext Books, London, 1970.

(*Paper received* 10*th August*, 1970.)

NOTE ADDED IN OCTOBER 1976

The six years since the presentation of this paper have brought great advances in the application of combustion theory to engineering processes. It is now commonplace to employ three-dimensional numerical models of steady-flow combustion processes; great advances have been made in the knowledge of chemical kinetics; and turbulence modelling has advanced to the point at which the hydrodynamic aspects of combustor flows can be well predicted. Relevant references may be found in the list starting on page T 1,8; they include:-Zuber (1972) and Patankar and Spalding (1973,1974).

Difficulties remain, especially in connexion with the quantitative understanding of the interactions of turbulence and chemistry. Reviews of the topic may be found in:

Murthy S N B (Ed.) "Turbulent mixing in non-reactive and reactive flows", Plenum Press, New York, 1975; and Spalding D B "The ESCIMO theory of turbulent combustion", Imperial College, Mechanical Engineering Department Report No. HTS/76/13, 1976. See also Bracco F V (Ed.) "Turbulent reactive flows", Special issue of Combustion Science and Technology, Vol. 13, Nos. 1-6, pages 1-275, 1976.

EXERCISES TO FACILITATE ABSORPTION OF THE MATERIAL OF CHAPTER 1

INSTRUCTIONS

Each of the following problems (1.1, 1.2, etc) is to be marked A, B, C, D or E.

For sentence-completion problems, the mark is to indicate which alternative completion is the best. For linked-statement problems the marks should be ascribed according to the following rule (S1 ≡ first statement, S2 ≡ second statement, A ≡ argument implied by the link-word, "because"):

If S1 true,	S2 true,	A true,	Mark	: A
If S1 true,	S2 true,	A false,	Mark	: B
If S1 true,	S2 false,		Mark	: C
If S1 false,	S2 true,		Mark	: D
If S1 false,	S2 false,		Mark	: E

1.1 The designer of a gasoline-engine combustion chamber attempts to reduce its tendency to "knock"
<u>because</u>
gasoline engines are major contributors to atmospheric pollution by oxides of nitrogen.

1.2 Diesel engines may produce exhaust smoke when producing maximum design torque
<u>because</u>
the fuel/air ratio in the cylinder is fuel-rich under these conditions.

1.3 Diesel engines are likely to increase their share of the automotive market
<u>because</u>
fuel prices are increasing.

1.4 The designer of the combustor of an aircraft gas turbine attempts to reduce all of the following except:
<u>A</u> the amount of smoke in the exhaust.

B the pressure drop across the combustor.
C the temperature of the combustor walls.
D the non-uniformity of temperature in the gases leaving the combustor.
E the range of fuel-air ratios within which the flame will stay alight.

1.5 The designer of an industrial furnace usually attempts to do all of the following except:
A ensure that the bulk of the fuel is burned at as short a distance from its point of injection as possible
B prevent any fuel from escaping from the furnace without burning.
C reduce the amount of noise generated by the flame.
D make the combustion system capable of operating successfully with fuels from more than one source of supply.
E ensure that the spatial distribution of heat transfer to the material to be heated corresponds to the needs of the process in question.

1.6 Combustion science is well able to meet the needs of the engineering designer
because
all relevant chemical-kinetic properties of materials are now known with sufficient quantitative accuracy.

1.7 The burning rate of a liquid-fuel droplet, suspended in still air, depends significantly upon all the following except:
A the chemical-kinetic properties of the fuel.
B the oxygen content of the air.
C the oxygen requirement of unit mass of fuel.
D the latent heat of vaporisation of the liquid.
E the thermal conductivity of the gaseous sheath around the droplet.

1.8 The height of a turbulent diffusion flame, burning in still air, depends upon all the following except:

A the diameter of the nozzle.

B the density of the fuel gas.

C the velocity of the fuel gas at the injection point.

D the viscosity of the fuel gas.

E the oxygen requirement of unit mass of fuel.

1.9 Dimensional analysis can be applied to combustion phenomena, for the generalisation of experimental data

because

these phenomena are expressions of the laws of thermodynamics, fluid mechanics, heat and mass transfer and chemical kinetics.

1.10 All the following chemical reactions are likely to play a part in the burning of hydrogen and air except:

A $O_2 + H \rightleftharpoons O + OH$

B $H_2 + OH \rightleftharpoons H_2O + H$

C $O_2 + N_2 \rightleftharpoons 2NO$

D $H_2O + O \rightleftharpoons 2OH$

E $2H + M \rightleftharpoons H_2 + M$

1.11 The stirred reactor ("Longwell bomb") is an apparatus characterised by all the following except:

A it operates in steady flow.

B when the flow rates of fuel and air are both sufficiently high, the flame is extinguished.

C chemical reaction rates can be deduced from the extinction flow rates, for various fuel-air ratios.

D These reaction rates are the same as are exhibited by the same fuel burning in air in a practical combustion chamber.

E The extinction flow rate increases more than in proportion to increases in pressure.

1.12 Measurement of the angle of the inner cone of a Bunsen burner can be used as the starting point of a determination of chemical reaction rates

<u>because</u>

all the fuel gas supplied to the burner is consumed in that cone.

1.13 Of the obstacles in the path of making quantitative predictions of practical combustion phenomena, the most serious is:

<u>A</u> inadequacy of current knowledge of chemical kinetics.

<u>B</u> the impossibility of solving the equations numerically for three-dimensional flow.

<u>C</u> inability to predict the detailed behaviour of turbulent fluids.

<u>D</u> transfer of heat by radiation is a non-linear phenomenon.

<u>E</u> there are too few reliable experimental data to permit the validity of the prediction methods to be established.

1.14 The time-average reaction rate in a turbulent flame is related to the time-average concentrations and temperatures at the point in question in the same way as are the instantaneous rate, concentrations and temperature

<u>because</u>

chemical reaction obeys linear laws.

1.15 A mathematical model of a combustion process usually:

<u>A</u> represents completely the behaviour of the practical process.

<u>B</u> provides a set of equations which partially simulate the process, but which are too difficult to solve.

<u>C</u> provides a soluble set of equations, the solution of which adequately represents the behaviour of the real process in practically important aspects.

<u>D</u> is useful for research workers, but not for design and development engineers.

<u>E</u> can completely eliminate the necessity for trial-and-error experimental development.

1.16 A mathematical method of performance prediction is of more use to a design engineer than knowledge of the performance of a large number of existing pieces of equipment, chiefly because:

A calculation gives more accurate results than experimental measurements.

B the results of calculations are usually easier to understand that compilations of experimental data.

C only by theoretical means can the performance be predicted of equipment that has not yet been built.

D mathematical methods are more quickly learned than experimental facts.

E there is no satisfactory terminating clause, because the premise is untrue.

The mathematical model of greatest relevance to:

1.17 Power-station furnaces,

1.18 The gasoline engine,

1.19 The open-hearth furnace,

1.20 The liquid propellant rocket,

1.21 The Bunsen burner,

is

A the free turbulent jet.

B the liquid-fuel droplet in stagnant surroundings

C the one-dimensional pre-mixed laminar flame.

D the single carbon particle in stagnant surroundings.

E Ignition by a spark of gas at rest.

The fuel used in:

1.22 A steam power plant for central electricity generation in the UK,

1.23 An aircraft gas turbine,

1.24 A spark-ignition engine for a road vehicle,

1.25 A compression-ignition engine for a road vehicle,

1.26 An open-hearth furnace for steel manufacture,

1.27 A gas turbine for use on land,
is usually:
A residual oil.
B gasoline.
C gas oil.
D kerosine.
E coal.

The main influence on the choice of fuel for:

1.28 Steam power plants for central power stations in the UK,

1.29 Aircraft gas turbines,

1.30 Spark-ignition engines,

1.31 Compression-ignition engines,

1.32 Open-hearth furnaces,

1.33 Land gas turbines,
is:
A politics and tradition.
B the fuel must be the cheapest one that is easily pumped, and must not generate solid or liquid ash.
C the fuel must vaporise at low temperatures, and burn without solid residue.
D the fuel must be cheap, and easily pumped.
E the fuel must vaporise fairly easily, but not so easily as to be set alight by stray sparks, it must burn without solid residue.

1.34 Coal tends every year to provide a smaller fraction of the fuel needs of the United Kingdom
because
its supply cannot keep pace with the demand.

1.35 Mathematical models are becoming less useful to combustion designers
because
digital computers can now cheaply perform very large numbers of arithmetic calculations.

1.36 The ignition of a pre-mixed gas by a spark is likely to be more difficult to predict mathematically than steady propagation of a plane laminar flame
because
it involves partial rather than ordinary differential equations.

The fundamental science which must be referred to in order to calculate:

1.37 The temperature difference between the inlet and the outlet of an adiabatic combustion chamber in which combustion is complete,

1.38 The temperature taken up by the water-cooled walls of a boiler furnace,

1.39 The ignition delay of a gaseous mixture of fuel and air, suddenly compressed by means of a piston,

1.40 The rate at which oxygen is transferred to the surface of a burning carbon particle,

1.41 The flow pattern of gas in a combustion chamber to which the fuel and air are supplied from jets,
is called:

A fluid mechanics.
B chemical kinetics.
C heat transfer.
D mass transfer.
E thermodynamics.

1.42 Natural gas is likely to supply an increasing fraction of the United Kingdom's fuel
because
the products of its combustion are comparatively clean.

ANSWERS TO MULTIPLE-CHOICE PROBLEMS

Answer	Problem number (1's omitted)									
A	3,	5,	7,	13,	19,	26,	28,	36,	41	
B	1,	9,	20,	24,	31,	33,	39,	42		
C	2,	10,	12,	15,	16,	21,	25,	27,	30,	34, 38
D	8,	11,	17,	22,	23,	32,	35,	40		
E	4,	6,	14,	18,	22,	29,	37			

DISCUSSION PROBLEMS

1.43 Of the problems to which you at first gave answers different from those in the table,

a) which did you subsequently feel you should have answered correctly, on the basis of information available to you?

b) which were those which you could not reasonably have been expected to answer correctly?

c) which were those to which you believe the answers in the table are either wrong, or not the only acceptable ones?

Explain your answer to question c, case by case.

1.44 The lecture "Combustion as Applied to Engineering" was prepared early in 1970. Explain in what ways the relation of combustion science to engineering has been altered, since that date, by at least two of:

a) changes in the fuel-supply situation.

b) developments in computational techniques.

c) improvements in knowledge of physics and chemistry.

d) improvements in methods of measurement.

e) development of new ways of using fuels.

CHAPTER 2

MASS TRANSFER I

2.1 NATURE AND OUTLINE OF SUBJECT

A) NATURE

Mass transfer is the science of the change of composition of mixtures by molecular and turbulent interdiffusion.

Examples of this interdiffusion include:

- gas from a tap spreads out into all parts of the room.
- carbon packed around iron in "case-hardening" diffuses into metal to make steel.
- fresh water from a river mixes with salt water from the ocean and dilutes it.
- sugar dissolves in tea (with change of phase).
- water condenses from a steam-air mixture.
- solid carbon dioxide (dry ice) sublimes and mixes with air.

B) OUTLINE

Study of mass transfer necessitates attention to the following topics:

- Definitions (of names of aspects of the phenomena, eg diffusion, concentration).
- Laws (ie empirically observed regularities).
- Material properties.
- Differential equations describing general processes.
- Boundary conditions and solutions relating to particular processes.
- Employment in analysis, prediction, design.

2.2 PRACTICAL IMPORTANCE OF SUBJECT

Among the mass-transfer processes of practical concern are:

- In Mechanical Engineering: combustion of fuels for power generation; oxygen-cutting of steel; corrosion of metals; condensation of steam.

- In Chemical Engineering: fractional distillation of petroleum to yield gasoline, kerosine, gas oil and residual oil; formation of ammonia from nitrogen and hydrogen on the surface of a catalyst; extraction of sulphur and other impurities in lubricants.

- In Metallurgical Processing: removal of carbon from cast iron to form steel; reduction of iron ore in a blast furnace.

- In Aeronautical Engineering: formation of ice on aircraft wings; the "ablation" of the heat shield of a re-entering space vehicle.

2.3 DEFINITIONS

- Density ≡ mass of mixture per unit volume ≡ ρ (kg/m³)

- Species ≡ chemically distinct substances, H_2O, H_2, H, O_2, etc.

- Partial density of A ≡ mass of chemical compound (species) A per unit volume ≡ ρ_A (kg/m³).

- Mass fraction of A ≡ $\rho_A/\rho \equiv m_A$. (-)

Note that the definitions imply:

$$\rho_A + \rho_B + \rho_C + \ldots\ldots\ldots = \rho \; ; \qquad (2.3\text{-}1)$$

and

$$m_A + m_B + m_C + \ldots\ldots\ldots = 1 \; . \qquad (2.3\text{-}2)$$

- Total mass velocity of mixture in a specified direction ≡ mass of mixture crossing unit area normal to that direction in unit time ≡ G_{tot} (kg/m²s), later called simply G.

- Total mass velocity of chemical species A in this specified direction ≡ mass of A ...... etc ≡ $G_{tot,A}$.
  This implies that:
  $$G_{tot,A} + G_{tot,B} + \ldots\ldots\ldots = G_{tot} \quad . \tag{2.3-3}$$
- Convective mass velocity of A in specified direction ≡ $m_A\ G_{tot}$ ≡ $G_{conv,A}$.
  This implies that:
  $$G_{conv,A} + G_{conv,B} + \ldots\ldots = G_{tot} \quad ; \tag{2.3-4}$$
  but, in general,
  $$G_{conv,A} \neq G_{tot,A} \quad . \tag{2.3-5}$$
- Diffusive mass velocity of A in specified direction ≡ $G_{tot,A} - G_{conv,A}$ ≡ $G_{diff,A}$. (2.3-6)
  This implies that:
  $$G_{diff,A} + G_{diff,B} + \ldots\ldots = 0 \quad . \tag{2.3-7}$$
- Velocity of mixture in the specified direction ≡ G_{tot}/ρ ≡ V.
  It is possible, but not especially useful, to define "velocities" for individual components; it will not be done here.
- Concentration: a word used loosely for either partial density or mass fraction (or mole fraction, partial pressure, etc).
- Composition of mixture ≡ the set of mass fractions.

2.4 EXPERIMENTAL LAW (THE BASIS OF MASS TRANSFER)

Fick's law of diffusion:

$$G_{diff,j} \equiv -\Gamma_j \frac{\partial m_j}{\partial x} \quad , \tag{2.4-1}$$

where: Γ_j ≡ "exchange coefficient of j in the mixture",
x ≡ distance in direction in which G is measured.

Remarks:

(a) This is similar to Fourier's and Ohm's laws; it is linear.

(b) The minus sign is a consequence of convention. It means that matter diffuses from high concentration to low.

(c) Γ_j is more usually written as $\mathcal{D}_j\rho$, where $\mathcal{D}_j$ is "diffusion coefficient of j in the mixture".

(d) Dimensions of Γ_j are kg/m s; those of $\mathcal{D}_j$ are m^2/s.

(e) In gases in laminar motion or at rest, Γ_j is of same order as μ; $\mathcal{D}_j$ is of same order as ν.
For liquids, $\Gamma_j \ll \mu$; for solids, Γ_j is very small indeed.
$\mu \equiv$ dynamic viscosity; $\gamma \equiv$ kinematic viscosity.

(f) For laminar gases, Γ_j is of the order of $\rho v_{mol}\lambda$ where v_{mol} is the molecular velocity, λ is the mean free path.
For turbulent liquids or gases, Γ_j is of order $\rho v_{turb}\,\ell$, where v_{turb} is a representative velocity of turbulence, and ℓ is the mean eddy size.

(g) The law is approximate: in some circumstances $G_{diff,j}$ can be finite when $\partial m_j/\partial x = 0$; for gradients of temperature, of pressure, and of mass fractions of <u>other</u> components, can provoke diffusion of j.

2.5 DIFFERENTIAL EQUATION FOR m_j IN ONE-DIMENSIONAL FLOW

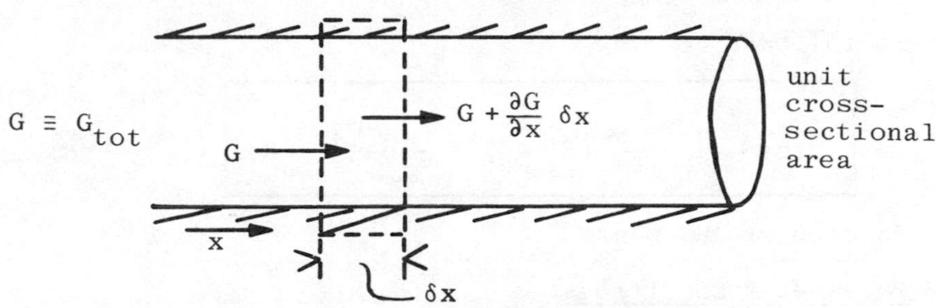

A) LAW OF MASS CONSERVATION (IRRESPECTIVE OF CHEMICAL SPECIES)

This law implies that:

	Inflow from left		$G\delta t$
-	Outflow to right	-	$(G + \frac{\partial G}{\partial x}\delta x)\,\delta t$
=	Increase of mass content	=	$\frac{\partial \rho}{\partial t}\,\delta t\,\delta x$.

Hence $$\frac{\partial G}{\partial x} = -\frac{\partial \rho}{\partial t} \quad . \qquad (2.5\text{-}1)$$

For steady flow (in which $\partial/\partial t$ terms are zero):

$$\frac{dG}{dx} = 0 \quad ; \qquad (2.5\text{-}2)$$

i.e. G is independent of x as well as t.

B) LAW OF CONSERVATION OF CHEMICAL SPECIES j

Similarly, $$\boxed{\frac{\partial G_{tot,j}}{\partial x} = -\frac{\partial \rho_j}{\partial t} + R_j} \quad , \qquad (2.5\text{-}3)$$

where $R_j \equiv$ rate of production of species j per unit volume, by chemical reaction.

C) DIFFERENTIAL EQUATION FOR m_j

Substitution of the appropriate definitions leads to:

$$m_j \frac{\partial G_{tot}}{\partial x} + G_{tot} \frac{\partial m_j}{\partial x} - \frac{\partial}{\partial x}\left(\Gamma_j \frac{\partial m_j}{\partial x}\right)$$
$$= -\left(m_j \frac{\partial \rho}{\partial t} + \rho \frac{\partial m_j}{\partial t}\right) + R_j \quad . \qquad (2.5\text{-}4)$$

Since $$\frac{\partial G_{tot}}{\partial x} = -\frac{\partial \rho}{\partial t} \quad , \qquad (2.5\text{-}1)$$

it follows:

$$\boxed{G_{tot} \frac{\partial m_j}{\partial x} - \frac{\partial}{\partial x}\left(\Gamma_j \frac{\partial m_j}{\partial x}\right) = -\rho \frac{\partial m_j}{\partial t} + R_j} \quad . \qquad (2.5\text{-}4)$$

Particular cases are:

- For steady conditions:

$$G_{tot} \frac{dm_j}{dx} - \frac{d}{dx}\left(\Gamma_j \frac{dm_j}{dx}\right) = R_j \quad . \qquad (2.5\text{-}5)$$

- For steady no-reaction conditions:

$$G_{tot} \frac{dm_j}{dx} - \frac{d}{dx}\left(\Gamma_j \frac{dm_j}{dx}\right) = 0 \quad . \qquad (2.5\text{-}6)$$

- <u>If also Γ_j = constant:</u>

$$G_{tot}\frac{dm_j}{dx} - \Gamma_j\frac{d^2m_j}{dx^2} = 0 \quad . \qquad (2.5\text{-}7)$$

- <u>The integral form of steady-state equation is:</u>

$$G_{tot}m_j - \Gamma_j\frac{dm_j}{dx} = \int R_j dx + \text{const} \quad . \qquad (2.5\text{-}8)$$

2.6 THE STEFAN-FLOW PROBLEM

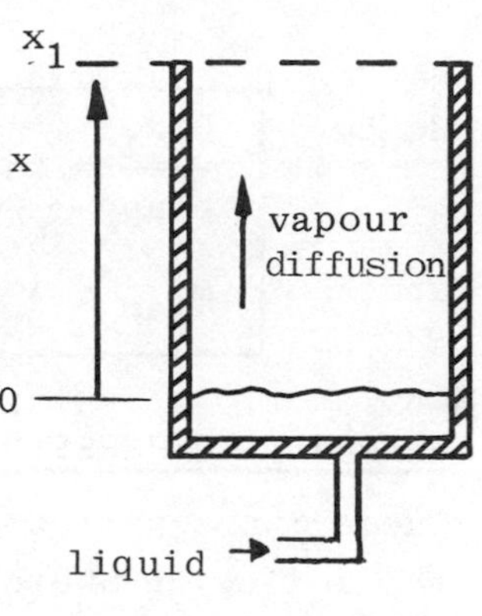

A) PHYSICAL SITUATION

- Steady state.
- Vapour diffuses upwards and escapes.
- Air does not dissolve in liquid.
- Γ_j = uniform.
- There is no reaction.

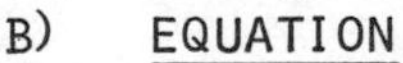

B) EQUATION

$$G\frac{dm_{vap}}{dx} - \Gamma_{vap}\frac{d^2m_{vap}}{dx^2} = 0 \quad . \qquad (2.6\text{-}1)$$

C) BOUNDARY CONDITIONS

$$x = 0 \;:\; m_{vap} = m_{vap,0} = m_{vap,sat} \quad , \qquad (2.6\text{-}2)$$

$$x = x_1 \;:\; m_{vap} = m_{vap,1} \quad . \qquad (2.6\text{-}3)$$

D) SOLUTION

- <u>First integration yields:</u>

$$G\,m_{vap} - \Gamma_{vap}\frac{dm_{vap}}{dx} = \text{const} = G_{tot,vap} \quad . \qquad (2.6\text{-}4)$$

Note that only the vapour moves; so

$$G_{tot,vap} = G \text{ in this case} \quad ; \qquad (2.6\text{-}5)$$

and $G_{tot,air} = 0$. (2.6-6)

- Second integration yields:

$$\Gamma_{vap} \frac{dm_{vap}}{dx} = G(m_{vap} - 1) \quad ; \qquad (2.6\text{-}7)$$

therefore

$$\ell n(m_{vap} - 1) = \frac{G}{\Gamma_{vap}} x + \text{const} \quad . \qquad (2.6\text{-}8)$$

- Insertion of boundary conditions now leads to:

$$\ell n\ (m_{vap,0} - 1) = \text{const} \quad ; \qquad (2.6\text{-}9)$$

and $$\ell n\ (m_{vap,1} - 1) = G\ x_1 / \Gamma_{vap} + \text{const} \quad . \qquad (2.6\text{-}10)$$

Hence $$\boxed{\frac{G\ x_1}{\Gamma_{vap}} = \ell n \left\{ 1 + \frac{m_{vap,0} - m_{vap,1}}{1 - m_{vap,0}} \right\}} \quad , \qquad (2.6\text{-}11)$$

and $$\boxed{m_{vap} = 1 - (1-m_{vap,0})\ \exp\ (Gx/\Gamma)} \quad . \qquad (2.6\text{-}12)$$

E) GRAPHICAL EXPRESSION

These two equations imply relations having the forms of the following sketches.

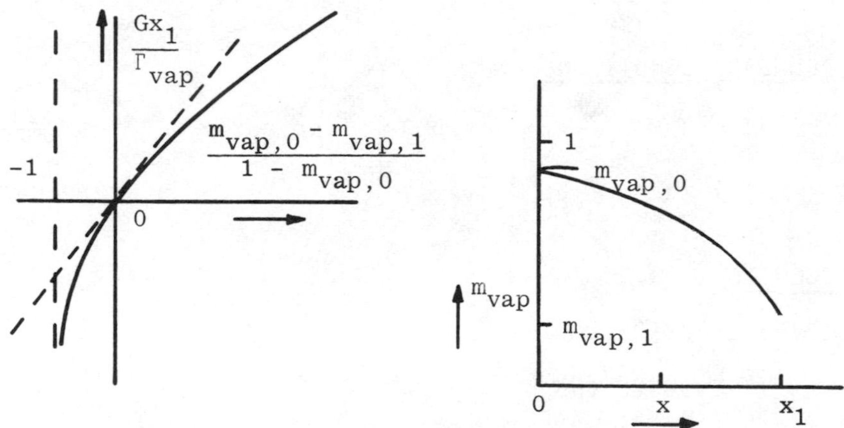

F) REMARKS

- The mass-transfer (vaporisation) rate increases with the magnitude of the concentration difference $m_{vap,0} - m_{vap,1}$.

- The rate is <u>proportional</u> to the difference only at small values.
- The diffusive and convective fluxes vary with x, their sum being constant.

2.7 REFERENCES

BIRD R B, STEWART W E and LIGHTFOOT E N
"Transport Phenomena"
Wiley, New York, 1960

HIRSCHFELDER J O, CURTIS C F and BIRD R B
"The Molecular Theory of Gases and Liquids"
Wiley, New York, 1956

KAYS W M
"Convective Heat and Mass Transfer"
McGraw Hill, New York, 1966

SHERWOOD T K and PIGFORD R L
"Absorption and Extraction"
McGraw Hill, New York, 1952

SPALDING D B
"Convective Mass Transfer"
Edward Arnold, London, 1963

WELTY J R, WICKS C E and WILSON R E
"Fundamentals of Momentum, Heat and Mass Transfer"
Wiley, New York, 1969

EXERCISES TO FACILITATE ABSORPTION OF THE MATERIAL OF CHAPTER 2

ANALYTICAL PROBLEMS

2.1 Derive the following differential equation for m_j, where x, y, and z are three Cartesian co-ordinate directions, and the mass velocities of the mixture in these directions are given the symbols G_x, G_y, G_z.

State whether the equation holds for steady or unsteady conditions.

$$G_x \frac{\partial m_j}{\partial x} + G_y \frac{\partial m_j}{\partial y} + G_z \frac{\partial m_j}{\partial z} - \frac{\partial}{\partial x}\left(\Gamma_j \frac{\partial m_j}{\partial x}\right) - \frac{\partial}{\partial y}\left(\Gamma_j \frac{\partial m_j}{\partial y}\right) - \frac{\partial}{\partial z}\left(\Gamma_j \frac{\partial m_j}{\partial z}\right) = R_j$$

2.2 When air blows over a warm-water surface, a boundary layer of retarded, warmed and moistened air forms above the surface. Early analysts of the vaporisation process treated the boundary layer as a "stagnant film" of gas adhering to the liquid surface; the rate of vaporisation was calculated using the Stefan-flow equations.

In a particular experimental situation, the rate of vaporisation from the water surface is found to be 4×10^{-4} kg/m s; the air immediately at the water surface has a water vapour content of .0218 kg H_2O/kg mixture, while in the bulk of the air stream the value of m_{vap} is .0052. If $\Gamma_{vap} = 2 \times 10^{-5}$ kg/m s at the temperature in question, what is the thickness of the equivalent "stagnant film" of air?

2.3 Show that, when Γ_{vap} varies with x, the mass-transfer rate in a Stefan-flow situation can be expressed by:

$$G = \frac{\ln\left[1 + \dfrac{m_{vap,0} - m_{vap,1}}{1 - m_{vap,0}}\right]}{\int_0^{x_1} \Gamma_{vap}^{-1}\, dx}$$

2.4 Derive a formula for the mass-transfer rate in a Stefan-flow situation for which $\Gamma_{vap} = a + b\, m_{vap}$, where a and b are constants.

ANSWERS TO ANALYTICAL PROBLEMS

2.1 -

2.2 0.00085 m

2.3 -

2.4
$$G = \frac{1}{x_1}\left[b(m_{vap,0} - m_{vap,1}) + (a+b)\,\ell n\left\{1 + \frac{m_{vap,0} - m_{vap,1}}{1 - m_{vap,0}}\right\}\right]$$

MULTIPLE-CHOICE PROBLEMS

Instructions as for exercises of Chapter 1.

2.5 All the following processes involve mass transfer except:

- A water is transformed into steam in a boiler;
- B water condenses from a steam-air mixture in a condenser;
- C oxygen and nitrogen are separated by the fractional distillation of liquid air;
- D steel is manufactured from molten cast-iron by the blowing through it of a gas containing oxygen;
- E a flame, ignited by a spark, spreads through a stoichiometric mixture of hydrogen and oxygen.

It is a consequence of the relevant definitions that the following expressions:

2.6 $\rho_A + \rho_B + \rho_C + \ldots\ldots\ldots\ldots\ldots\ldots\ldots$ (all components)

2.7 $G_{diff,A} + G_{diff,B} + \ldots\ldots\ldots\ldots\ldots$ (all components)

2.8 $G_{tot,A} + G_{tot,B} + \ldots\ldots\ldots\ldots\ldots\ldots$ (all components)

2.9 $m_A + m_B + \ldots\ldots\ldots\ldots\ldots\ldots\ldots\ldots$ (all components)

2.10 $-G_{tot} + G_{conv,A} + G_{conv,B} + \ldots\ldots\ldots$ (all components)

equal:

A 1;

B -1;

C 0;

D a positive quantity, not necessarily 1;

E a quantity which may be positive or negative.

2.11 All the following statements about Fick's law of diffusion are true except:

A its mathematical form is: $G_{diff,j} = \Gamma_j \frac{\partial m_j}{\partial x}$

B Γ_j is not necessarily independent of the direction of the diffusion flux.

C In gases, Γ_j is of the same order as the viscosity.

D In non-turbulent liquids, Γ_j is very much smaller than the viscosity.

E In non-turbulent liquids, Γ_j is very much smaller than the thermal conductivity divided by the specific heat.

In the equation:

$$G_{tot} \frac{\partial m_j}{\partial x} - \frac{\partial}{\partial x}\left(\Gamma_j \frac{\partial m_j}{\partial x}\right) = -\rho \frac{\partial m_j}{\partial t} + R_j$$

2.12 the first term,

2.13 the second term,

2.14 the third term,

2.15 the fourth term,

2.16 the whole left-hand side,

represents:

A the source of j, eg by chemical reaction.

B the diffusion term.

C the gradient of the total flux of j.

D the gradient of the convective flux of j.

E the transient term.

Of the following equations relating to the Stefan-flow problem, the equation:

2.17 $m_{vap} = 1 - (1 - m_{vap,0}) \exp (Gx/\Gamma)$,

2.18 $\Gamma_{vap} \dfrac{dm_{vap}}{dx} = G\,(m_{vap} - 1)$,

2.19 $G \dfrac{dm_{vap}}{dx} + \Gamma_{vap} \dfrac{d^2 m_{vap}}{dx^2} = 0$,

2.20 $\dfrac{Gx_1}{\Gamma_{vap}} = \ell n \left\{ 1 + \dfrac{m_{vap,0} - m_{vap,1}}{1 - m_{vap,0}} \right\}$,

2.21 $x = 0{:}\quad m_{vap} = m_{vap,sat}$.

<u>represents</u>:

<u>A</u> an invalid statement.

<u>B</u> an expression for the vapour-concentration profile.

<u>C</u> a differential equation expressing the constancy of the total flux.

<u>D</u> a formula from which the mass-transfer rate can be determined.

<u>E</u> a boundary condition.

Classify the following statements, all of which relate to a gaseous mixture of two substances, A and B, according to whether they represent:

<u>A</u> definitions.

<u>B</u> correct logical deductions from definitions.

<u>C</u> incorrect logical deductions from definitions.

<u>D</u> statements of experimental findings having a high degree of truth.

<u>E</u> statements of experimental findings having only a moderate degree of truth.

2.22 $G_{diff,A} = G_{tot,A} - G_{conv,B}$.

2.23 $G_{tot,A} = m_A(G_{tot,A} + G_{tot,B}) - G_{diff,A}$.

2.24 $G_{diff,A} = - \Gamma_A \dfrac{\partial m_A}{\partial x}$.

2.25 $G_{conv,A} = m_A(G_{tot,A} + G_{tot,B})$.

2.26 Γ_A = viscosity of gaseous mixture.

2.27 $G_{diff,A} + G_{diff,B} = 0$

2.28 $\Gamma_A = \mathcal{D}_A \rho$

ANSWERS TO MULTIPLE-CHOICE PROBLEMS

Answer	Problem number (2's omitted)						
A	5,	9,	11,	15,	19,	25,	28
B	13,	17,	22,	27			
C	7,	10,	16,	18,	23		
D	6,	12,	20,	24			
E	8,	14,	21,	26			

DISCUSSION PROBLEMS

2.29 As for exercise 1.43.

2.30 Equation (2.6-7) appears to imply that the diffusion rate is equal to:

$$\frac{-\Gamma_{vap}}{1-m_{vap}} \cdot \frac{dm_{vap}}{dx} ,$$

which differs from the Fick's law expression by the presence of $(1-m_{vap})$ in the denominator. Explain this contradiction.

2.31 Supply a nomenclature list for Chapter 2, defining all novel quantities. (Note: This exercise is recommended to the reader in respect of every chapter; but it will not be specifically mentioned again.)

CHAPTER 3
DROPLET VAPORISATION I

3.1 OUTLINE OF PROBLEM

A) THE PHENOMENON CONSIDERED

A small sphere of liquid in an infinite gaseous atmosphere vaporises and finally disappears.

B) PRACTICAL IMPORTANCE

- Vaporisation is a prerequisite for burning liquid fuel in: gasoline, diesel and gas-turbine engines; in furnaces, industrial and (many kinds of) domestic heaters.
- It occurs in the drying of milk powder.
- It occurs in the de-superheaters of steam plant.
- Fires may be extinguished by a cloud of droplets.
- Fog dispersal may involve vaporisation.

C) WHAT IS TO BE PREDICTED

Mathematical expressions are required which express the quantitative influence, on the time of vaporisation, of the properties of the liquid, vapour and atmosphere.

3.2 MATHEMATICAL MODEL

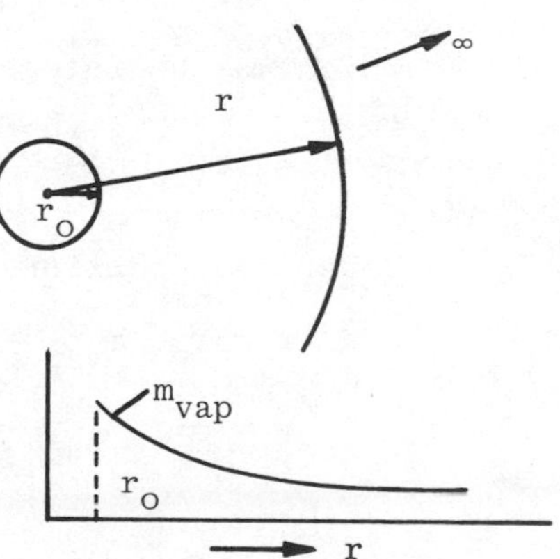

A) DESCRIPTION AND ASSUMPTIONS

The model is defined as follows:

- spherical symmetry (non-radial motion neglected);
- (quasi-) steady state in gas;
- Γ_{vap} independent of radius;
- large distance between droplets;
- no chemical reaction.

B) CALCULATION OF THE DISTRIBUTION OF VAPOUR CONCENTRATION, m_{vap}, IN THE GAS

The analysis proceeds as follows:

(i) Mass Conservation

$$Gr^2 = G_o r_o^2 , \quad (3.2\text{-}1)$$

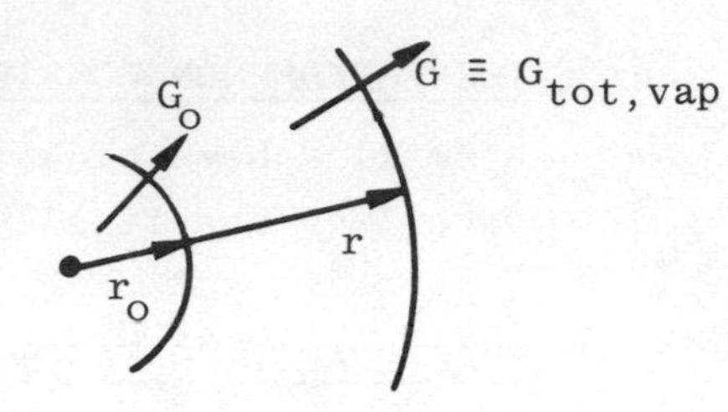

where

$G_o \equiv$ rate of phase change of liquid per unit surface area

$r_o \equiv$ radius of droplet surface.

(ii) Conservation of Vapour

Because there is no loss of vapour by chemical reaction:

$$G_{tot,vap} r^2 = G_{tot,vap,o} r_o^2$$

$$= G_o r_o^2 \text{ in this case} \quad . \quad (3.2\text{-}2)$$

Therefore from Fick's Law:

$$\left(m_{vap} G - \Gamma_{vap} \frac{dm_{vap}}{dr} \right) r^2 = G_o r_o^2 \quad , \quad (3.2\text{-}3)$$

ie

$$\Gamma_{vap} \frac{dm_{vap}}{dr} r^2 = G_o r_o^2 (m_{vap} - 1) \quad . \quad (3.2\text{-}4)$$

C) SOLUTION

- Rearrangement leads to:

$$\frac{dm_{vap}}{m_{vap} - 1} = \frac{G_o r_o^2}{\Gamma_{vap}} \cdot \frac{dr}{r^2} \quad , \quad (3.2\text{-}5)$$

which may be integrated to:

$$\ln(m_{vap} - 1) = \frac{G_o r_o^2}{\Gamma_{vap}} \left(\frac{-1}{r} \right) + \text{const} \quad . \quad (3.2\text{-}6)$$

- Boundary conditions are:

$$r = r_o : m_{vap} = m_{vap,o} \quad , \quad (3.2\text{-}7)$$

$$r = \infty : m_{vap} = m_{vap,\infty} \quad , \quad (3.2\text{-}8)$$

The solution for the vaporisation rate is therefore:

$$\frac{G_o r_o}{\Gamma_{vap}} = \ell n \left(1 + \frac{m_{vap,o} - m_{vap,\infty}}{1 - m_{vap,o}} \right) \quad , \qquad (3.2\text{-}9)$$

while that for the vapour distribution is:

$$\left(\frac{1 - m_{vap}}{1 - m_{vap,\infty}} \right) = \left(\frac{1 - m_{vap,o}}{1 - m_{vap,\infty}} \right)^{(r_o/r)} . \qquad (3.2\text{-}10)$$

Obviously, the qualitative form of the distribution is as shown.

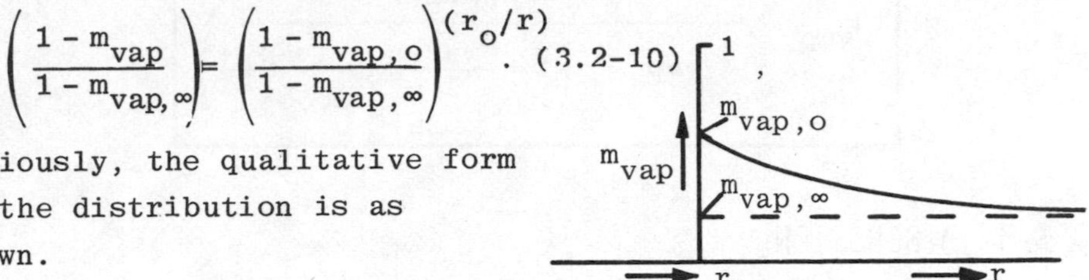

D) VARIATION OF DROPLET DIAMETER WITH TIME

(i) Differential Equation

Because vapour is formed at the expense of the liquid, the rate of droplet-radius change is proportional to G_o; thus:

$$\frac{dr_o}{dt} = - \frac{G_o}{\rho_{liq}} \quad , \qquad (3.2\text{-}11)$$

where t stands for time and

ρ_{liq} for the density of the liquid;

or, in terms of diameter D,

$$\frac{dD}{dt} = -2 \frac{G_o}{\rho_{liq}} \quad . \qquad (3.2\text{-}12)$$

Hence:

$$\frac{dD}{dt} = -4 \frac{\Gamma_{vap}}{D\rho_{liq}} \ell n \left(1 + \frac{m_{vap,o} - m_{vap,\infty}}{1 - m_{vap,o}} \right) . \qquad (3.2\text{-}13)$$

(ii) Solution

If Γ_{vap}, $m_{vap,o}$ and $m_{vap,\infty}$ are all independent of time, the $D \sim t$ variation is:

$$D_o^2 - D^2 = 8 t \frac{\Gamma_{vap}}{\rho_{liq}} \ell n \left(1 + \frac{m_{vap,o} - m_{vap,\infty}}{1 - m_{vap,o}} \right) , \qquad (3.2\text{-}14)$$

where $D_o \equiv D$ when $t = 0$.

(iii) Vaporisation time

The time at which the diameter falls to zero, t_{vap}, is therefore given by:

$$t_{vap} = \frac{D_o^2 \rho_{liq}}{8\,\Gamma_{vap}\,\ell n\left(1 + \dfrac{m_{vap,o} - m_{vap,\infty}}{1 - m_{vap,o}}\right)} \quad . \qquad (3.2\text{-}15)$$

3.3 DISCUSSION

A) PRACTICAL IMPLICATIONS

- $t_{vap} \propto D_o^2$. Therefore, if t_{vap} must be small, so must D_o. This is the reason for the use of "atomisers" (single-jet, rotating-cup, impingement, swirl, steam-assisted, etc).
- t_{vap} decreases as $m_{vap,o}$ increases; so volatile fuels vaporise fast.

B) LIMITATIONS OF UTILITY OF MODEL

- $m_{vap,o}$ has a strong influence, but is not usually known. It depends on temperature; means will have to be found of calculating this in the next two lectures.
- Relative motion of droplet and air influences (augments) the vaporisation rate by causing departures from symmetry, which steepen the average concentration gradient.
- The vapour fields of neighbouring droplets interact.
- $m_{vap,\infty}$ and $m_{vap,o}$ may both vary with time.
- Γ_{vap} usually depends somewhat on temperature and composition.
- Often the initial droplet diameter is only approximately known.

C) CONCLUSIONS

- The model provides useful insight into an important physical process.

It will need many refinements in order to be useful in engineering design.

Most of the refinements can be made, and some will be introduced below.

The most troublesome of all the obstacles to prediction is imprecision of knowledge of initial droplet size.

3.4 PRACTICAL MEANS OF PRODUCING STREAMS OF SMALL DROPLETS

A) SWIRL "ATOMISERS"

The liquid is supplied in a tangential fashion to a cylindrical chamber having a small-diameter hole in one end. The liquid flows over the lip of this hole like water over a broad- crested weir; the thickness of the film δ may therefore be estimated from:

- Weir theory:

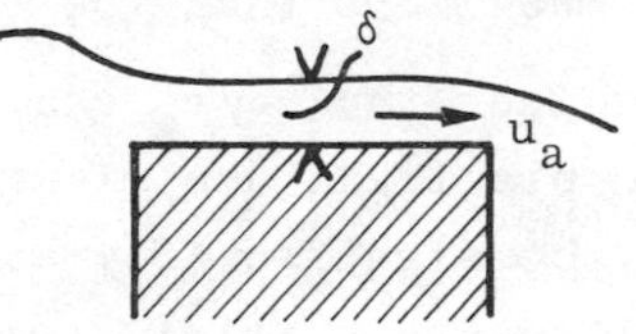

$$\frac{g\,\delta}{u_a^2} \approx 1, \quad (3.4\text{-}1)$$

where g is the gravitational acceleration.

- Mass conservation:

$$\dot{m}_{fuel} = 2\pi r_a u \delta\, \rho_{liq} \quad .(3.4\text{-}2)$$

- Conservation of angular momentum:

$$\frac{v_a}{v_c} = \frac{r_c}{r_a} \quad , \quad (3.4\text{-}3)$$

where v is the swirl component of the liquid velocity, v_a is v in the aperture, v_c is the swirl velocity on entry at chamber-wall radius.

- Replacement of gravitational acceleration by centripetal acceleration in the weir formula:

$$\frac{v_a^2}{r_a}\,\frac{\delta}{u_a^2} \approx 1 \qquad . \qquad (3.4\text{-}4)$$

- Relation of v_c to $\dot{m}_{fu}$:

$$v_c = \dot{m}_{fu}/(\rho_{liq}A_{in}) \qquad , \qquad (3.4\text{-}5)$$

where A_{in} stands for area of tangential inlet duct.

- Resulting expression relating thickness of film to geometrical factors:

$$\frac{\delta}{r_a} = \left[\frac{1}{2}\cdot\frac{A_{in}}{\pi r^2{}_a}\cdot\frac{r_a}{r_c}\right]^{2/3} \qquad . \qquad (3.4\text{-}6)$$

- Evidently the film thickness will be much smaller than the aperture radius,provided that the inlet-duct area is much less than the aperture area, and the aperture radius r_a is much less than the chamber radius r_c.
- The fact that δ/r_a can be much less than unity explains the practical use of swirl atomisers; for δ can in this way be significantly smaller than normal machining tolerances. Swirl atomisers thus provide ways of producing thin liquid films without drilling very small holes or machining very thin slots.
- Such small holes and slots would, incidentally, become clogged with impurities unless the fuel were very well filtered. Fine filtering is expensive.
- However, the pressure drop in the fuel line across the atomiser may be considerable. The simple theory implies that it is given by:

$$\Delta p = \frac{\rho_{liq}}{2}\,(u_a{}^2 + v_a{}^2) \ ,$$

ie

$$\frac{\Delta p}{\frac{1}{2}\rho_{liq}\left[\dot{m}_{fu}/(\pi r^2{}_a\rho_{liq})\right]^2} = \frac{1}{2}\,\frac{r_a}{\delta}\left(1 + \frac{r_a}{\delta}\right) \qquad . \qquad (3.4\text{-}7)$$

The right-hand side would of course equal unity, if the liquid were supplied without any tangential velocity component, and the aperture therefore "ran full".

- In practice, the small dimensions of the chamber allow viscous effects to play a significant part; therefore film thicknesses may be several times as large as the above analysis indicates; and, whereas the analysis implies that the thickness is independent of flow rate, practical devices exhibit larger thicknesses at lower flow rates.
- At very low flow rates, surface-tension effects also become significant.
- The film leaving the lip of the aperture spreads out as a conical sheet, the angle of which is given, according to the above theory, by:

$$\tan(\text{angle}) = v_a/u_a \approx (r_a/\delta)^{\frac{1}{2}} \quad . \qquad (3.4\text{-}8)$$

- Once again, viscous effects cause departures from this formula in practice.
- To preserve continuity, the thickness of the film diminishes as the distance from the aperture increases. Eventually surface-tension forces, triggered by aero-dynamic disturbances, cause the film to break up and become a cloud of droplets, the diameter of which is of the same order as the film thickness. This completes the "atomisation" process.

B) <u>ROTATING-CUP ATOMISERS</u>

The liquid fuel is poured on to the inner surface of a rotating cup and acquires its swirl velocity.

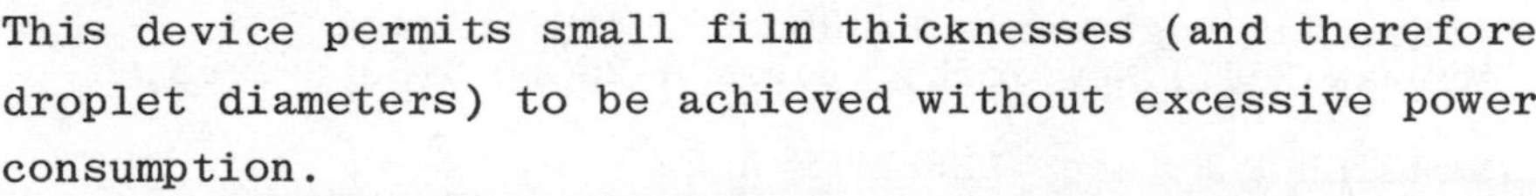

This device permits small film thicknesses (and therefore droplet diameters) to be achieved without excessive power consumption.

It is not used for gas-turbine or diesel engines, because

of its size, weight and comparative complexity. However, a few gas turbines do atomise the fuel by supplying it to the rim of a spinning disc which is attached to the shaft of the engine.

c) OTHER DEVICES

In some industrial furnaces, the fuel is caused to flow through an orifice through which also flows steam or compressed air. The frictional forces generated by the high relative velocities causc the liquid to be broken into small sheets and filaments, which then assume droplet form. Such atomisers are especially useful for high-viscosity fuels, eg residual oil.

In diesel engines, the liquid is forced at extremely high pressures through small orifices. Atomisation is effected by friction between the emerging jet of liquid and the high-pressure gas into which it is injected. Because the flow is not continuous, there is no need for the whole fuel-supply line to be subjected to the high pressure, which is generated locally, and in pulses.

In rocket motors, the atomisation is often effected by causing jets to impinge. This device is used because, even though the atomisation may not be as fine as is desirable, the quantity of liquid to be injected through the propellant-supply assembly is so enormous that extreme geometrical simplicity is necessary.

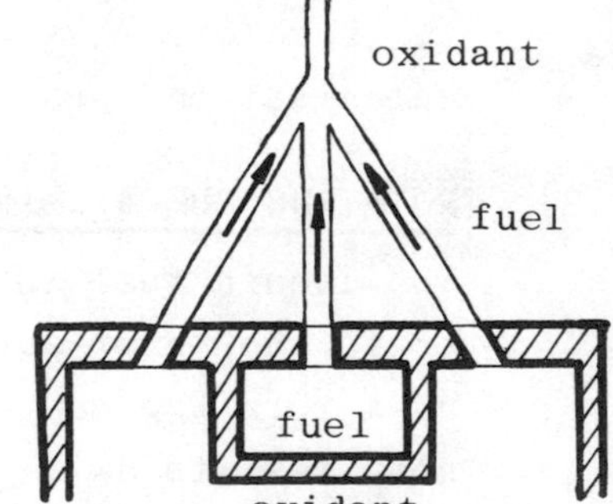

4 REFERENCES

PERRY J H (Editor)
"Chemical Engineer's Handbook" 4th Edition
McGraw Hill, New York, pages 9-25 to 9-30, 1963.

NORSTER E R and LEFEBVRE A H
"Effects of Fuel-Injection Method on Gas-Turbine Combustor Emissions" in "Emissions from Continuous Combustion Systems" Cornelius W and Agnew W G (Editors)
Plenum Press, New York, pp 255-278, 1972.

EXERCISES TO FACILITATE ABSORPTION OF THE MATERIAL OF CHAPTER 3

ANALYTICAL PROBLEMS

3.1 The droplet-vaporisation model in the lecture is one of point symmetry. Make a corresponding analysis of vaporisation from a cylindrical surface of fuel, presuming axial symmetry, and show that the vaporisation rate is zero when the atmosphere is infinite in extent. Decide whether you consider this implication to be realistic; if you do not, to what is the lack of realism due?

3.2 A water droplet at 10^{0}C vaporises into dry air at 1 atmosphere pressure. Its initial diameter is 0.1 mm. Calculate the time of vaporisation.

Data: m_{H_2O} = .0075 for saturated air at 10^{0}C and 1 atm.

$\Gamma_{H_2O} \approx 2.6 \times 10^{-5}$ kg/m s under these conditions.

$\rho_{liq} \approx 10^3$ kg/m^3.

Answer: 6.43 s.

MULTIPLE-CHOICE PROBLEMS

Mark the following items with A, B, C, D or E by reference to equation (3.2-10) according to whether the change in conditions indicated in the item, other things being equal, would:

A increase the time of vaporisation;

B decrease the time of vaporisation;

C not affect the time of vaporisation;

D probably increase the vaporisation time, but this is not implied in the equation;

E probably decrease the vaporisation time, but this is not implied in the equation.

3.3 The initial diameter increases.

3.4 The pressure of the atmosphere increases.

3.5 The droplet is falling at a significant velocity through the atmosphere.

3.6 The temperature of the atmosphere increases.

3.7 The liquid is replaced by one which is less volatile.

3.8 Both $m_{vap,o}$ and $m_{vap,\infty}$ are increased by equal amounts.

3.9 The specific heat of the liquid increases.

3.10 The droplet is one of a cloud, and the distance btween it and its nearest neighbours is decreased.

3.11 Swirl atomisers are used in aircraft gas turbines
<u>because</u>
the fuel has a high viscosity.

3.12 Steam-atomising burners are used in high-performance liquid-propellant rocket motors
<u>because</u>
there is insufficient space for a multiplicity of swirl atomisers.

3.13 Rotating-cup atomisers are not used in industrial furnaces
<u>because</u>
the fuel has to be supplied to them at high pressure.

3.14 Swirl, impinging-jet and rotating-cup atomisers are all employed in practice
<u>because</u>
they create liquid films which are appreciably thinner than the smallest tolerances involved in their manufacture.

3.15 Fuel is supplied to the atomiser of a diesel engine at very high pressure
<u>because</u>
a finite ignition delay ensues before the first injected fuel starts to burn.

Mark each of the following items, which all relate to the mathematical model of a vaporising droplet:

- A if the item is valid only for uniform Γ_{vap};
- B if it is valid only for very low concentrations of vapour;
- C if it is never valid;
- D if it is valid for an axi-symmetrical model (Exercise 3.1) but not for a point-symmetrical one;
- E if it is valid even if the diffusion flux depends on gradients of other mixture components in addition to that of m_{vap}.

3.16 $G_{tot,vap} r^2 = G_o r_o^2$

3.17 $\ell n(m_{vap} - 1) = -\dfrac{G_o r_o^2}{\Gamma_{vap} r} + \text{const}$

3.18 $\ell n(1 - m_{vap}) = \dfrac{G_o r_o^2}{\Gamma_{vap} r} - \text{const}$

3.19 $G_{tot,vap} = -\Gamma_{vap} \dfrac{dm_{vap}}{dr}$

3.20 $\Gamma_{vap} \dfrac{dm_{vap}}{d(\ell n r)} = G_o r_o (m_{vap} - 1)$

ANSWERS TO MULTIPLE-CHOICE PROBLEMS

Answer	Problem number (3's omitted)			
A	3,	7,	14,	17
B	6,	8,	15,	19
C	4,	9,	11,	18
D	12,	20,		
E	5,	10,	13,	16

DISCUSSION PROBLEMS

3.21 As for exercise 1.43

3.22 List and explain the main factors governing the choice of fuel-atomising devices for power plants and industrial furnaces, and the practical uses to which the various types of device are put.

CHAPTER 4

MASS TRANSFER II

4.1 OUTLINE OF LECTURE

A) MOTIVATION

- It is necessary to know temperatures in order to calculate droplet-vaporisation times; for the mass fraction of vapour adjacent to the liquid surface depends upon the liquid temperature. Such interactions of diffusional and thermal processes are common. Means must be provided for computing them.
- Although thermodynamics and heat transfer provide much of the relevant knowledge, their application to diffusing systems is not obvious; and, even when their application has been correctly made, some forms of the equations are more easily handled than others. So a formal presentation will be worth looking at.

B) THE PROBLEM

This lecture is therefore concerned with how to compute distributions of temperature and of energy flux in diffusing media, perhaps with chemical reaction.

C) OUTLINE OF LECTURE

There are three main topics:

- The expression for the energy flux.
- The differential equation for one-dimensional flow (several forms).
- The application to the Stefan-flow problem.

4.2 THE ENERGY FLUX

A) DEFINITIONS

- $h_j \equiv$ partial enthalpy of component j in a mixture.

  Note that, for ideal mixtures which are usually in question in combustion processes :

  $$h_j = \int_{T_o}^{T} c_j \, dT + h_{j,o} \quad , \qquad (4.2\text{-}1)$$

  where: $c_j \equiv$ specific heat of j at constant pressure, independent of anything but temperature;

  $h_{j,o} \equiv$ value of h_j at T_o, in part arbitrary.

- $h \equiv$ specific enthalpy of mixture

  $$= \sum_{\text{all } j} m_j h_j \quad . \qquad (4.2\text{-}2)$$

- $c \equiv$ mean specific heat of mixture at constant pressure

  $$= \sum_{\text{all } j} m_j c_j \quad . \qquad (4.2\text{-}3)$$

- $\tilde{h} \equiv$ stagnation enthalpy of mixture

  $$= h + V^2/2 \quad , \qquad (4.2\text{-}4)$$

  where $V \equiv$ mixture velocity

- $Q \equiv$ heat flux per unit area (caused by temperature gradient)

  $$= -\lambda \frac{dT}{dx} \quad , \qquad (4.2\text{-}5)$$

  where $\lambda \equiv$ thermal conductivity of mixture. This is Fourier's law.

B) THE ENERGY FLUX

The total energy flux E across a surface, per unit area, can be regarded as the algebraic sum of individual components as follows:

$$
\begin{aligned}
E \equiv\ & Q && , \text{ heat flux,} \\
& + W_s && , \text{ shear work,} \\
& + G\, V^2/2 && , \text{ kinetic energy,} \\
& + \sum_{\text{all } j} h_j\, G_{tot,j} && , \text{ enthalpy flux.}
\end{aligned} \qquad (4.2\text{-}6)
$$

4.3 THE DIFFERENTIAL EQUATION FOR ONE-DIMENSIONAL STEADY FLOW

A) THE FIRST LAW OF THERMODYNAMICS APPLIED TO A CONTROL VOLUME

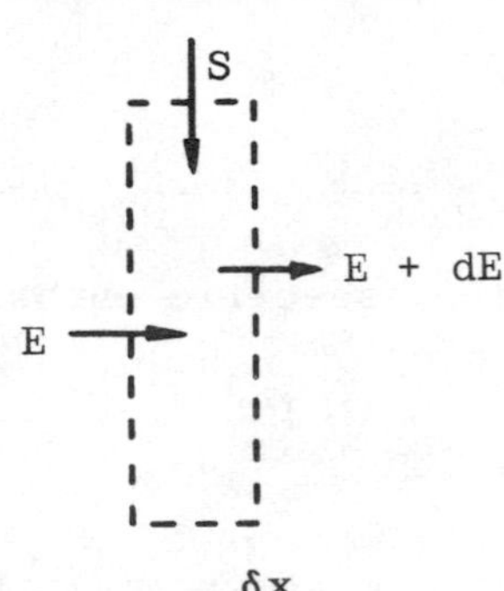

This law may be expressed as:

$$S = \frac{dE}{dx} \quad , \quad (4.3\text{-}1)$$

where S ≡ source from outside, eg by radiation (<u>not</u> chemical reaction). Note that it is being supposed that the only variations are in the x direction.

B) FIRST FORM OF DIFFERENTIAL EQUATION FOR TEMPERATURE

- From above :

$$S = -\frac{d}{dx}\left(\lambda \frac{dT}{dx}\right) + \frac{dW_s}{dx} + G\,\frac{d(V^2/2)}{dx} + \sum\left[h_j \frac{dG_{tot,j}}{dx} + G_{tot,j}\,\frac{dh_j}{dx}\right] \quad . \quad (4.3\text{-}2)$$

- Here G is the sum of the $G_{tot,j}$'s; and use has already been made of $\frac{dG}{dx} = 0$.

- The generation rate of species j per unit volume is R_j. Therefore:

$$\frac{d\,G_{tot,j}}{dx} = R_j \quad ; \quad (4.3\text{-}3)$$

and so the following term appearing in (4.3-2) can be written as :

$$\sum_{\text{all } j} h_j\, \frac{d\,G_{tot,j}}{dx} = \sum_{\text{all } j} h_j\, R_j \quad . \quad (4.3\text{-}4)$$

Note that, as a matter of definition:

$$\sum_{\text{all } j} R_j = 0 \qquad . \qquad (4.3\text{-}5)$$

- Because the $h_{j,o}$'s in the definition of h_j, equation (4.2-1), are constants, the last terms of (4.3-2) can be written as :

$$G_{tot,j} \frac{dh_j}{dx} = G_{tot,j}\, c_j \frac{dT}{dx} \qquad . \qquad (4.3\text{-}6)$$

- Hence, the first form of the differential equation is:

$$\boxed{S = \frac{-d}{dx}\left(\lambda \frac{dT}{dx}\right) + \frac{dW_s}{dx} + G\,\frac{d(V^2/2)}{dx} + \sum h_j R_j + \left[\sum G_{tot,j} c_j\right] \frac{dT}{dx}}$$

$$. \qquad (4.3\text{-}7)$$

- Note that a flow-rate-average specific heat <u>could</u> be defined, as follows :

$$\bar{\bar{c}} \equiv \sum G_{tot,j}\, c_j \;/\; G \qquad . \qquad (4.3\text{-}8)$$

Then the last term would be $G\,\bar{\bar{c}}\, dT/dx$. Of course, $\bar{\bar{c}}$ is not in general equal to c.

c) SECOND FORM OF THE DIFFERENTIAL EQUATION FOR T

- The relevant definitions imply :

$$G_{tot,j} \equiv m_j G + G_{diff,j} \qquad . \qquad (4.3\text{-}9)$$

So $$c_j\, G_{tot,j} = c_j m_j G + c_j\, G_{diff,j} \qquad , \qquad (4.3\text{-}10)$$

and $$\sum_{\text{all } j} c_j\, G_{tot,j} = c\,G + \sum_j c_j\, G_{diff,j} \qquad . \qquad (4.3\text{-}11)$$

Since $\sum_j G_{diff,j} = 0$, it is useful to write the last term as:

$$\sum_j c_j\, G_{diff,j} = \underbrace{\sum_j c\, G_{diff,j}}_{0} + \sum_j (c_j - c)\, G_{diff,j}$$

$$. \qquad (4.3\text{-}12)$$

- Hence a second form of the differential equation is obtained, namely :

$$S = \frac{-d}{dx}\left(\lambda\frac{dT}{dx}\right) + \frac{dW_s}{dx} + G\,\frac{d(V^2/2)}{dx} + \sum_j h_j R_j + cG\,\frac{dT}{dx} + \sum_j (c_j - c)\, G_{diff,j}\frac{dT}{dx} \quad . \qquad (4.3\text{-}13)$$

- Obviously, if $(c_j - c) = 0$ or $G_{diff,j} = 0$ for all j, the last term vanishes.

- If the reaction rates of all species can be linked to that of a single species, the following substitution can be made :

$$\sum_{all\ j} h_j R_j \equiv -H_{fu} R_{fu} \quad , \qquad (4.3\text{-}14)$$

where H_{fu} is the heat of combustion of the fuel.

4.4 THE STEFAN-FLOW PROBLEM

A) STATEMENT OF CONDITIONS

- Only vapour has finite G_{tot}; so G is equal to $G_{tot,vap}$.
- It is supposed that chemical reaction will be absent.
- Shear work and kinetic energy will be supposed negligible.
- There are no sources of energy; therefore S equals zero.
- The fluid properties λ and c_{vap} are presumed independent of T.

B) DIFFERENTIAL EQUATION

From (4.3-7):

$$0 = -\frac{d}{dx}\left(\lambda\frac{dT}{dx}\right) + G\,c_{vap}\,\frac{dT}{dx} \quad . \qquad (4.4\text{-}1)$$

C) SOLUTION

- Integration leads to:

$$\lambda\,\frac{dT}{dx} - G\,c_{vap}\,T = -\,Q_o - c_{vap}\,T_o\,G \quad , \qquad (4.4\text{-}2)$$

ie $\frac{dT}{dx} = \frac{G\, c_{vap}\,(T - T_o) - Q_o}{\lambda}$. (4.4-3)

- Further integration yields:

$$\ell n \left[T - T_o - \frac{Q_o}{G\, c_{vap}} \right] = \frac{G\, c_{vap}\, x}{\lambda} + \text{const} \, . \qquad (4.4\text{-}4)$$

- Insertion of the boundary conditions leads to:

$$\boxed{\ell n \left[1 + \frac{c_{vap}\,(T_o - T_1)}{Q_o/G} \right] = \frac{G\, c_{vap}\, x_1}{\lambda}} \, . \qquad (4.4\text{-}5)$$

D) <u>DISCUSSION</u>

(i) When the argument of ℓn is close to 1, the equation tends to:

$$\frac{G c_{vap}\,(T_o - T_1)}{Q_o/G} = \frac{G\, c_{vap}\, x_1}{\lambda}$$

ie $- Q_o = \lambda(T_1 - T_o)/x_1$. (4.4-6)

So only heat conduction is important.

(ii) In general, the heat flux to the surface can be expressed as :

$$- Q_o = G\, c_{vap}(T_1 - T_o) / \left[e^{\frac{G\, c_{vap}\, x_1}{\lambda}} - 1 \right] . \qquad (4.4\text{-}7)$$

So a positive G reduces the rate of heat transfer at the liquid surface. The effect is one of the reasons for the use of "transpiration cooling".

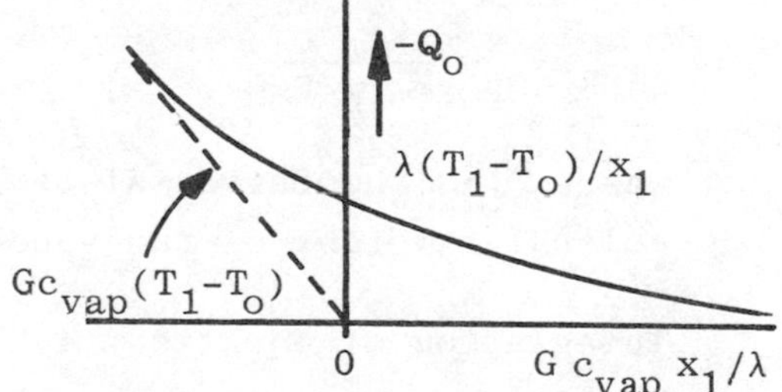

(iii) If there is no external heat supply to the liquid, and the liquid temperature is steady, it can be supposed that :

$$- Q_o = GL \qquad , \qquad (4.4\text{-}8)$$

where L is the latent heat of vaporisation of the liquid. Then the vaporisation rate is fixed by:

$$G = \frac{\lambda}{x_1\ c_{vap}}\ \ell n \left[1 + \frac{c_{vap}\ (T_1 - T_o)}{L} \right] \quad . \quad (4.4\text{-}9)$$

It should be noted that the specific heat of the air, c_{air}, plays no part.

(iv) Equation (4.4-9) connects the vaporisation rate with the temperature at the surface, T_o. Equation (2.6-11) connects it with the mass fraction of vapour in the gas phase there, $m_{vap,o}$. These two quantities are also linked by a further relation expressing the existence of thermodynamic equilibrium between the liquid and gas phases:

$$m_{vap,o} = m_{vap,o}(T_o,\ \text{pressure}) \quad . \quad (4.4\text{-}10)$$

This relation is usually as shown in the sketch. It may be deduced from:-

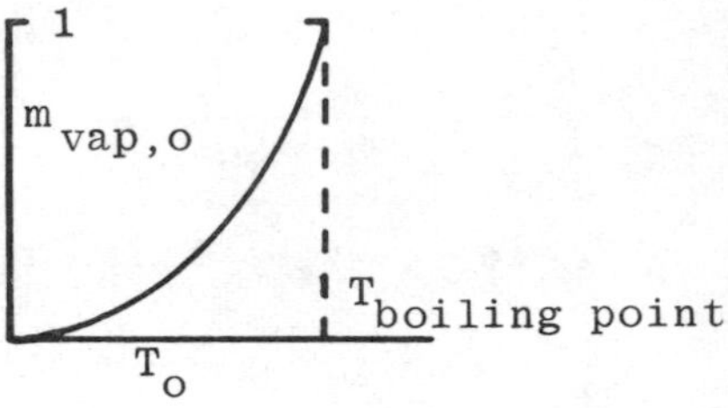

(1) the Clausius-Clapeyron relation:

$$\frac{dp_{vap,o}}{dT_o} = \frac{L}{v_o\ T_o} \quad , \quad (4.4\text{-}11)$$

where $p_{vap,o}$ is the vapour pressure at the surface, and v_o is the specific volume of the vapour given by:

$$v_o = \frac{\mathcal{R}\ T_o}{M_{vap}\ p_o} \quad , \quad (4.4\text{-}12)$$

where $\mathcal{R}$ is the Universal Gas Constant and M_{vap} is the molecular weight of the vapour,

(2) integration of equation (4.4-11) from the specified boiling point at the prevailing pressure p, permitting $p_{vap,o}$ to be determined as a function of T_o; and

(3) the relation between $p_{vap,o}$ and $m_{vap,o}$, which can be determined from the following relations, valid

for mixtures of Ideal Gases, such as are usually in question in combustion processes:

$$\frac{m_j}{M_j} = \frac{p_j}{p} \cdot \frac{1}{\bar{M}} \quad ; \quad (4.4\text{-}13)$$

$$\bar{M} = \sum \frac{p_j M_j}{p} = 1/\sum \frac{m_j}{M_j} \quad . \quad (4.4\text{-}14)$$

4.5 REFERENCES

SPALDING D B
"Convective Mass Transfer"
Edward Arnold, London, 1963

This book builds up the theory of convective heat and mass transfer on the basis of an alternative model of the elementary processes, namely the "Reynolds Flux". Apart from the replacement of Γ_{vap}/x_1 and $\lambda/(c_{vap}\ x_1)$ by "conductances", g, having the dimensions kg/m^{2}s, the formulae are identical.
The book may therefore be consulted as a source of "driving-force" expressions, generalising those encountered in Chapters 2 and 4, namely:

$$\frac{m_{vap,o} - m_{vap,1}}{1 - m_{vap,o}} \quad \text{and} \quad \frac{c_{vap}\ (T_1 - T_o)}{(Q_o/G)}$$

EXERCISES TO FACILITATE ABSORPTION OF THE MATERIAL OF CHAPTER 4

ANALYTICAL PROBLEMS

4.1 Derive the following differential form of the Steady-Flow Energy Equation, and state the conditions for which it is valid:

$$G_x \frac{\partial h}{\partial x} + G_y \frac{\partial h}{\partial y} + G_z \frac{\partial h}{\partial z} - \frac{\partial}{\partial x}\left(\lambda \frac{\partial T}{\partial x}\right) - \frac{\partial}{\partial y}\left(\lambda \frac{\partial T}{\partial y}\right) - \frac{\partial}{\partial z}\left(\lambda \frac{\partial T}{\partial z}\right) + \sum_j h_j R_j$$

$$+ \sum_j (c_j - \bar{c}) \left[G_{diff,j,x} \frac{\partial T}{\partial x} + G_{diff,j,y} \frac{\partial T}{\partial y} + G_{diff,j,z} \frac{\partial T}{\partial z} \right] = 0$$

4.2 Determine by trial and error the temperature that would be taken up by water vaporising through a stagnant gaseous film into air, under the following conditions:

$T_1 = 26.43^0C$, $m_{H_2O,1} = .003$,

$c_{H_2O} = 2000\ J/kg^0C$, $\lambda = 0.5 \times 10^{-2}\ J/m\ s^0C$,

$\Gamma_{H_2O} = 2 \times 10^{-5}\ kg/m\ s$, $L = 2.29 \times 10^6\ J/kg$.

The heat-transfer rate to the water surface is just sufficient to supply the latent heat of vaporisation. The following table contains the necessary information about $m_{H_2O,o}$.

T_o	15.6	16.7	17.8	18.9	20.0	21.1	22.2	23.35
$m_{H_2O,o}$	.0109	.0117	.0126	.0135	.0145	.0155	.0166	.0178

Answer: 20.0^0C.

4.3 Suppose that heat is supplied electrically to the water reservoir of the last problem at a steadily increasing rate. Sketch the variations with time of G, Q_o, T_o and $m_{H_2O,o}$, and check that your sketches are compatible with the governing equations. (Pay special attention to the behaviour at $G \to \infty$).

MULTIPLE-CHOICE PROBLEMS

Classify the following statements according to whether they represent:

A definitions.

B correct logical deductions from definitions.

C incorrect logical deductions from definitions.

D statements of experimental findings having a high degree of truth.

E statements of experimental findings having only a moderate degree of truth.

4.4 $Q = -\lambda\frac{dT}{dx}$.

4.5 $h_j = \int_{T_o}^{T} c_j dT + h_{j,o}$.

4.6 $E = Q + W_s + GV^2/2 + \sum h_j G_{tot,j}$.

4.7 $\sum R_j = 0$.

4.8 c is independent of temperature.

4.9 $\lambda = \sum m_j c_j \Gamma_j$.

4.10 $c = \sum m_j c_j$.

4.11 $\frac{dh_j}{dx} = c_j \frac{dT}{dx}$.

4.12 $\sum c_j G_{tot,j} = c\ G - \sum c_j\ G_{diff,j}$.

4.13 The quantity $\lambda/(x_1 c_{vap})$ can be regarded as a conductance <u>because</u>
it has the typical units: kg/m²s.

4.14 The quantity $B_h \equiv c_{vap} (T_1 - T_o)/L$ can be regarded as a "driving force" for mass transfer
because
the mass-transfer rate is proportional to B_h when $|B_h| \ll 1$.

4.15 Equation (4.4-5) is not valid for condensation of vapour
because
Q_o is essentially positive.

4.16 Equation (4.4-5) is valid only for uniform thermal conductivity
because
equation (4.4-3) is valid only for uniform thermal conductivity.

4.17 When G tends to minus infinity, Q_o tends to $G\, c_{vap}(T_1 - T_o)$
because
$\ell n(1 - x)$ tends to minus infinity as x tends to minus infinity.

4.18 The Stefan problem is worthy of study, in a course on mass transfer,
because
it represents an easily-analysable idealisation of the problems of mass transfer through boundary layers and other more complex flows.

4.19 The Clausius-Clapeyron equation may be used to calculate the relation between $m_{vap,o}$ and T_o
because
this relation does not depend upon the magnitude of the partial pressures of other components of the mixture.

4.20 The composition of a mixture of Ideal Gases can be characterised just as well in terms of the partial pressures of the individual components as in terms of the mass fractions
because
$m_j = p_j / \sum p_j$.

ANSWERS TO MULTIPLE-CHOICE PROBLEMS

Answer	Problem number (4's omitted)			
A	5,	10,	14,	18
B	7,	11,	13,	
C	12,	16,	20,	
D	4,	6,	17,	19
E	8,	9,	15,	

DISCUSSION PROBLEMS

4.21 As for exercise 1.43.

4.22 In the book cited in the references, the central formula for the calculation of mass-transfer rates is:

$$\dot{m}'' = g\, B$$

<u>where</u>: <u>$\dot{m}''$</u> stands for the rate of mass transfer with the phase under consideration, (kg/m s);

<u>g</u> stands for the "conductance" for mass transfer (kg/m s);

<u>B</u> stands for the dimensionless driving force, defined as:

$$B \equiv \frac{\phi_G - \phi_S}{\phi_S - \phi_T}$$

<u>where</u>: ϕ stands for a "conserved property" (eg enthalpy in a flow free from energy sources; or mass fraction of a chemically inert material), and the subscripts have the following meanings:

G represents the state of the considered phase at a position remote from the interface;

S represents the state of the considered phase at the interface;

T represents the state of the "transferred substance", ie that material the transfer of which across the interface would be equivalent to the actual mass and heat transfers there.

Compare the formulae derived for the Stefan Flows of Chapters 2 and 4, and the droplet model of Chapter 3, with the above formulae, and explain the similarities and differences.

CHAPTER 5

DROPLET VAPORISATION II

5.1 OUTLINE OF PROBLEM

A) STATEMENT

In chapter 2, droplet vaporisation was considered from the point of view of Fick's law of diffusion. The resulting formula connected the vaporisation time with the vapour concentrations. However, the concentration at the surface is often not known.

In this chapter, droplet vaporisation is considered from the thermal point of view. A vaporisation-time formula is arrived at which, in conjunction with that of chapter 2, and with certain thermodynamic data, permits the vaporisation time finally to be determined.

The task is :-

given :
- droplet diameter,
- transport properties of vapour,
- thermodynamic properties of liquid, vapour and atmosphere,
- initial droplet temperature (presumed uniform),

calculate: variations with time of
- droplet diameter,
- droplet temperature,
- vaporisation rate,
- mass fraction of vapour in the gas phase at the droplet surface.

B) EXPECTED BEHAVIOUR

(i) When a cold droplet is injected into a hot gas, its temperature can be expected to rise, and the value of $m_{vap,o}$ also, as shown in the sketch.

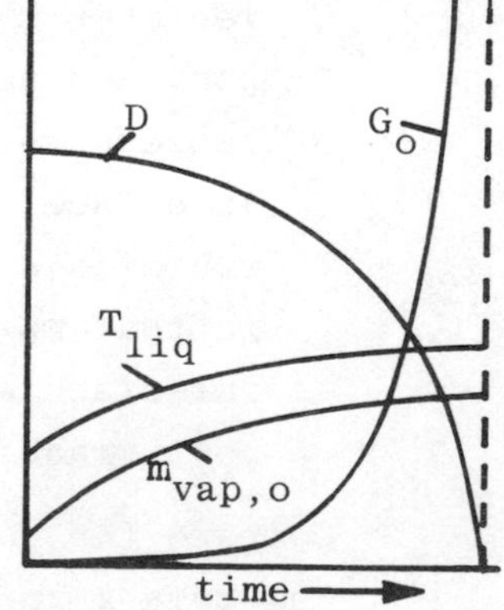

(ii) When the liquid is hot at first, and the atmosphere cool, the behaviour is likely to be as shown in this sketch.

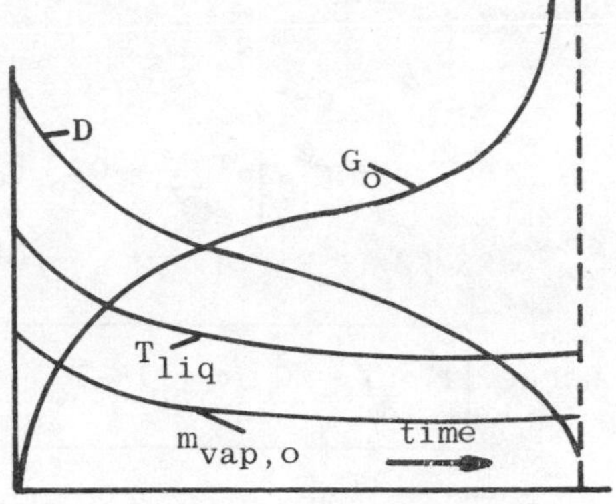

(iii) If the initial liquid temperature has a specific intermediate temperature (which we shall be able to calculate), T_{liq} and $m_{vap,o}$ do not change; and D follows a quadratic curve.

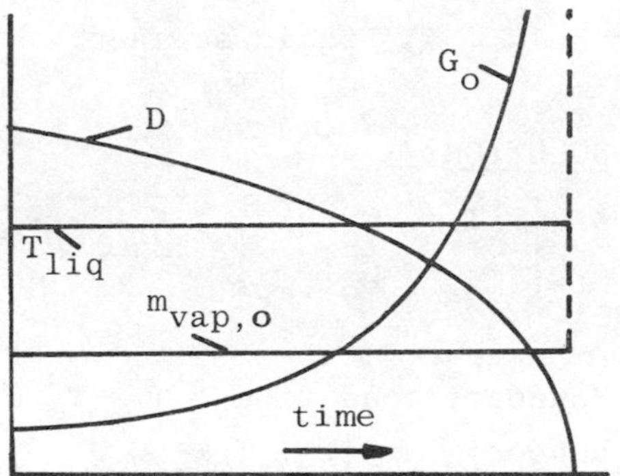

5.2 CALCULATION OF THE VAPORISATION RATE IN TERMS OF TEMPERATURE

A) ASSUMPTIONS

- spherical symmetry.
- quasi-steady state.
- λ and c_{vap} uniform.
- large distance between droplets.
- no chemical reaction.
- no radiation.
- shear work and kinetic energy negligible.

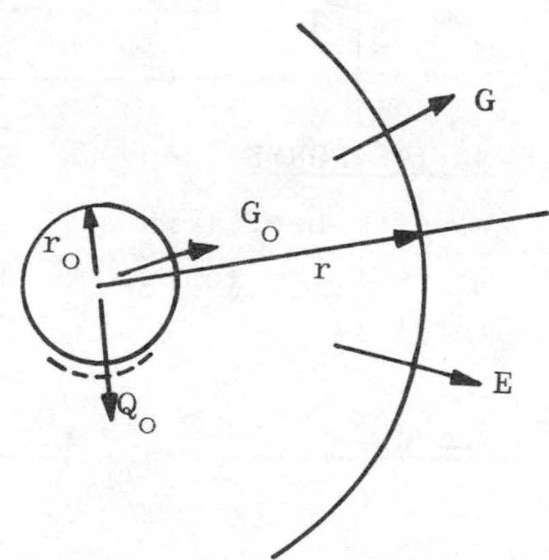

B) DIFFERENTIAL EQUATIONS

- Mass conservation:

$$G = G_{tot,vap}, \quad (5.2\text{-}1)$$

$$Gr^2 = G_o r_o^{\;2} \quad . \quad (5.2\text{-}2)$$

Energy :

$$E_o r_o^2 = E r^2$$
$$= r^2 \left[-\lambda \frac{dT}{dr} + G_{tot,vap} \left\{ c_{vap}(T-T_o) + h_{vap,o} \right\} \right]$$
$$= r_o^2 \left[-\left(\lambda \frac{dT}{dr} \right)_o + G_{tot,vap}\, h_{vap,o} \right] \qquad .(5.2\text{-}3)$$

Hence
$$\boxed{r^2 \lambda \frac{dT}{dr} = G_o \left\{ c_{vap}(T-T_o) - Q_o/G_o \right\} r_o^2} \qquad ,(5.2\text{-}4)$$

where Q_o ≡ heat flux through gas phase close to the liquid surface.

C) <u>SOLUTION</u>

$$\ell n \left[T - T_o - \frac{Q_o}{G_o c_{vap}} \right] = \frac{G_o c_{vap} r_o^2}{\lambda} \left[-\frac{1}{r} \right] + \text{const.} \qquad .(5.2\text{-}5)$$

Boundary condition : $T = T_\infty$ at $r \to \infty$.(5.2-6)

Hence :

$$\ell n \left[\frac{T_\infty - T_o - \frac{Q_o}{G_o c_{vap}}}{\left(-\frac{Q_o}{G_o c_{vap}} \right)} \right] = \frac{G_o c_{vap} r_o}{\lambda} \qquad .(5.2\text{-}7)$$

and so :

$$\boxed{\ell n \left[1 + \frac{c_{vap}(T_\infty - T_o)}{(-Q_o/G_o)} \right] = \frac{G_o c_{vap} r_o}{\lambda}} \qquad .(5.2\text{-}8)$$

D) <u>ALTERNATIVE FORMS</u>

The above is useful when T_o and $(-Q_o/G_o)$ are <u>known</u>. When Q_o is <u>to be found</u>, the following alternative forms are useful :-

$$1 + \frac{c_{vap} G_o (T_\infty - T_o)}{-Q_o} = \exp \left(\frac{G_o c_{vap} r_o}{\lambda} \right) \qquad ,(5.2\text{-}9)$$

i.e. $$-Q_o = \frac{c_{vap} \ G_o(T_\infty - T_o)}{\exp\left(\frac{G_o c_{vap} r_o}{\lambda}\right) - 1} \qquad ,(5.2\text{-}10)$$

and so : $$\boxed{\frac{-Q_o r_o}{\lambda(T_\infty - T_o)} = \frac{G_o c_{vap} r_o/\lambda}{\exp(G_o c_{vap} r_o/\lambda) - 1}} \qquad .(5.2\text{-}11)$$

This may be expressed graphically as follows :

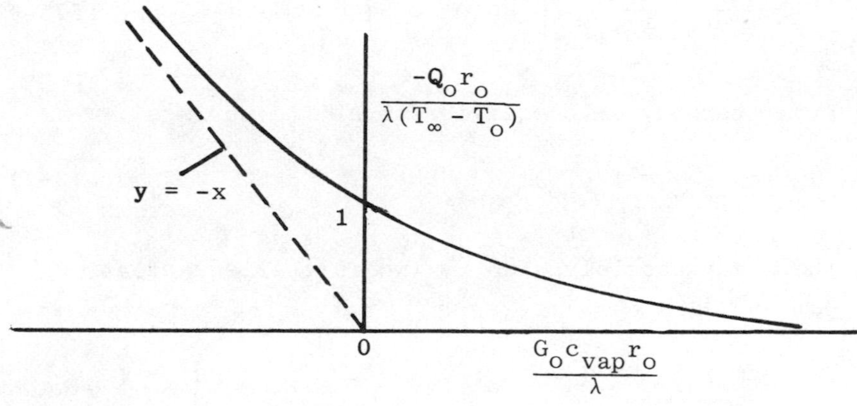

5.3 EQUATIONS FOR DROPLET BEHAVIOUR

A) ASSEMBLY OF EQUATIONS

- From chapter 3 :

$$G_o = \frac{\Gamma_{vap}}{r_o} \ \ln\left(1 + \frac{m_{vap,o} - m_{vap,\infty}}{1 - m_{vap,o}}\right) \qquad .(5.3\text{-}1)$$

- Because the vaporisation reduces the droplet radius :

$$\frac{dr_o}{dt} = - \frac{G_o}{\rho_{\ell iq}} \qquad .(5.3\text{-}2)$$

- From section 5.2B :

$$-Q_o = \frac{c_{vap}\, G_o\, (T_\infty - T_o)}{\exp\left(\frac{G_o c_{vap} r_o}{\lambda}\right) - 1} \qquad .(5.3\text{-}3)$$

- From a heat balance on droplet :

$$-Q_o 4\pi r_o{}^2 = \frac{4}{3}\pi r_o{}^3 c_{\ell iq} \frac{dT_o}{dt} + G_o L\, 4\pi r_o{}^2 \qquad ,$$

i.e. $$\frac{dT_o}{dt} = \frac{3}{r_o c_{\ell iq}} \left(-Q_o - G_o L \right) \qquad .\ (5.3\text{-}4)$$

- From thermodynamic equilibrium at the surface :

$$m_{vap,o} = f\,(T_o) \qquad .(5.3\text{-}5)$$

- There may possibly also be two further equations, namely :

$$\left.\begin{aligned} \frac{dm_{vap,\infty}}{dt} &= \ldots. \\ \text{and} \quad \frac{dT_\infty}{dt} &= \ldots. \end{aligned}\right\} \qquad (5.3\text{-}6)$$

The equations account for the saturation of the atmosphere with vapour, and for its change of temperature. Here $m_{vap,\infty}$ and T_∞ will be regarded as constants for simplicity.

B) LINKAGE OF EQUATIONS

The equations and unknowns of section 5.3A are linked together as shown by the following diagram, in which (1) $\equiv$ eq. 5.3-1, etc. :

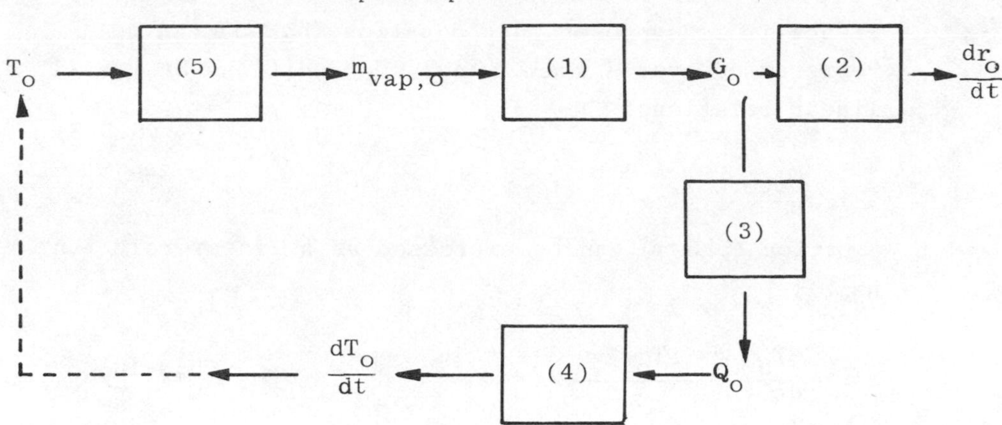

C) REMARKS ABOUT THE EQUATIONS

- These equations form a complete set; i.e. if the initial values of the variables T_o and r_o are given, and the auxiliary quantities $\Gamma_{vap,}$ λ etc. are known, the equations can be solved for all values of time t.

- The equations form a non-linear set through the logarithm and the $m_{vap,o} \sim T_o$ relation . They will therefore require numerical solution on a computer.

- The problem is a very simple one for a computer : there are just two simultaneous first-order differential equations to be solved.

D) APPROXIMATE SOLUTION PROCEDURE

- Approximate solutions can be obtained by way of linearisation. Thus, for small differences of concentration and temperature, equation (5.3-1) becomes :

$$G_o \approx \Gamma_{vap}(m_{vap,o} - m_{vap,\infty})/r_o \quad ; (5.3\text{-}7)$$

and equation (5.3-3) becomes :

$$-Q_o \approx \lambda(T_\infty - T_o)/r_o \quad . (5.3\text{-}8)$$

- If now it is supposed that equation (5.3-5) can be represented over a small range of conditions by the linear relation :

$$m_{vap,o} \approx a + b\,T_o \qquad ,(5.3\text{-}9)$$

equation (5.3-4) can be expressed as a linear relation, namely :

$$\frac{dT_o}{dt} \approx \frac{pT_o + q}{r_o^2} \qquad ,(5.3\text{-}10)$$

where :

$$p \equiv \frac{-3(\lambda + bL\Gamma_{vap})}{c_{\ell iq}} \qquad ,(5.3\text{-}11)$$

and,

$$q \equiv 3\,\frac{\left\{\lambda T_\infty + (m_{vap,\infty} - a)L\Gamma_{vap}\right\}}{c_{\ell iq}} \qquad .(5.3\text{-}12)$$

- From equations (5.3-9) and (5.3-2), however, it can be concluded :

$$\frac{dr_o^2}{dt} \approx kT_o + \ell \qquad ,(5.3\text{-}13)$$

where :

$$k \equiv -\,2b\Gamma_{vap} \qquad ,(5.3\text{-}14)$$

and

$$\ell \equiv 2(m_{vap,\infty} - a)\,\Gamma_{vap} \qquad .(5.3\text{-}15)$$

- Combination of the two differential equations (5.3-10)

and (5.3-13) yields :

$$\frac{dT_o}{dr_o^2} \approx \frac{pT_o + q}{kT_o + \ell} \cdot \frac{1}{r_o^2} \qquad ,(5.3\text{-}16)$$

which can be solved analytically.

- Finally, insertion of $T_o(r_o^2)$ from (5.3-16) into (5.3-13) permits a solution for the time of vaporisation t_{vap} in terms of the quadrature :

$$\boxed{t_{vap} \approx \int_{r_{o,start}}^{o} (kT_o + \ell)^{-1}\, dr_o^2} \qquad .(5.3\text{-}17)$$

- Further analysis is left to the reader. It would confirm, and give quantitative expression to, the expectations displayed in the graphs of section 5.1B.

5.4 EQUILIBRIUM VAPORISATION

A) THE PROBLEM

- Equilibrium vaporisation is defined as the process which ensues when the droplet is injected at such a temperature that the heat transfer to the droplet surface from the gas exactly equals the vaporisation rate times the latent heat of vaporisation.

$$-Q_o = G_o L \qquad .(5.4\text{-}1)$$

- This implies, according to equation (5.3-4), that T_o is invariant with time.

- The task is now to determine the values of T_o, $m_{vap,o}$, and G_o and t_{vap} in these circumstances.

B) DETERMINATION OF T_o AND $m_{vap,o}$

- The following equations are relevant :

(3.2-9): $$G_o = \frac{\Gamma_{vap}}{r_o} \ln\left(1 + \frac{m_{vap,o} - m_{vap,\infty}}{1 - m_{vap,o}}\right) \qquad ,(5.4\text{-}2)$$

(5.2-8): $$G_o = \frac{\lambda}{c_{vap} r_o} \ln\left(1 + \frac{c_{vap}(T_\infty - T_o)}{L}\right) \qquad ,(5.4\text{-}3)$$

(5.3-5): $$m_{vap} - f(T_o) \qquad .(5.4\text{-}4)$$

- Elimination of G_o gives the following relation between the vapour concentration and the temperature at the interface :

$$1 + \frac{(m_{vap,o} - m_{vap,\infty})}{(1 - m_{vap,o})} = \left[1 + \frac{c_{vap}(T_\infty - T_o)}{L}\right]^{\frac{\lambda}{c_{vap}\Gamma_{vap}}} \qquad .(5.4\text{-}5)$$

It should here be noted that :

(1) The quantity $\lambda/(c_{vap}\Gamma_{vap})$ is dimensionless.

(2) For gaseous mixtures of nearly uniform molecular weight, its value is close to unity.

(3) When the molecular weight of the vapour is greater than that of the atmosphere, its value may somewhat exceed unity.

- It is convenient to express $m_{vap,o}$ explicitly in terms of T_o, from (5.4-5), with the result :

$$m_{vap,o} = 1 - \frac{(1 - m_{vap,\infty})}{\left[1 + \frac{c_{vap}(T_\infty - T_o)}{L}\right]^{\{\lambda/(c_{vap}\Gamma_{vap})\}}} \qquad .(5.4\text{-}6)$$

Obviously, when T_∞ tends to infinity or L tends to zero, $m_{vap,o}$ tends to unity.

- The values of $m_{vap,o}$ and T_o which actually prevail can be determined by simultaneous solution of (5.4-4) and (5.4-6).

Both are non-linear. The following graph illustrates the situation.

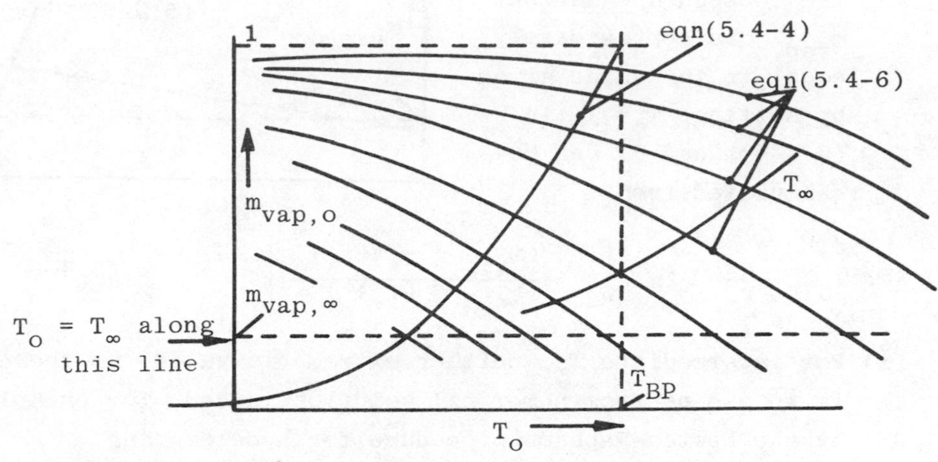

c) CONCLUSIONS

Inspection of the diagram, and of the equations which it expresses, permits the following conclusions to be drawn:-

(i) When T_∞ is much greater than the boiling-point temperature $T_{B.P.}$, equations (5.4-4) and (5.4-6) are solved by a value of $m_{vap,o}$ close to 1 and of T_o close to $T_{B.P.}$.

Then the vaporisation rate is best calculated from :

$$G_o \approx \frac{\lambda}{c_{vap} r_o} \ln \left(1 + \frac{c_{vap}(T_\infty - T_{B.P.})}{L} \right) \qquad .(5.4\text{-}7)$$

(ii) When T_∞ is low, and $m_{vap,\infty}$ is close to zero, T_o is close to T_∞; for, under these circumstances, the slopes of the two curves are very different :

$$\left| \frac{\partial m_{vap,o}}{\partial T_o} \right|_{(5.4\text{-}6)} \gg \left| \frac{\partial m_{vap,o}}{\partial T_o} \right|_{(5.4\text{-}4)} \qquad .(5.4\text{-}8)$$

That this implies $T_o \approx T_\infty$ is shown by the following sketch :

The consequence is that $m_{vap,o}$ is approximately equal to the value given by setting: $T_o = T_\infty$ in (5.4-4); and G_o can be calculated from :

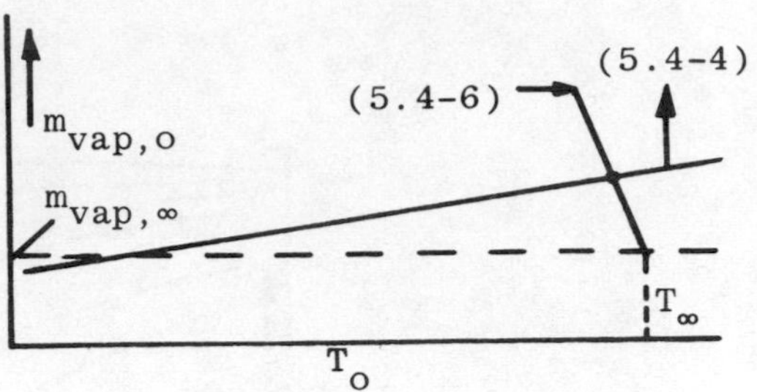

$$G_o \approx \frac{\Gamma_{vap}}{r_o} \ln \left[1 + \frac{(m_{vap,\infty} - f(T_\infty))}{(1 - f(T_\infty))}\right] \quad .(5.4\text{-}9)$$

(iii) For intermediate T_∞, neither extreme is valid; so there is no escape from numerical solution, aided, for example, by the Newton-Raphson procedure for accelerating convergence.

D) VAPORISATION TIME

The time for the complete disappearance of the droplet can be deduced either from :

$$t_{vap} = \frac{D_o^2 \rho_{\ell iq}}{8(\lambda/c_{vap}) \ln \left[1 + \frac{c_{vap}(T_\infty - T_o)}{L}\right]} \quad ,(5.4\text{-}10)$$

or from :

$$t_{vap} = \frac{D_o^2 \rho_{\ell iq}}{8\Gamma_{vap} \ln \left[1 + \frac{m_{vap,\infty} - m_{vap,o}}{1 - m_{vap,o}}\right]} \quad .(5.4\text{-}11)$$

Both are valid; but which is to be used depends on whether T_o or $m_{vap,o}$ is the easier to estimate.

5.5 REFERENCES

Godsave G A E (1953)

"Studies of the combustion of drops in a fuel spray - the burning of single drops of fuel".

Fourth Symposium on Combustion, p. 818, Williams and Wilkins, Baltimore.

EXERCISES TO FACILITATE ABSORPTION OF MATERIAL OF CHAPTER 5

ANALYTICAL PROBLEM

5.1 Develop the linearised theory of section 5.3D to show that the droplet radius r_o and the droplet temperature T_o are related to their initial values $r_{o,i}$ and $T_{o,i}$, by :

$$T_o - T_{o,i} + \left(\frac{\ell}{k} - \frac{q}{p}\right)\ell n \left(\frac{\frac{q}{p} + T_o}{\frac{q}{p} + T_{o,i}}\right) = \frac{p}{k}\,\ell n \left(\frac{r_o^2}{r_{o,i}^2}\right)$$

where the symbols have the meanings defined in the text.

MULTIPLE-CHOICE PROBLEMS

Mark the following items with A, B, C, D or E according to whether the theory of droplet vaporisation under equilibrium conditions implies that the change in conditions indicated in the item,

A. would increase T_o and $m_{vap,o}$.

B. would decrease T_o and $m_{vap,o}$.

C. would leave T_o and $m_{vap,o}$ unchanged.

D. would cause T_o to rise while $m_{vap,o}$ fell.

E. would cause T_o fo fall while $m_{vap,o}$ rose.

5.2 r_o diminishes.

5.3 pressure increases.

5.4 thermal conductivity increases.

5.5 pressure diminishes.

5.6 diffusion coefficient of vapour diminishes.

5.7 L diminishes.

5.8 $m_{vap,\infty}$ increases.

5.9 T_∞ diminishes.

5.10 $m_{vap,\infty}$ and T_∞ both diminish.

5.11 thermal conductivity and diffusion coefficient are both increased by the same factor.

5.12 A droplet of liquid at its boiling point, when injected into a gaseous atmosphere at the same temperature will tend to become cooler

<u>because</u>

the vaporisation rate must be infinite when $m_{vap,o}$ equals unity.

5.13 A droplet injected into its own saturated vapour cannot vaporise

<u>because</u>

when $m_{vap,o}$ equals $m_{vap,\infty}$ the driving force for mass transfer is zero.

5.14 When steam condenses on to droplets of cold water injected into it in a spray condenser, the rate of condensation depends upon the rate of heat conduction within the liquid

<u>because</u>

the thermal conductivity of the gas is much greater than that of the liquid.

5.15 When a droplet is injected into very hot gas, its equilibrium temperature is close to its boiling temperature at the prevailing pressure

<u>because</u>

small differences in $m_{vap,o}$ in the neighbourhood of unity make large differences in the rate of vaporisation.

5.16 A rain-drop vaporises into atmospheric air at a rate depending upon the difference between the humidity of the air and the humidity which saturated air would have at the prevailing atmospheric temperature

because

the value of $\lambda/(c_{vap}\Gamma_{vap})$ is close to unity for water vapour in air.

Mark the following equations A, B, C, D or E according to whether they represent, for a steady-state system of point symmetry :

A the total-mass conservation equation.

B the energy conservation equation.

C the vapour-conservation equation.

D a relation expressing thermodynamic relation.

E an incorrect equation.

5.17 $G_o c_{vap} r_o/\lambda = \ln\left[1 + c_{vap}(T_\infty - T_o)/(-Q_o/G_o)\right]$

5.18 $Gr_o^2 = G_o r^2$

5.19 $\dfrac{dp_{vap,o}}{dT_o} = \dfrac{L}{v_o T_o}$

5.20 $G_o = \dfrac{\Gamma_{vap}}{r_o} \ln\left[1 + \dfrac{(m_{vap,\infty} - m_{vap,o})}{(1 - m_{vap,o})}\right]$

5.21 $r^2 \lambda \dfrac{dT}{dr} = \{G_o c_{vap}(T - T_o) - Q_o\}\, r_o^2$

5.22 $m_{vap,o} = f(T_o)$

5.23 $m_{vap,o} = 1 - \dfrac{(1 - m_{vap,\infty})}{\left[1 + \dfrac{c_{vap}(T_\infty - T_o)}{L}\right]^{c_{vap}\Gamma_{vap}/\lambda}}$

When a volatile liquid is in contact with a mixture of its own vapour and a permanent gas, and its temperature is :

5.24 far below the boiling point (B.P.),

5.25 equal to the B.P.,

5.26 just below the B.P.,

5.27 just above the B.P.,

the laws of thermodynamics imply that :

A thermodynamic equilibrium cannot prevail.

B $m_{vap,o}$ is close to zero.

C $m_{vap,o}$ exceeds unity.

D $m_{vap,o}$ is just below unity.

E $m_{vap,o}$ equals unity.

5.28 Of the following possible features of a vaporising-droplet model :

(i) Spherical symmetry,

(ii) Steady state,

(iii) Uniform gas density,

(iv) $\Gamma_{vap} = \lambda/C_{vap}$

(v) Uniform liquid temperature,

(vi) Uniform gas temperature,

(vii) Thermodynamic equilibrium at the surface,

(viii) $m_{vap} << 1$,

the model discussed in Chapter 5 employs :

A all.

B all except (vi) and (viii).

C (i), (ii), (iii), (vii) only.

D (v), and some others, but not (vi).

E (vii), (viii) and some others.

5.29 In the gas-phase region adjoining a vaporising liquid-fuel surface, the ratio (vapour transfer by convection)/(net vapour transfer) is given by :

A $m_{vap,o}$

B $m_{vap,o}/(1-m_{vap,o})$

C $-m_{vap,o}$

D $-(\Gamma_{vap} dm_{vap}/dr)_o/G_o$

E None of the above.

5.30 All the following statements are true except :

A the temperature T_S, which is taken up by a droplet vaporising into a vapour-gas mixture, is known as the "wet-bulb" temperature of the mixture.

B the wet-bulb temperature is always lower than both the mixture temperature and the boiling point of the liquid at the prevailing pressure.

C the concentration of vapour at the droplet surface exceeds that in the bulk of the gas.

D the wet-bulb temperature is lower than the dew-point of the mixture.

E the convection and diffusion contributions to the total vapour transfer have the same sign.

The equation: $\ln\left(1 + \frac{m_{vap,o} - m_{vap,\infty}}{1 - m_{vap,o}}\right) = \frac{G_o r_o}{\Gamma_{vap}}$, implies

that, when :

5.31 $m_{vap,\infty} = 0$ and $m_{vap,o} \neq 0$,

5.32 $m_{vap,\infty} = m_{vap,o} \neq 1$,

5.33 $m_{vap,\infty} = m_{vap,o} = 1$,

the rate of vaporisation per unit area, G_o :

A equals zero.

B must be determined by some other consideration.

C becomes infinite.

D is positive and finite.

E is negative and finite.

The vaporisation time of a liquid-fuel droplet immersed in the gaseous products of combustion, is likely to increase in proportion to :

5.34 The first power,

5.35 The square,

5.36 The power - 1,

of :

A the pressure of the gas.

B the thermal conductivity of the mixture.

C the density of the liquid.

D the latent heat of vaporisation of the liquid.

E the initial diameter of the droplet.

5.37 It is possible to construct mathematical models, more complex than that described in the lectures, to take quantitative account of all of the following facts except :

A the fuel droplets may enter the gas at a temperature below that of the wet bulb.

B the gas temperature and composition change during the course of vaporisation.

C the Reynolds number is finite.

D some fuels "crack" in the liquid phase.

E most fuels are mixtures of components having different volatilities.

5.38 The rate of vaporisation rises to a finite maximum as $m_{vap,o}$ increases

<u>because</u>

it is not possible for $m_{vap,o}$ to exceed unity.

5.39 If the diffusion coefficient varies with m_{vap}, the vaporisation time will still be proportional to the square of the initial radius,

because

the thermal conductivity will probably vary in proportion.

5.40 If a droplet is injected at a temperature below its wet-bulb temperature, its vaporisation time will exceed that which is implied by equilibrium-vaporisation theory

because

less heat will be required to vaporise it.

5.41 It is possible for a droplet to vaporise at a finite rate, even when $m_{vap,o}$ equals $m_{vap,\infty}$

because

$m_{vap,o}$ may equal unity.

ANSWERS TO MULTIPLE-CHOICE PROBLEMS

Answer	Problem number (5's omitted)
A	6, 8, 15, 27, 32, 41
B	4, 7, 9, 10, 12, 16, 17, 21, 24, 33, 36, 39
C	2, 11, 14, 20, 34, 40
D	3, 19, 22, 26, 28, 29, 30, 31, 37, 38
E	5, 13, 18, 23, 25, 36

DISCUSSION PROBLEMS

5.42 As for problem 1.43.

5.43 In some rocket motors, the pressure in the combustion chamber exceeds the critical pressure of the propellant, which is however injected far below its critical temperature. Consider whether the vaporisation-time formulae are valid. If they are, is it better to use the formula based on concentrations or that based on temperatures? If they are not valid, what is the reason?

CHAPTER 6

MASS TRANSFER III

6.1 INTRODUCTION

A) MOTIVATION

So far chemical reaction has been specifically excluded from the problems considered; yet the aim of the lectures is to show how combustion phenomena can be predicted. The present lecture advances in this direction.

Real combustion phenomena involve hundreds of interacting chemical reactions; yet their overall effects are often quite simply described : fuel and oxygen disappear; carbon dioxide and steam take their places; the temperature rises and/or heat is produced. It is therefore useful to focus attention on a model of combustion which accords with reality in respect of the overall effects, but which suppresses the distracting intermediate details.

The purpose of so doing is to generate quantitative predictions of combustion phenomena which are easy to make and understand, which fit reality in its main features, which are without positively misleading auxiliary implications, and which can be refined when necessary.

B) THE MAIN IDEA

The model to be introduced is the Simple Chemically-Reacting System (SCRS), defined in three parts having differing levels of importance. The SCRS refers just to three substances : fuel, oxidant and product. It also involves an inessential but convenient assumption about their thermodynamic properties, and a more far-reaching assumption about their transport properties.

A major consequence of the last-mentioned

assumption is that the state of the gaseous mixture at any point in a flame can be characterised by just two parameters : one expresses the result of mixing, whether molecular (i.e. by diffusion and heat conduction) or turbulent; the other expresses the result of chemical reaction.

Even greater simplicity is introduced when the latter parameter can be taken as having one or other of two extreme values, corresponding either to complete lack of reaction (as when a combustible gas flows into the atmosphere without being ignited), or to reaction completed to the fullest possible extent (as when the mixture composition is in thermodynamic equilibrium at all points). The latter extreme is nearly attained in many circumstances.

C) PRACTICAL RELEVANCE

With the aid of the ideas now to be introduced, it will be possible :-

(i) to calculate the burning rates of liquid-fuel droplets.

(ii) to predict the length and shape of laminar and turbulent diffusion flames.

(iii) to calculate the burning times of solid-fuel particles.

(iv) to lay a foundation for the later treatment of combustion phenomena in which chemical kinetics plays a significant part, for example :- spark ignition, flame propagation through a combustible mixture, stabilisation by re-circulation of combustion products.

6.2 THE SIMPLE CHEMICALLY REACTING SYSTEM

A) DEFINITION

(i) Part 1. The SCRS involves a reaction between two reactants (fuel and oxidant) in which these combine, in fixed proportions by mass, to produce a unique product :

fuel + oxidant → product

1 kg skg 1 + s kg

Note that, in real reactions by contrast;

- there may be more than one combining proportion, e.g.

$$C + O_2 \rightarrow CO_2 \quad ,$$

or $$C + \tfrac{1}{2}O_2 \rightarrow CO \quad ;$$

- the reactions may proceed by more than one step, e.g. H_2 and O_2 proceed to H_2O via H, OH, O, HO_2, etc.; and hydrocarbons break down into innumerable intermediates.

(ii) Part 2. The specific heats of all mixture components (fuel, oxidant, product, diluent) are equal, and independent of temperature, i.e.

$$c_j = c \quad . \quad (6.2\text{-}1)$$

Note that this Part is not necessary; but, while making big simplifications in the chemistry, one might as well make such minor ones in the thermodynamics as bring algebraic ease at small cost in reality.

(iii) Part 3. All the transport properties, Γ_{fu}, Γ_{ox}, Γ_{prod}, Γ_{dil} and λ/c are equal at any point in the mixture. They need not be uniform, however (i.e. they can have different values from point to point).

(iv) Example; the oxidation of methane.

- Methane (CH_4), when it burns with air, is believed to proceed by way of the following reactions, in which M stands for any molecule :

$CH_4 + M$	$\rightleftharpoons$	$CH_3 + H + M$
$CH_4 + O$	$\rightleftharpoons$	$CH_3 + OH$
$CH_4 + H$	$\rightleftharpoons$	$CH_3 + H_2$
$CH_4 + OH$	$\rightleftharpoons$	$CH_3 + H_2O$
$CH_3 + O$	$\rightleftharpoons$	$HCHO + H$
$CH_3 + O_2$	$\rightleftharpoons$	$HCHO + OH$
$CH_3 + O_2$	$\rightleftharpoons$	$HCO + H_2O$
$HCO + OH$	$\rightleftharpoons$	$CO + H_2O$
$HCHO + OH$	$\rightleftharpoons$	$H + CO + H_2O$
$CO + OH$	$\rightleftharpoons$	$CO_2 + H$
$H + O_2$	$\rightleftharpoons$	$OH + O$
$O + H_2$	$\rightleftharpoons$	$H + OH$
$O + H_2O$	$\rightleftharpoons$	$2\ OH$
$H + H_2O$	$\rightleftharpoons$	$H_2 + OH$
$H_2O + M$	$\rightleftharpoons$	$H + OH + M$
$HCO + M$	$\rightleftharpoons$	$H + CO + M$

- In the SCRS version of methane combustion, however, it is supposed that this complex set of reactions can be replaced by the following :

$$CH_4 + 2O_2 \rightleftharpoons CO_2 + 2H_2O.$$

The implication is that the intermediate species (H_2, H, O, OH, CO, HCO, HCHO, CH_3), though they may play vital roles in bringing the combustion about, do not themselves attain concentrations of the same order as those of the main reactants (CH_4 and O_2) and the main products (CO_2 and H_2O).

- The mixture having two molecules of H_2O to one of CO_2 is regarded as a single substance, i.e. the product.

- Ordinarily the nitrogen of the air is present; and this enters in practice into additional chemical reactions with oxygen and hydrogen of which the most important are believed to be :

$$\begin{array}{lcl} O + N_2 & \rightleftharpoons & NO + N \\ N + O_2 & \rightleftharpoons & NO + O \\ N_2 + O_2 & \rightleftharpoons & N_2O + O \\ N_2O + O & \rightleftharpoons & 2NO \\ N_2 + O_2 & \rightleftharpoons & 2NO \\ N + OH & \rightleftharpoons & NO + H \\ N_2 + OH & \rightleftharpoons & N_2O + H \end{array}$$

In the SCRS version however, N_2 is supposed to be a simple diluent, entering no chemical reaction at all.

- Since the molecular weights of methane and oxygen are respectively 16 and 32 (to a sufficient accuracy for our purposes), the value of the stoichiometric ratio s of the SCRS definition is 4.0, i.e. 4 kg of O_2 are required for the complete combustion of 1 kg of CH_4.
- The composition of air is given, with sufficient accuracy for our purposes, by the following table :

Gas	Molecular wt.	% by mass	% by volume
O_2	32	23.2	21.0
N_2	28	76.8	79.0
Air	29	100	100

It follows that 1 kg of CH_4 requires 4 ÷ 0.232 i.e. 17.24 kg of air for its complete combustion.

- The specific heats of methane, oxygen, nitrogen, carbon dioxide and steam are all different; and they vary with temperature. However, for the purposes of the SCRS, all the specific heats can be taken as equal to each other and to a constant value, say 1100 J/kg C.

Because of the actual variation of the specific heats with temperature, the heat of combustion of methane (i.e. the enthalpy decrease on reaction for isothermal constant-pressure reaction) depends upon the temperature level at which the reaction occurs. Within the framework of the SCRS, however, the heat of combustion can be given a single value, say 4.0 x 10^7 J/kg.

B) THE DEFINITION OF ENTHALPY FOR A SCRS

- The general definition of specific enthalpy for an ideal-gas mixture, and its relation to specific heats and temperature, are provided by equations (4.2-1),(4.2-2) and (4.2-3). Introduction of (6.2-1) from Part 2 of the SCRS definition yields the result :

$$h = c(T - T_o) + \Sigma\, m_j h_{j,o} \quad , \quad (6.2\text{-}2)$$

where : T_o stands for an arbitrary reference temperature, $h_{j,o}$ is the specific enthalpy of species j at that reference temperature, and Σ denotes summation over all components of the mixture.

- Some of the $h_{j,o}$'s can be set equal to zero, but not all simultaneously. Specifically, the $h_{j,o}$'s must obey a relation which ensures that, when 1 kg of fuel and s kg of oxidant disappear to form (1 + s) kg of product, the decrease of enthalpy at constant temperature must equal H, the heat of combustion of the fuel. This implies :

$$h_{fu,o} + s\, h_{ox,o} = (1 + s)h_{prod,o} + H \quad . \quad (6.2\text{-}3)$$

- Innumerable choices are possible within this limitation, of which the following alternatives are convenient :-

(i) Choose :

$$T_o = 0 \quad ; \quad (6.2\text{-}4)$$

$$h_{ox,o} = h_{prod,o} = h_{dil,o} = 0 \quad . \quad (6.2\text{-}5)$$

Then, from (6.2-3) :

$$h_{fu,o} = H \quad , (6.2\text{-}6)$$

and :

$$\boxed{h = c\,T + m_{fu}\,H} \quad . (6.2\text{-}7)$$

(ii) Choose T_o as before and :

$$h_{fu,o} = h_{prod,o} = h_{dil,o} = 0 \quad . (6.2\text{-}8)$$

Then there follows :

$$h_{ox,o} = H/s \quad , (6.2\text{-}9)$$

and :

$$\boxed{h = c\,T + m_{ox}\,H/s} \quad . (6.2\text{-}10)$$

Both these choices will be used, at different points, in later developments.

c) THE TWO-STREAM MIXING PROCESS

An idealisation of many combustion-chamber flows is represented by the diagram, according to which fluid from a "first" stream, F, flows in at the rate f kg/s,

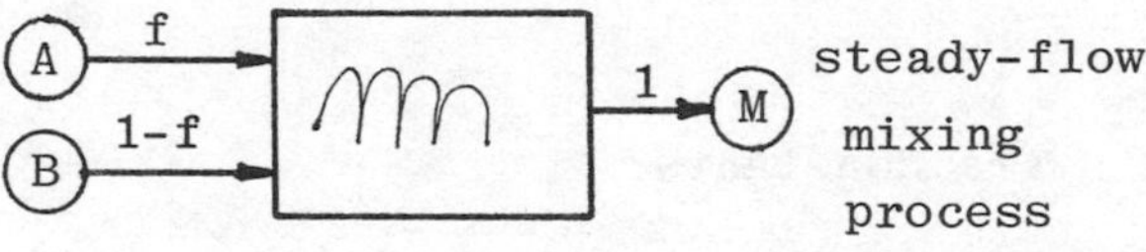

and fluid from an "auxiliary" stream, A, flows in at the rate (1-f) kg/s, to form a "mixture", M, which flows out at the rate of 1 kg/s. (Note that it will sometimes, but not always, be useful to regard F as standing for "fuel" and A as standing for "air"; for often the fuel enters in a pure state, while the oxidant is oxygen, which enters mixed in atmospheric proportions with nitrogen.)

- Any extensive property of a fluid ϕ, which is free from sources and sinks, obeys the following relation for such a two-stream mixing process, as a matter of definition :

$$f\phi_F + (1 - f)\phi_A = \phi_M \quad , (6.2\text{-}11)$$

which leads to :

$$f = \frac{\phi_M - \phi_A}{\phi_F - \phi_A} \quad . (6.2\text{-}12)$$

Such a property can be referred to conveniently by calling it a "conserved property".

- It is easy to prove (see tutorial exercises) that the following properties are conserved properties in this sense :

m_{in},	the mass fraction of a chemically inert mixture component;
m_{dil},	since a diluent is by definition inert;
$m_{fu}-m_{ox}/s$,	a "composite" mass fraction;
$m_{fu}+m_{prod}/(1+s)$,	another;
$m_{ox}/s+m_{prod}/(1+s)$,	yet another;

together with any linear combination of such conserved properties, i.e.

$$a_1\phi_1 + a_2\phi_2 + \ldots . + b,$$

where a_1, a_2, ... and b are constants.

- Thus, to take one particular ϕ, the first composite one, we conclude that, in a two-stream mixing process, the fuel and oxidant mass fractions are linked with f as follows :

$$f = \frac{(m_{fu} - m_{ox}/s)_M - (m_{fu} - m_{ox}/s)_A}{(m_{fu} - m_{ox}/s)_F - (m_{fu} - m_{ox}/s)_A} \quad . (6.2\text{-}13)$$

- In the particular case in which the F stream contains only fuel, and the A stream contains oxygen but no fuel (this is common in combustion practice), it follows :

$$m_{fu,A} = 0 \qquad , (6.2\text{-}14)$$

$$m_{fu,F} = 1,\ m_{ox,F} = 0 \qquad , (6.2\text{-}15)$$

$$f = \frac{(m_{fu} - m_{ox}/s)_M + m_{ox,A}/s}{1 + m_{ox,A}/s} \qquad . (6.2\text{-}16)$$

- If chemical reaction is complete within the mixing chamber, either fuel or oxidant will have zero concentration in the M state. The first case occurs when f is smaller than f_{stoich}, the "stoichiometric" value, the second when f exceeds f_{stoich}. These two cases, and the intermediate one, are represented by the following deductions from (6.2-16) :

$$f < f_{stoich} : \quad f = \frac{-m_{ox,M}/s + m_{ox,A}/s}{1 + m_{ox,A}/s} \qquad ; (6.2\text{-}17)$$

$$f < f_{stoich} : \quad f = \frac{m_{fu,M} + m_{ox,A}/s}{1 + m_{ox,A}/s} \qquad ; (6.2\text{-}18)$$

$$f = f_{stoich} : \quad f = \frac{m_{ox,A}/s}{1 + m_{ox,A}/s} \qquad . (6.2\text{-}19)$$

The last of these relations can serve as a definition of f_{stoich}.

- f is sometimes called the "mixture fraction". It is to be distinguished quite clearly from the mass fraction of fuel, m_{fu}, and the fuel/air ratio. The tutorial exercises will aid the comprehension of the distinction.

D) THE TWO-STREAM ADIABATIC MIXING PROCESS

- If the steady-flow mixing process is adiabatic, and free from work input, the stagnation enthalpy $\tilde{h}$ is a conserved property. $\tilde{h}$ is defined by :

$$\tilde{h} \equiv h + V^2/2 \qquad . \quad (6.2\text{-}20)$$

As a consequence, f and h are related by :

$$f = \frac{\tilde{h}_M - \tilde{h}_A}{\tilde{h}_F - \tilde{h}_A} \qquad . \quad (6.2\text{-}21)$$

- If the kinetic-energy term can be neglected, of course we have :

$$f = \frac{h_M - h_A}{h_F - h_A} \qquad . \quad (6.2\text{-}22)$$

- If further the enthalpy is defined by (6.2-7), and the F stream consists of pure fuel while there is none in the A stream, we conclude :

$$f = \frac{(cT + m_{fu}H)_M - cT_A}{cT_F + H - c\,T_A} \qquad . \quad (6.2\text{-}23)$$

- Finally, if the fuel reacts to the fullest possible extent, one has the following set of $f \sim T_M$ relations to accompany the $f \sim m_{fu,M}$ and $f \sim m_{ox,M}$ relations (6.2-17),(6.2-18) and (6.2-19) :

$$f < f_{stoich} \; : \; f = \frac{T_M - T_A}{T_F - T_A + H/c} \qquad ; \quad (6.2\text{-}24)$$

$$f > f_{stoich} \; : \; f = \frac{T_M - T_A - \dfrac{m_{ox,A}}{s}\dfrac{H}{c}}{T_F - T_A - \dfrac{m_{ox,A}}{s}\dfrac{H}{c}} \qquad ; \quad (6.2\text{-}25)$$

$$f = f_{stoich} \; : \; T_M = T_A + f_{stoich}\,(T_F - T_A + H/c) \qquad . \quad (6.2\text{-}26)$$

- Here it should be observed that, in order to provide relations of maximum utility, equation (6.2-18) was invoked to eliminate $m_{fu,M}$ from (6.2-23). Alternatively, of course, one might have started from the different definition of h, equation (6.2-10).

- The last equation, (6.2-26), is a formula for the adiabatic flame temperature of a Simple Chemically-Reacting System.

6.3 DIFFERENTIAL EQUATIONS OF TRANSPORT OF MATTER AND ENERGY

A) MOTIVATION

The equations governing the conservation of species and energy contain the reaction rate terms R_j. Often these are not known with any precision.

Fortunately there are many processes, namely the "diffusion-controlled" flames, for which precise knowledge of the R_j's is not necessary.

These will be brought into focus by derivation of special differential equations from which the R_j's are absent, even for situations for which chemical reaction is present.

Attention will be confined to the one-dimensional steady-state situation for simplicity.

B) EQUATIONS FOR FUEL AND OXIDANT

- Equation (2.5-5) can be written, first for fuel and then for oxidant, as :

$$G \frac{dm_{fu}}{dx} - \frac{d}{dx}\left(\Gamma_{fu} \frac{dm_{fu}}{dx}\right) = R_{fu} \qquad , (6.3\text{-}1)$$

and :

$$G \frac{dm_{ox}}{dx} - \frac{d}{dx}\left(\Gamma_{ox} \frac{dm_{ox}}{dx}\right) = R_{ox} \qquad . (6.3\text{-}2)$$

- Because the fuel and oxidant engage in a simple chemical reaction, with a fixed stoichiometric ratio s, it follows that the reaction rates are related by :

$$R_{ox} = s\, R_{fu} \qquad . (6.3\text{-}3)$$

- Moreover, part 3 of the definition of the SCRS states :

$$\Gamma_{ox} = \Gamma_{fu} \equiv \Gamma_{fuox} \text{ , say} \qquad . (6.3\text{-}4)$$

- Combination of the above four equations leads to :

$$\boxed{G \frac{d}{dx}\left(m_{fu} - \frac{m_{ox}}{s}\right) - \frac{d}{dx}\left\{\Gamma_{fuox}\frac{d}{dx}\left(m_{fu} - \frac{m_{ox}}{s}\right)\right\} = 0} \qquad .(6.3\text{-}5)$$

- Thus the composite variable $(m_{fu} - m_{ox}/s)$, which in section 6.2 C was revealed as a "conserved property", obeys the same differential equation as does the mass fraction of a chemically inert material. The source term has disappeared.

- Although proved for 1D steady flow, this result is generally true for an SCRS.

- It is just as easy to prove that the variables $\{m_{fu} + m_{prod}/(1 + s)\}$ and $\{m_{ox}/s + m_{prod}/(1 + s)\}$ obey equations of the same kind. These were of course also revealed to be conserved properties in Section 6.2C.

c) THE ENERGY EQUATION

- Equation (4.3-13) can be written, by reason of (4.3-14), in the following form for an SCRS from which shear work, kinetic-energy effects, and external energy sources are absent :

$$c\,G\frac{dT}{dx} - \frac{d}{dx}\left(\lambda\frac{dT}{dx}\right) = H\,R_{fu} \qquad . (6.3\text{-}6)$$

- If equation (6.3-1) is multiplied by H and added to (6.3-6), and if it is recalled that, by reason of part 3 of the definition of the SCRS, the following relation holds :

$$\lambda/c = \Gamma_{fu} \equiv \Gamma_{hfu}\text{ , say} \qquad , (6.3\text{-}7)$$

there results :

$$G\frac{d}{dx}(cT + Hm_{fu}) - \frac{d}{dx}\left\{\Gamma_{hfu}\frac{d}{dx}(cT + Hm_{fu})\right\} = 0 \qquad . (6.3\text{-}8)$$

- Since the "composite" dependent variable of this equation is one of the possible definitions of h, it follows

$$\boxed{G\frac{dh}{dx} - \frac{d}{dx}\left(\Gamma_{hfu}\frac{dh}{dx}\right) = 0} \quad . \quad (6.3\text{-}9)$$

Therefore enthalpy also obeys a differential equation, in these circumstances, which lacks a source term.

- It is not difficult to show that equation (6.3-9) holds for any other definition of h, for example (6.2-10).

D) FURTHER EQUATIONS

- It is a generally valid theorem that, if ϕ_I and ϕ_{II} obey identical homogeneous second-order differential equations, any linear combination of these, e.g.

$$\phi_{III} \equiv a\phi_I + b\phi_{II} + c \quad , \quad (6.3\text{-}10)$$

also obeys an identical equation. This is easily proved in terms of the equation we are concerned with, as follows :

$$G\frac{d\phi_{III}}{dx} - \frac{d}{dx}\left(\frac{\Gamma d\phi_{III}}{dx}\right)$$

$$= a\left\{\frac{Gd\phi_I}{dx} - \frac{d}{dx}\left(\frac{\Gamma d\phi_I}{dx}\right)\right\}$$

$$+ b\left\{\frac{Gd\phi_{II}}{dx} - \frac{d}{dx}\left(\frac{\Gamma d\phi_{II}}{dx}\right)\right\} = 0 \quad . \quad (6.3\text{-}11)$$

- This being so, if the equations for two ϕ's are valid in the same integration domain, their solutions must be related by :

$$\frac{\phi_I - \phi_{I,1}}{\phi_{I,2} - \phi_{I,1}} = \frac{\phi_{II} - \phi_{II,1}}{\phi_{II,2} - \phi_{II,2}} \quad , \quad (6.3\text{-}12)$$

where the subscripts 1 and 2 denote the values at the two particular locations which define the integration constants. This can be proved by defining :

$$\phi_{III} \equiv \frac{\phi_I - \phi_{I,1}}{\phi_{I,2} - \phi_{I,1}} - \frac{\phi_{II} - \phi_{II,1}}{\phi_{II,2} - \phi_{II,1}} \quad ; \quad (6.3\text{-}13)$$

this is a quantity which must be zero at the two defining locations; and, since the differential equation is homogeneous (i.e. source-free), the only possible solution is :

$$\phi_{III} = 0, \text{ throughout} \quad . \quad (6.3\text{-}14)$$

- If a quantity f is defined by :

$$f \equiv (h - h_A)/(h_F - h_A) \quad , \quad (6.3\text{-}15)$$

or by :

$$f \equiv \frac{(m_{fu} - m_{ox}/s) - (m_{fu} - m_{ox}/s)_A}{(m_{fu} - m_{ox}/s)_F - (m_{fu} - m_{ox}/s)_A} \quad , \quad (6.3\text{-}16)$$

this quantity will obey the same differential equations as do h and $(m_{fu} - m_{ox}/s)$, because their equations are second-order, and source-free. Then f obeys the differential equations, for an SCRS :

$$\boxed{G\frac{df}{dx} - \frac{d}{dx}\left(\Gamma\frac{df}{dx}\right) = 0} \quad . \quad (6.3\text{-}17)$$

- If f is defined in either of these ways, in view of the relations derived in section 6.2, it can usefully be thought of as being the mixture fraction for a 2-stream mixing process which <u>could have</u> produced gas of the local condition, even if this was actually produced by a different process involving diffusion, heat conduction and chemical reaction.

E) DISCUSSION

- The differential equations (6.3-5), (6.3-9) and (6.3-17) are extremely useful in flame theory, not least because they do not contain chemical-kinetic terms (sources).
- They are the one-dimensional steady-state versions of more general equations.
- They are especially useful when the reaction can be presumed

to go to completion. The next section concerns this.

6.4 THE PROPERTIES OF FULLY REACTED GASES

A) PRACTICAL RELEVANCE

- In many flames the fuel and oxidant are supplied separately, and the mixing and burning take place close at hand. Examples are :

(i) The Bunsen burner with closed air inlet.

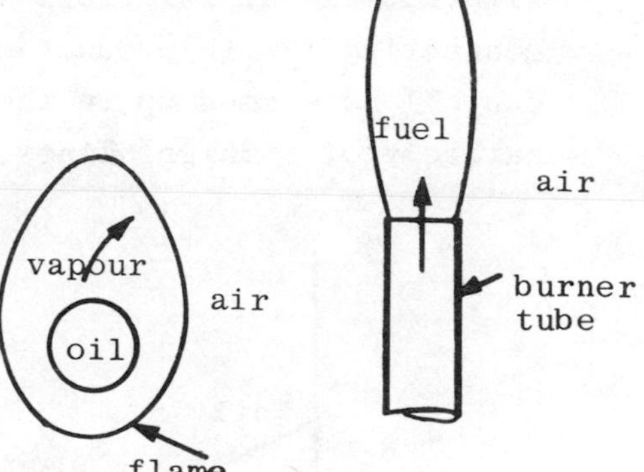

(ii) The oil droplet.

(iii) The industrial furnace, in which a steam-air mixture is injected at high velocity, surrounded by a stream of air, into the furnace space.

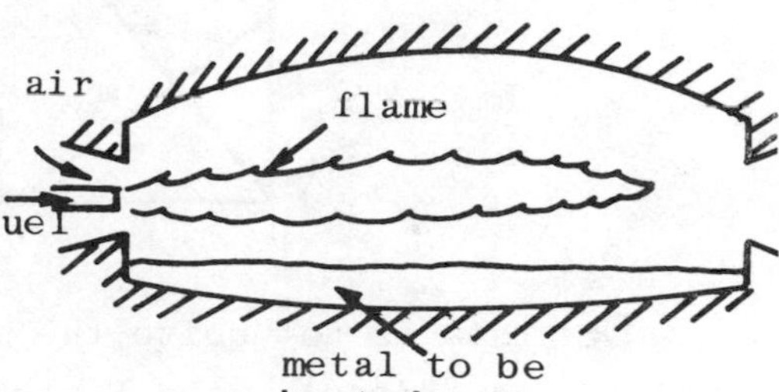

- In such flames, provided that the SCRS assumptions are obeyed, it is meaningful to characterise the gases at any point in the flame by f, defined by (6.3-15) or (6.3-16); it does not matter which. Thus the mixing pattern can be described by a series of constant-f surfaces as indicated.

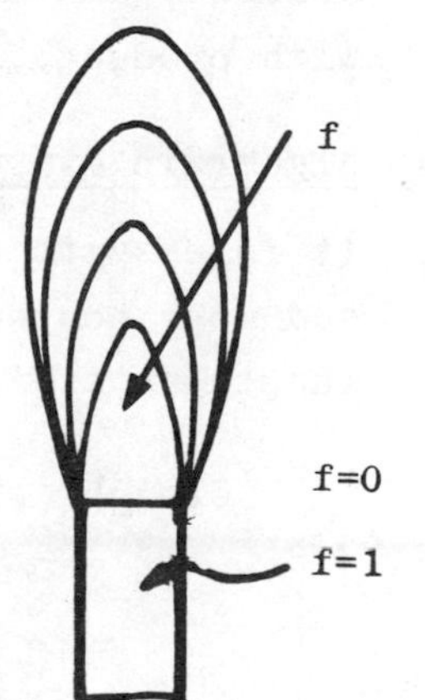

- Usually, the chemical-reaction rate is so fast that the gases are almost fully reacted at every point. To neglect the "almost" is therefore a convenient simplification.

B) CONSEQUENCES FOR RELATIONS BETWEEN CONCENTRATIONS

(i) Graphical expression

The relations (6.2-17) (6.2-18) and (6.2-19) together with additional relations which can be derived connecting f with product and diluent concentrations, can all be summed up in the following graph, consisting entirely of straight lines.

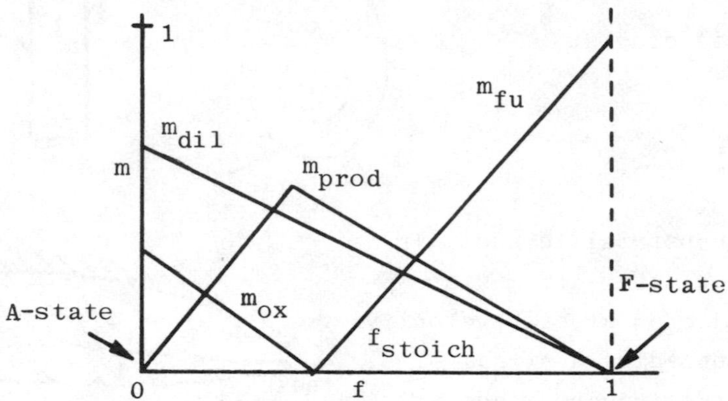

This is not quite the most general form (nor were the equations) : the F stream could, in general have some diluent; and both streams might already be contaminated with product.

(ii) Algebraic expression

It is easy to derive equations for the various lines, and most convenient to do so in terms of linear functions of f and f_{stoich}. For example :

$$f < f_{stoich} : m_{fu} = 0 \qquad , \quad (6.4\text{-}1)$$

$$m_{ox} = m_{ox,A} \frac{(f_{stoich} - f)}{f_{stoich}} \quad ; \quad (6.4\text{-}2)$$

$$f > f_{stoich} : m_{ox} = 0 \quad , \quad (6.4\text{-}3)$$

$$m_{fu} = \frac{f - f_{stoich}}{1 - f_{stoich}} \quad ; \quad (6.4\text{-}4)$$

$$\text{any } f : m_{dil} = m_{dil,A} (1 - f) \quad , \quad (6.4\text{-}5)$$

$$m_{prod} = 1 - m_{dil} - m_{ox} - m_{fu} \quad . \quad (6.4\text{-}6)$$

The value of f_{stoich}, of course, can be deduced from equation (6.2-19).

c) CONSEQUENCES FOR RELATIONS INVOLVING TEMPERATURE

(i) Graphical expression

Provided that there are no sources of energy, and that kinetic heating is negligible, (6.3-15) shows that f and h are linearly related. Further consequences are shown by the straight-line diagram.

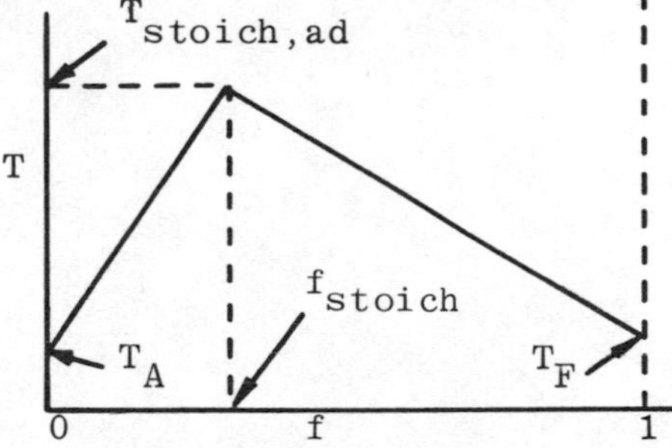

(ii) Algebraic expressions

Equation (6.2-26) implies that the temperature in the stoichiometric adiabatic mixture is given by :

$$T_{stoich,ad} = T_A + (T_F - T_A + H/c) f_{stoich} \quad . \quad (6.4\text{-}7)$$

The equations for the two branches of the temperature profile can then be deduced from :

$$f < f_{stoich} : T = T_A + \frac{f}{f_{stoich}} (T_{stoich,ad} - T_A) \quad , \quad (6.4\text{-}8)$$

$$f > f_{stoich} : T = T_F + \frac{(1-f)}{\left(1-f_{stoich}\right)}(T_{stoich,ad}-T_F) \quad .(6.4\text{-}9)$$

c) CLOSURE

The results derived in this chapter are of great importance for the understanding of flame phenomena. The diagrams of sections 6.4 B and 6.4 C should be memorised.

EXERCISES TO FACILITATE ABSORPTION OF MATERIAL OF CHAPTER 6

MULTIPLE-CHOICE PROBLEMS

6.1 The Simple Chemically Reacting System is defined in such a way as to permit the fuel and oxidant to combine in variable proportions

because

most fuels contain carbon, which is capable of forming two distinct oxides, CO and CO_2.

6.2 Uniformity of specific heats is made one of the features of the SCRS

because

the specific heats of all common gases appearing in flames differ from one another by only a few per cent.

6.3 Equality of the transport properties (Γ's, λ/c) is made a feature of the SCRS

because

then composite fluid properties can be defined which obey source-free differential equations.

6.4 The existence of the species O, OH, HCO, etc., is ignored in the SCRS version of the methane $\sim$ oxygen reaction

because

their concentrations are usually much smaller than those of CH_4, O_2, H_2O and CO_2.

6.5 Oxides of nitrogen do feature in the SCRS version of methane-air combustion

because

the oxide NO usually has as high a concentration in the exhaust gases as does CO_2.

6.6 The mass fraction of oxygen in air is numerically greater than the mole fraction of oxygen in air,

<u>because</u>

the molecular weight of oxygen is greater than that of nitrogen.

6.7 The concepts of "mixture fraction" and "two-stream mixing process" are useful aids to description and understanding

<u>because</u>

the composition and temperature of the gases within most steady-flow combustion chambers vary little from one location to another.

6.8 A conserved property ϕ obeys the equation

$$f\phi_F + (1 - f)\phi_A = \phi_M$$

<u>because</u>

that is how conserved properties are defined.

6.9 The properties m_{in}, m_{dil}, m_{fu} and m_{ox} can all be regarded as conserved properties for an SCRS

<u>because</u>

inert substances and diluents take no part in the reaction, and fuel and oxygen combine in constant proportions.

6.10 $\{m_{ox}/s + m_{prod}/(1 + s)\}$ is a conserved property in an SCRS

<u>because</u>

it is a linear combination of m_{ox} and m_{prod}.

6.11 If m_{ox}/s is greater than m_{fu}, f is greater than f_{stoich}

<u>because</u>

m_{ox}/s equals m_{fu} in a stoichiometric mixture.

6.12 For a SCRS in a steady one-dimensional system, the equation

$$G\frac{d}{dx}\left(m_{fu} - \frac{m_{ox}}{s}\right) + \frac{d}{dx}\left\{\Gamma_{fuox}\frac{d}{dx}\left(m_{fu} - \frac{m_{ox}}{s}\right)\right\} = 0$$

is valid

because

$$R_{fu} = sR_{ox}$$

6.13 If ϕ_I and ϕ_{II} both obey the equation

$$G\frac{d\phi}{dx} - \frac{d}{dx}\left(\Gamma\frac{d\phi}{dx}\right) = \phi$$

so does ϕ_{III} defined as a $\phi_I + b\phi_{II} + c$ where a, b and c are any constants

because

the source term on the right-hand side is linear.

6.14 In an adiabatic SCRS, in a one-dimensional steady-flow system with diffusion, the relations between $\tilde{h}$, $m_{fu} - m_{ox}/s$ and other conserved properties are the same as in an adiabatic two-stream mixing process

because

each obeys an identical second-order differential equation.

6.15 The fully-reacted SCRS is a useful theoretical model for many practically-important diffusion-controlled flames

because

chemical reaction rates are much lower than mixing rates in this kind of flame.

In the following diagram, which is valid for a fully-reacted adiabatic SCRS with pure fuel in the F stream, the lines marked 6.16, 6.17, 6.18, etc., could represent :

A the oxygen mass fraction.

B the product mass fraction

C the temperature.

D the fuel mass fraction.

E the enthalpy.

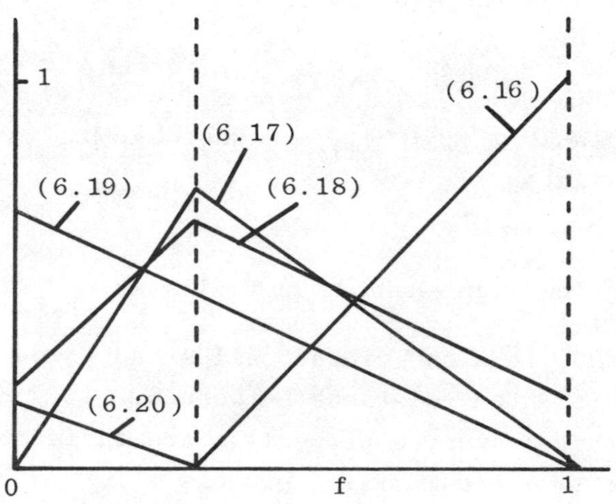

In the diagram on the right which represents a steady laminar diffusion flame, the locations marked (6.21), (6.22) etc. are probably characterised by :

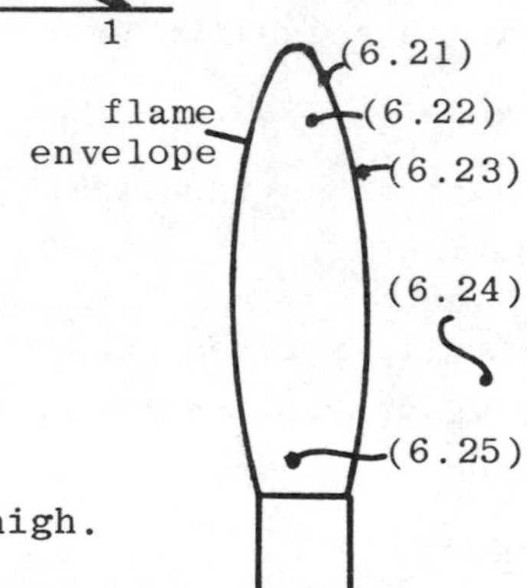

A $f > f_{stoich}$ and temperature high.

B $f > f_{stoich}$ and temperature low.

C $f = f_{stoich}$.

D $f < f_{stoich}$ and temperature high.

E $f < f_{stoich}$ and temperature low.

Carbon monoxide burns in a turbulent diffusion flame, for which the oxidant is atmospheric air. Helium is added as a tracer to the carbon monoxide stream, in which its concentration is 0.1% by mass. Samples of the gas in the flame are extracted from various locations, and analysed for helium content. Where the mass fraction of helium is found to be :

6.26 10^{-3}

6.27 $.406 \times 10^{-3}$

6.28 $.289 \times 10^{-3}$

6.29 $.203 \times 10^{-3}$

6.30 0.

the other gases probably consist of :

A air only.

B CO only.

C CO, CO_2 and N_2 but no O_2.

D CO_2, N_2 and O_2 but no CO.

E CO_2 and N_2 only.

ANSWERS TO MULTIPLE-CHOICE PROBLEMS

Answer	Problem number (6's omitted)
A	3, 6, 8, 14, 20, 22, 30
B	4, 10, 17, 25, 26, 27
C	2, 7, 11, 15, 18, 21, 27, 28
D	1, 9, 13, 16, 23, 29
E	5, 12, 19, 24, 28, 30

DISCUSSION PROBLEMS

6.32 As for 1.43.

6.33 "Some of the $h_{j,o}$'s can be set equal to zero, but not all simultaneously". Explain this statement.

6.34 The only differential equations which have been considered in the text have been for one-dimensional steady situations. Demonstrate that the results about the relation between the conserved properties in an SCRS are valid also for two-and three-dimensional steady processes.

6.35 Consider whether the derived relations between the conserved properties of an SCRS are valid also for one-dimensional <u>un</u>steady situations.

CHAPTER 7
DROPLET COMBUSTION

7.1 INTRODUCTION

A) THE PHENOMENON IN QUESTION

The following sketch shows what happens when a fuel droplet burns in an oxidising atmosphere. There is usually some relative motion between the droplet and the nearby gas, either because of the momentum with which the droplet is injected, or because of natural-convection processes. This relative motion causes the flame to be unsymmetrical.

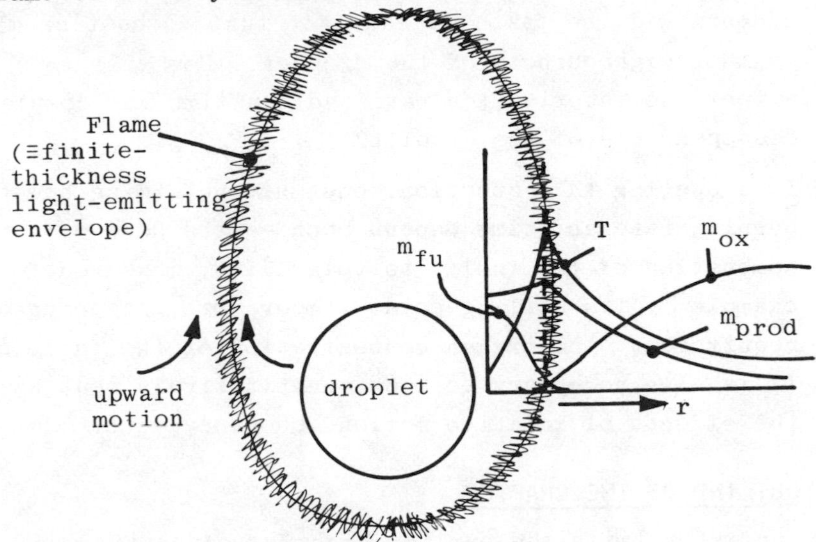

The flame of a candle, and that around the head of a matchstick soon after ignition, are familiar examples of flames having similar characteristics.

When the droplet is very small, the viscosity of the gas, operating over the short distances involved, allows only slight relative motion. Then the droplet becomes very nearly spherical.

In the course of time, the diameter of the droplet diminishes, finally to zero. The temperature of the liquid may change during this burning process.

B) PRACTICAL IMPORTANCE

In most power-producing devices and furnaces, liquid fuels are injected into the combustion space in the form of droplets. The size of the combustion chamber is in part dependent on the time taken for vaporisation and burning.

The burning of liquid fuels from wicks and pools obeys similar laws.

Chapter 5 was concerned with the vaporisation of droplets of liquid into an infinite atmosphere of hot gas; now account will be taken of the fact that combustion may occur in the neighbourhood of the droplet. How will this affect the vaporisation rate and the time for complete disappearance of the droplet?

In answering this question, one must determine how the burning rate and time depend upon:- the heat of combustion of the fuel; its volatility, measured for example by its boiling-point temperature; its oxygen requirement; the oxygen concentration of the fuel; etc. It is also necessary to know quantitatively what are the effects of relative motion, buoyancy, etc.

C) OUTLINE OF THE CHAPTER

Consistent with the earlier practice of setting up simple mathematical models, which are realistic enough to be reliable predictive tools but simple enough to permit analytical study, attention will be concentrated on a spherically-symmetrical model. The Simple Chemically-Reacting System of Chapter 6 will also be incorporated; and it will be supposed that equilibrium vaporisation prevails.

The result will be a formula for the burning time which is very similar to that for the vaporisation time, namely :

$$t_b = D_o^2 \rho_{liq} / \{8\Gamma \, \ell n \, (1 + B)\} \qquad , (7.1\text{-}1)$$

wherein the "driving force", B, differs from its earlier forms (see equations 5.4-10 and 5.4-11), by the appearance of the quantities H, s and $m_{ox,\infty}$.

Of course, the analysis is unrealistic in some respects. The nature of the unrealism, and how it may be repaired, are discussed at the end of the chapter.

7.2 MATHEMATICAL MODEL

A) DESCRIPTION

- Spherical symmetry is presumed. This means that all non-radial motion in the gas is neglected.
- Quasi-steady conditions prevail in the gas. This is what has been tacitly assumed for droplet vaporisation; it is valid provided that the gas densities are much smaller than the liquid densities, so that the amount of fuel "stored" in the gas phase at any moment is much less than that in the liquid droplet.
- The distances between the droplets are much larger than the droplet diameters.
- The combustion process is of SCRS type (Chapter 6).
- The transport properties (Γ's) are independent of radius. (N.B. In view of the large temperature variations, this is not very realistic; however, it can be presumed that suitable "average" values can be defined.)
- The chemical-kinetic constants are such as to allow neither fuel vapour nor oxygen to penetrate the reaction zone in significant amounts.

B) DIFFERENTIAL EQUATIONS

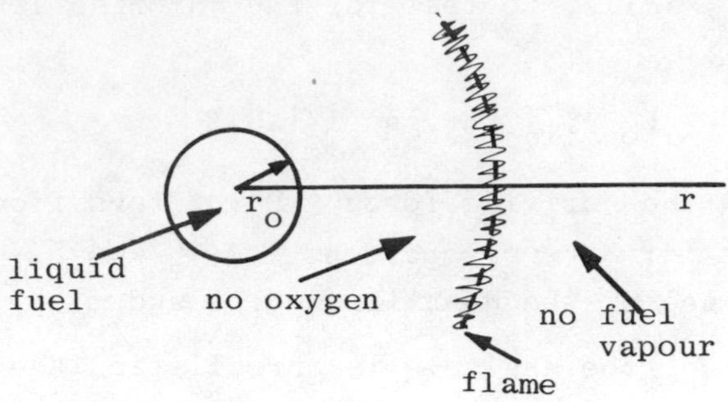

(i) Mass conservation

As for earlier droplet models :

$$Gr^2 = G_o r_o^{\ 2} \qquad . \quad (7.2\text{-}1)$$

Note that, in the gas space enclosed by the flame, the total mass flux G is given by :

$$G = G_{tot,fu} \qquad ; \quad (7.2\text{-}2)$$

and outside the flame, the following equation holds :

$$G = G_{tot,prod} + G_{tot,ox} \qquad . \quad (7.2\text{-}3)$$

Of course $G_{tot,ox}$ is negative, because oxygen has to be transferred inwards in order to sustain chemical reactions.

(ii) Conservation of any conserved property ϕ

The next flux of ϕ from the droplet, divided by 4π, is equal to the sum of the convective and diffusive fluxes at any radius, including $r = r_o$. Hence :

$$\left(G\phi - \Gamma_\phi \frac{d\phi}{dr}\right) r^2 = \left\{G_o \phi_o - \Gamma_\phi \left(\frac{d\phi}{dr}\right)_o\right\} r_o^{\ 2} \qquad . \quad (7.2\text{-}4)$$

Here ϕ, as explained in Chapter 6, can stand for : m_{in}, $m_{fu} - m_{ox}/s$, h, etc. Γ_ϕ is the appropriate exchange coefficient.

C) SOLUTION FOR THE VAPORISATION RATE

Equation (7.2-4) can be integrated, because it is presumed that Γ_ϕ is a constant. Insertion of the boundary condition for $r \to \infty$ then leads to :

$$\frac{G_o r_o}{\Gamma_\phi} = \ln\left[1 + \frac{(\phi_o - \phi_\infty)}{-\Gamma_\phi (d\phi/dr)_o/G_o}\right] \qquad . \quad (7.2\text{-}5)$$

By interpretation of ϕ in the various possible ways, the following useful results are obtained :-

(i) <u>If $\phi \equiv m_{fu} - m_{ox}/s$:</u>

- $$\phi_\infty = - m_{ox,\infty}/s \qquad , \quad (7.2\text{-}6)$$

  because there is no fuel vapour in the bulk of the gas (the fast-kinetics presumption precludes this);

- $$\phi_o = m_{fu,o} \qquad , \quad (7.2\text{-}7)$$

  because no oxygen can penetrate to the liquid surface;

- $$\begin{aligned} -\Gamma_\phi \left(\frac{d\phi}{dr}\right)_o &= -\Gamma_{fuox}\left(\frac{dm_{fu}}{dr}\right)_o \\ &= G_{diff,fu,o} \\ &= G_{tot,fu,o} - G_{conv,fu,o} \\ &= G_o - m_{fu}\, G_o \end{aligned} \qquad . \quad (7.2\text{-}8)$$

Hence, from equation (7.2-5) :

$$\boxed{\frac{G_o r_o}{\Gamma_{fuox}} = \ln\left[1 + \frac{m_{fu,o} + m_{ox,\infty}/s}{1 - m_{fu,o}}\right]} \qquad . \quad (7.2\text{-}9)$$

(ii) <u>If $\phi \equiv cT + H\, m_{fu}$:</u>

- $$\phi_\infty = cT_\infty \qquad ; \quad (7.2\text{-}10)$$
- $$\phi_o = cT_o + H\, m_{fu} \qquad ; \quad (7.2\text{-}11)$$
- $$-\Gamma_\phi \left(\frac{d\phi}{dr}\right)_o = Q_o + H\,(1 - m_{fu})\, G_o \qquad . \quad (7.2\text{-}12)$$

Hence, insertion of these expressions into equation (7.2-5)

leads to :

$$\frac{G_o r_o}{\Gamma_{hfu}} = \ell n \left[1 + \frac{c(T_o - T_\infty) + H\, m_{fu,o}}{Q_o/G_o + H\,(1 - m_{fu,o})} \right] \quad . \quad (7.2\text{-}13)$$

(iii) If $\phi \equiv cT + H\, m_{ox}/s$:

- $$\phi_\infty = cT_\infty + H\, m_{ox,\infty}/s \quad ; \quad (7.2\text{-}14)$$
- $$\phi_o = cT_o \quad ; \quad (7.2\text{-}15)$$
- $$-\Gamma_\phi \left(\frac{d\phi}{dr}\right)_o = Q_o \quad ; \quad (7.2\text{-}16)$$

Hence the expression for the burning rate becomes :

$$\frac{G_o r_o}{\Gamma_{hox}} = \ell n \left[1 + \frac{\{c(T_\infty - T_o) + H\, m_{ox,\infty}/s\}}{- Q_o/G_o} \right] \quad . \quad (7.2\text{-}17)$$

(iv) Other ϕ's

Whatever the definition of ϕ, provided it represents a conserved property having a unique exchange coefficient, a corresponding burning-rate expression can be derived. For example, if ϕ is defined as the mass fraction of a chemically inert substance, the relation becomes :

$$\frac{G_o r_o}{\Gamma_{in}} = \ell n \left[1 + \frac{m_{in,\infty} - m_{in,o}}{m_{in,o}} \right] \quad . \quad (7.2\text{-}18)$$

The derivation of further expressions of this kind is left to the interested reader.

D) THE SELECTION OF A PRACTICALLY USEFUL BURNING-RATE FORMULA

- A formula is useful for evaluation only if all but one of the quantities appearing in it is known.

In the present case $m_{ox,\infty}$ and T_∞ are likely to be known; but conditions at the surface are less certain. However, it is known that $m_{fu,o}$ and T_o are linked by the thermodynamic-equilibrium relation; and T_o cannot exceed the boiling-point temperature which usually is known.

- If the burning rate is to be determined precisely, equations (7.2-9) and (7.2-17) must be combined to yield (with equal Γ's):

$$\frac{m_{fu,o} + m_{ox,\infty}/s}{1 - m_{fu,o}} = \frac{c(T_\infty - T_o) H\, m_{ox,\infty}/s}{-Q_o/G_o} \quad ; \quad (7.2\text{-}19)$$

then this relation between $m_{fu,o}$ and T_o can be solved simultaneously with the thermodynamic-equilibrium relation in the same way as for pure vaporisation in Chapter 5.

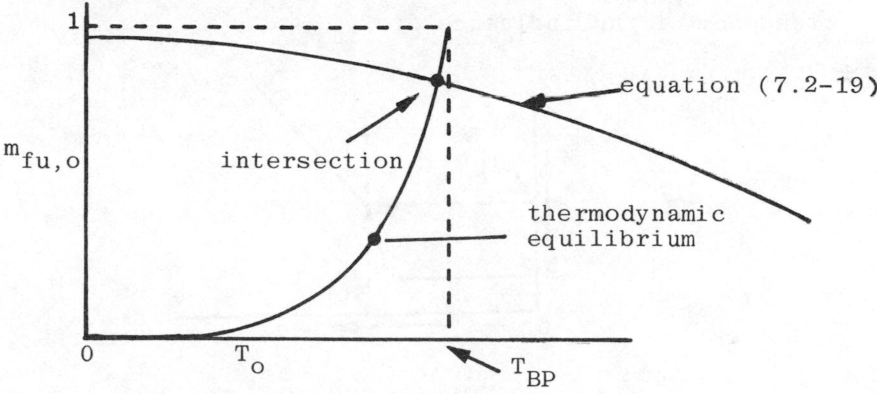

- The intersection of the two curves always occurs at a temperature just below the boiling point. Hence, the most useful practical formula is :

$$\frac{G_o r_o}{h_{ox}} \approx \ln\left[1 + \frac{\{c(T_\infty - T_{BP}) + H\, m_{ox,\infty}/s\}}{-Q_o/G_o}\right] \quad . \quad (7.2\text{-}20)$$

- For equilibrium vaporisation, i.e. that for which the droplets are injected already at a temperature close to their final temperature, $-Q_o/G_o$ can be set equal to the latent heat of vaporisation L. Somewhat more generally, the following equation can be employed :

$$-Q_o/G_o \approx L + c_{liq}\,(T_{B.P.} - T_{inj}) \qquad , (7.2\text{-}21)$$

where c_{liq} is the specific heat of the liquid and T_{inj} is the temperature with which the liquid enters the combustion space.

Use of equation (7.2-21) implies that the thermal conductivity of the droplet is small, so that the temperature distribution within it is as shown in the sketch below, throughout the burning process. Although perhaps not very realistic quantitatively, the formula certainly exhibits the correct tendency of the influence of T_{inj}.

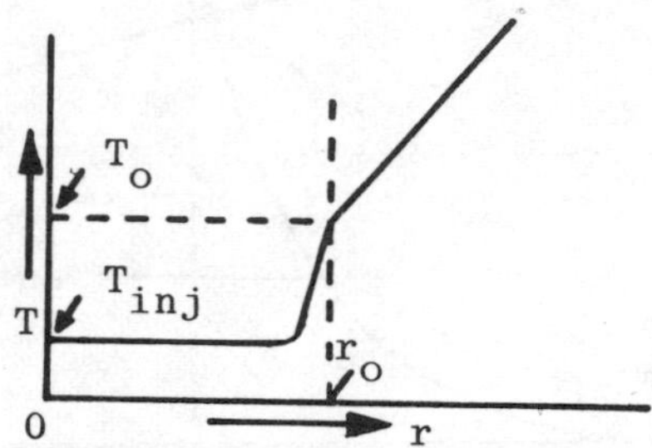

E) CALCULATION OF THE BURNING TIME

- By equating the mass-transfer rate at the surface, G_o, to the rate of diminution of droplet radius, as in earlier chapters, and by presuming that T_∞, $-Q_o/G$, etc. remain constant, it can be concluded that the burning time t_b is given by :

$$t_b \approx \frac{D_o^{\,2}\,\rho_{liq}}{8\Gamma_{hox}\,\ell n\left[1 + \dfrac{\{c(T_\infty - T_{BP}) + H\,m_{ox,\infty}/s\}}{-Q_o/G_o}\right]} \qquad , (7.2\text{-}22)$$

where D_o is the initial diameter of the droplet.

- Because Γ_{hox} is independent of pressure, and varies little from one gas to another, and because most fuels have similar densities and the logarithm varies much less rapidly than its argument, the quantity D_o^2/t_b exhibits no very wide variations.

  D_o^2/t_b is approximately equal to $10^{-6} m^2/s$, for hydrocarbons burning in air, for which the argument of the logarithm usually lies between 6 and 16.

- Experimental measurements are in good agreement with predictions, provided that these incorporate an average $\Gamma_{h,fu}$, defined as :

$$\Gamma_{h,fu} \equiv \left[r_o \int_o^\infty (\Gamma_{h,fu}\, r^2)^{-1}\, dr \right]^{-1} \quad . \quad (7.2\text{-}23)$$

7.3 DISCUSSION OF THE SOLUTION FOR THE SPHERICALLY-SYMMETRICAL MODEL

A) THE DRIVING FORCE FOR MASS TRANSFER

(i) Definition

Comparison of the burning-time formulae (7.1-1) and (7.2-22) shows that one may adopt the following definition of the mass-transfer "driving force", B :

$$B \equiv \frac{c(T_\infty - T_{BP}) + H\, m_{ox,\infty}/s}{(-Q_o/G_o)} \quad . \quad (7.3\text{-}1)$$

This is a dimensionless quantity, representing the ratio of the enthalpy excess of the bulk of the gas over the gas adjacent the surface to the enthalpy increase suffered by the fuel as it enters the gaseous phase.

(ii) Actual values

The following table (Spalding, 1955) gives an indication of the values attained by B in common circumstances. The fuels are supposed to be burning in atmospheric air, and $(-Q_o/G_o)$ is taken either as the latent heat of vaporisation L or as that quantity plus the enthalpy increase suffered by the liquid in rising from atmospheric temperature to the boiling point.

Fuel	$-Q_o/G_o=L$	$-Q_o/G_o=L+c_{liq}(T_{BP}-T_{inj})$
	Values of B	
C_5H_{12}	8.23	7.78
C_6H_{14}	9.00	6.39
C_7H_{16}	9.15	5.45
C_8H_{18}	9.70	5.02
$C_{10}H_{22}$	10.02	4.11
C_6H_6	7.74	6.09
$CH_3C_6H_5$	8.35	5.85
$(CH_2)_6$	8.25	6.22
CH_3OH	2.67	2.37
C_2H_5OH	3.50	2.95

Values of B appropriate to burning in pure oxygen are approximately four times as large; for $m_{ox,\infty}$ equals .232 for atmospheric air, and the H term is always the major one in the numerator of B.

(iii) Sensitivity of the burning rate to changes in conditions

Because one has :

$$\frac{d}{dB} \ln (1 + B) = \frac{1}{1+B} \quad , (7.3\text{-}2)$$

$\ln(1 + B)$ increases much less rapidly than B when the latter has a typical practical value, e.g. 9.0.

Consequently, the burning rates of liquid fuels are not strongly influenced by changes in $m_{ox,\infty}$, T_∞, etc. Even the four-fold increase resulting from a change from air to pure oxygen is likely to increase the burning rate only by about 40%. Therefore the differences of performance between a volatile fuel like gasoline and an involatile one like diesel oil will be slight, when both start from the same initial droplet size.

The viscosity of the fuel, which affects the atomisation behaviour and hence D_o, usually has a much greater effect on burning time than any other fuel property. This is one reason why otherwise highly-viscous fuels are heated before they reach the atomiser.

B) OTHER FEATURES OF THE SOLUTION

(i) Profiles of concentration and temperature

The variations with radius of gas properties around the droplet are easily computed from the relation :

$$\frac{G_o r_o^2}{\Gamma_\phi}\left(\frac{1}{r_o} - \frac{1}{r}\right) = \ln\left[1 + \frac{(\phi_o - \phi)}{\left(-\Gamma_\phi\left(\frac{d\phi}{dr}\right)_o / G_o\right)}\right] \quad , (7.3\text{-}3)$$

where ϕ can of course be given any appropriate form. However, it is also necessary, in order to complete the prediction, to know or assume the degree of combustion of the gases at each point. The assumption of complete reaction is close to the truth; then the prediction must be of the form shown in the sketch of Section 7.1 A.

This result follows immediately from the considerations of chapter 6, namely that the mixture fraction f is a conserved property, i.e. a ϕ, and that all the concentrations, and the temperature may be linked together by way of the diagrams shown in section 6.4 B and C.

(ii) The location of the flame

The radius of the flame can be related to the radius of the droplet by inserting :

$$r = r_{fl} \; : \; \phi = \phi_{stoich} \qquad , \quad (7.3\text{-}4)$$

into (7.3-4). There results :

$$\frac{G_o r_o}{\Gamma_\phi}\left(1 - \frac{r_o}{r_{fl}}\right) = \ln\left[1 + \frac{B(\phi_o - \phi_{stoich})}{(\phi_o - \phi_\infty)}\right] \quad . \quad (7.3\text{-}5)$$

Let ϕ now be defined as $(m_{fu} - m_{ox}/s)$, so that ϕ_{stoich} equals zero. Then there easily follows :

$$\frac{r_{fl}}{r_o} = \left\{1 - \frac{\ln\left[1 + Bm_{fu,o}/(m_{fu,o} + m_{ox,\infty}/s)\right]}{\ln(1 + B)}\right\}^{-1} . \quad (7.3\text{-}6)$$

Since, as is known, $m_{fu,o}$ approaches unity quite nearly, and $m_{ox,\infty}$ will usually be of the order of 0.06, while B is of the order of 8, this formula implies that r_{fl}/r_o is much greater than unity : the diameter of the flame is very much larger than the diameter of the droplet. Photographs of burning droplets confirm this.

7.4 SOME FEATURES OMITTED FROM THE MODEL

A) RELATIVE MOTION OF DROPLET AND GAS

- The relative motion associated with forced or natural convection always results in an increase in the burning rate and a decrease in the burning time as compared with the values predicted for the spherically symmetrical model.
- The reason is that the flame ceases to be a concentric spherical surface. Part of it approaches the liquid; and the correspondingly steepened gradients cause an increase in the rates of heat conduction and diffusion.
- The effect can be calculated by the solution of the appropriate partial differential equations. However, because droplets cannot remain for long in motion relative to the

surrounding gas, the total effect can scarcely ever change the average burning rate by a factor of more than 2. This is usually negligible, in view of the uncertainties about the initial droplet sizes produced by practical atomising devices.

B) OTHER FACTORS

- Real fuels are not usually pure substances having a distinct boiling point and latent heat of vaporisation. **However**, because of the comparative insensitivity of B to T_{BP} and of $\ell n\ (1 + B)$ to B, these uncertainities have rather little effect on the reliability of burning-time calculations; for averages can be guessed.
- When the pressure of the gas becomes high, and the critical pressure of the fuel is approached, the quasi-steady assumption breaks down, as indeed does the whole theory. At the critical point, for example, the latent heat of vaporisation vanishes, which appears to make the burning rate infinite. An unsteady-state analysis is required, if the burning time is to be properly predicted.
- It has been presumed that the sole mechanism of heat transfer is that of conduction. In reality, radiation plays a significant part, especially when soot is formed in the gas. Its effect is to accelerate vaporisation and combustion.
- Some "heavy" fuels cannot vaporise completely. Instead they form solid residues which remain when all the volatile matter has vaporised. These residues, often looking like "skeletons" of the original spherical droplet, are called "cenospheres". These too may burn, but more in the fashion of solid-carbon particles (Chapter 20).
- Chemical-kinetic effects can, in principle, intervene to cause extinction of the flame just before final droplet disappearance; the reason is that the necessary reaction

rate per unit volume must increase as r_o diminishes; and therefore a chemical-kinetic limit must be reached.

In practice, droplets are nearly always burned in volumes of gas which, as all the droplets vaporise, rise in temperature to a value near the adiabatic stoichiometric value. This diminishes the tendency to extinction, and renders the process unimportant.

7.5 REFERENCES

1. CANADA G S & FAETH G M (1973)
"Fuel droplet burning rates at high pressure".
Fourteenth Symposium (International) on Combustion. The Combustion Institute, Pittsburgh, pp 1345 - 1354.

2. GODSAVE G A E (1953)
"Studies of the combustion of drops in a fuel spray - the burning of single drops of fuel".
Fourth Symposium on Combustion, Williams and Wilkins, Baltimore, p 818.

3. HALL A R & DIEDERICHSEN J (1953)
"An experimental study of the burning of single drops of fuel in air at pressures up to twenty atmospheres".
Fourth Symposium on Combustion, Williams and Wilkins, Baltimore, p 837.

4. SPALDING D B (1953)
"The combustion of liquid fuels".
Fourth Symposium on Combustion, Williams and Wilkins, Baltimore, p 847.

5. SPALDING D B (1955)
"Some fundamentals of combustion".
Butterworth's, London.

6. SPALDING D B (1959)
"Theory of particle combustion at high pressures".
ARS Journal, Vol. 29, pp 828 - 835.

EXERCISES TO FACILITATE ABSORPTION OF MATERIAL OF CHAPTER 7

ANALYTICAL PROBLEMS

7.1 Evaluate B for n-octane (C_8H_{18}) burning in an atmosphere of the products of decomposition of hydrogen peroxide (H_2O_2) at 500 °C and 2 x 10^6 N/m^2, given the following data :

$H = 4.43 \times 10^7$ J/kg

$T_{BP} = 281\,°C$

$L = 1.2 \times 10^5$ J/kg

$c = 2.09 \times 10^3$ J/kg °C

Note that $2H_2O_2$ splits into $2H_2O$ (steam) plus O_2.

7.2 Evaluate the burning time of a droplet of n-octane burning as above, given that its initial diameter is 2 x 10^{-4}m, and that ρ_{liq} = 706 kg/m^3 and λ = 0.0729 J/m s °C.

7.3 What is the ratio of the flame radius to the droplet radius in the above case?

ANSWERS TO ANALYTICAL PROBLEMS

Problem number :	7.1	7.2	7.3
Answer :	54.3,	0.0254s,	20

MULTIPLE-CHOICE PROBLEMS

Indicate whether the following statements, 7.4 to 7.10, are :

A always true,

B true only for Simple Chemically Reacting Systems,

C true only for SCRS's with very fast chemical kinetics,

D true only for SCRS's with very fast chemical kinetics and without significant radiative transfer or other energy source,

E untrue even then.

7.4 It is easy to predict the maximum temperature in the flame around a vaporising droplet.

7.5 The vaporisation rate at the surface of a droplet is proportional to the reciprocal of the droplet radius.

7.6 The lifetime of a droplet at rest in an oxidising atmosphere and constantly at its equilibrium temperature is proportional to the reciprocal of the square of the initial diameter.

7.7 The radius of a spherically symmetrical flame can be deduced by solving the equation for the variation with radius of $m_{fu} - m_{ox}/s$.

7.8 It is possible to predict the vaporisation with radius of an inert diluent, in a spherically-symmetric flame around a droplet, knowing only : the concentration of this diluent in the atmosphere, the droplet radius, and the value of Γ (presumed constant).

7.9 The vaporisation rate of a burning liquid droplet can be calculated, in conditions of spherical symmetry, from knowledge of r_o, Γ, and the values at the surface and at infinity of the quantity

$$\left\{ \frac{m_{ox}}{s} + \frac{m_{prod}}{(1+s)} \right\} .$$

7.10 The effect of relative motion between a burning droplet and the atmosphere surrounding it is to reduce the burning rate as compared with that for a spherically-symmetrical model.

Mark the following statements 7.11 to 7.28, in accordance with the usual code (see page E 1.1).

7.11 It is of little practical value to develop a theory of droplet combustion which is accurate to better than 10%
<u>because</u>
initial droplet sizes are known only to much lower accuracies.

7.12 The temperature of a burning droplet can attain a temperature in excess of its boiling-point temperature
<u>because</u>
heats of combustion are commonly much greater than latent heats of vaporisation.

7.13 The equation : $G r^2 = G_o r_o^2$ is invalid for the droplet-burning model
<u>because</u>
fuel is consumed by chemical reaction at the reaction zone.

7.14 The relation $G_{tot,prod} > G_{tot,ox}$ is valid for the droplet-burning model only outside the reaction zone
<u>because</u>
the concentrations of both product and oxygen are negligible in the space between the liquid surface and the reaction zone.

7.15 Of the various driving-force expressions for burning liquids given in the text, the most useful in practice is

$$B \approx \frac{c(T_\infty - T_{BP}) + H\, m_{ox,\infty}/s}{L + c_{liq}(T_{BP} - T_{inj})}$$

<u>because</u>
all the quantities on the right-hand side are usually known.

7.16 If there is an inert mixture constituent in the atmosphere around a burning droplet, the value of $m_{in,o}$ must be finite
<u>because</u>
the consideration of the equations connecting $m_{fu,o}$ with T_o shows that $m_{fu,o}$ cannot attain the value unity.

7.17 The driving-force expression

$$B = \frac{m_{fu,o} + m_{ox,\infty}/s}{1 - m_{fu,o}}$$

is nearly always valid for a burning droplet, according to the SCRS assumptions,

because

chemical reactions are nearly always fast enough to prevent oxygen from penetrating to the liquid surface.

7.18 The appropriate average value of the exchange coefficient, accounting for the variation of Γ with radius, is :

$$\bar{\Gamma} = \left[\int_{r_o}^{\infty} (\Gamma r)^{-1} \, dr \right]^{-1}$$

because

Γ usually increases with temperature.

7.19 The burning rate of a given fuel droplet in air is only about one quarter of that in pure oxygen

because

the mass fraction of oxygen in air is around 23%.

7.20 The burning time of a fuel droplet is likely to increase as the pressure of the atmosphere surrounding it is raised

because

the resulting suppression of dissociation will slightly raise the maximum temperature in the flame.

7.21 A volatile fuel such as gasoline burns significantly more rapidly in a gas-turbine combustion chamber than a less volatile one such as kerosine, for equal sizes of injected droplets,

because

the expression

$$\frac{m_{fu,o} + m_{ox,\infty}/s}{1 - m_{fu,o}}$$

shows that the driving force tends to infinity as $m_{fu,o}$ approaches unity.

7.22 If λ/c could be made significantly greater than Γ_{fu} and Γ_{ox}, the burning rate of a droplet would still not be significantly affected, under equilibrium-vaporisation conditions,

because

under these conditions the heat conducted to the liquid surface has no effect on the vaporisation rate.

7.23 Fuel oils are usually heated before supply to the injection device of a furnace

because

the burning times of droplets of given size are significantly reduced by the raising of their injection temperature.

7.24 Increasing the oxygen content of the atmosphere around a burning droplet increases the ratio of the flame diameter to the droplet diameter

because

the increase reduces the value of the stoichiometric coefficient s.

7.25 It can be expected that the flame around an isolated burning droplet in a cold atmosphere will be extinguished before the droplet has disappeared completely

because

as the size diminishes the chemical reaction rate per unit volume rises, according to the spherically-symmetrical theory, to infinity.

7.26 Droplet-burning theory is of little relevance to gasoline-engine combustion

because

there is little prospect of producing fuels with low latent heats of vaporisation at an economic price.

7.27 Droplet-burning theory permits the precise prediction of the combustion-efficiency of gas-turbine engines

because

the droplet-size distributions produced by swirl atomisers are well known.

7.28 Droplet-burning theory is likely to be useful for the design of liquid-propellant rocket motors
because
in such motors the relative velocity of the injected droplets and the hot gases is very small.

ANSWERS TO MULTIPLE-CHOICE PROBLEMS

Answer	Problem number (7's omitted)					
A	5,	8,	11,	15,	17,	25
B	9,	16,	20,	26		
C	7,	14,	23,	28		
D	4,	12,	13,	18,	19,	21
E	6,	10,	22,	24,	27	

CHAPTER 8

LIQUID-PROPELLANT ROCKET

8.1 INTRODUCTION

A) TYPICAL CONFIGURATION

The diagram indicates, schematically, the main elements of a liquid-propellant rocket motor, namely :-

- injector manifold, supplying both fuel and oxidizer;
- cylindrical combustion space;
- convergent-divergent exhaust nozzle;
- cooling jacket, supplied with fuel used as coolant.

The designer's aim is to ensure that the propellants are completely mixed and burned before they enter the exhaust nozzle.

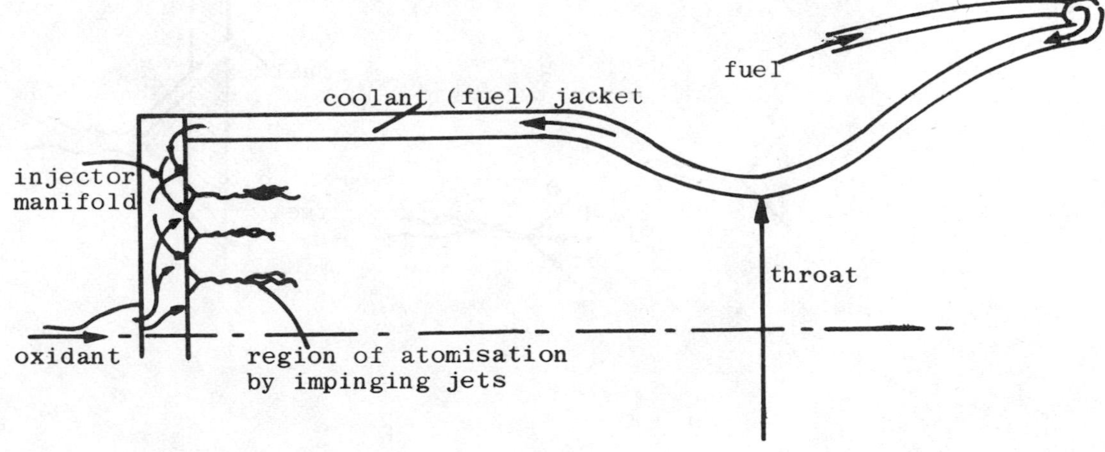

B) PROPELLANT SYSTEMS

The following three types are employed :-

(i) Two liquids,

e.g. kerosine and liquid O_2;
liquid H_2 and liquid O_2;
aniline and nitric acid;
hydrazine hydrate and hydrogen peroxide.

This is the most common type.

(ii) A single liquid (called a "monopropellant"),

e.g. hydrogen peroxide (+ catalyst);
nitromethane.

These have been rarely used because they either have a low energy content (specific impulse), or are prone to explosion in storage. However their advantages of simplicity ensure continued attention.

(iii) Liquid plus gas

e.g. Kerosine + decomposed H_2O_2;
gaseous H_2 + liquid O_2.

c) INJECTOR TYPES

These include the following :-

(i) Impinging jets.

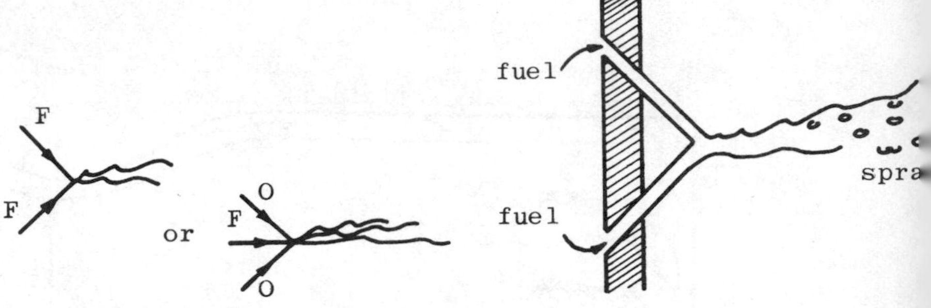

This is by far the most popular in practice.

(ii) Target plate + jet.

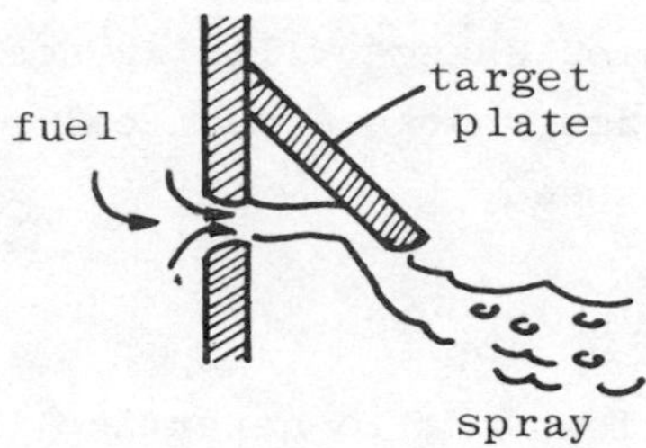

This type is suitable for small motors.

(iii) Swirl atomisers as for gas turbines. They are no longer used because they are too bulky and heavy.

8.2 FUNDAMENTAL CONSIDERATIONS

A) THERMODYNAMICS

(i) At the high pressures and temperatures typical of most rockets, the equilibrium products of H_2 and O_2, for example, contain, as well as H_2O : H_2, O_2, H, O, OH, in significant proportions.

Thermodynamics allows the compositions and temperatures to be evaluated, so that the maximum possible performance (thrust) can be determined; this maximum is achieved when the gases are in equilibrium throughout. It is used as a standard of reference, with which actual performance can be compared, and also for selecting propellant pairs which are likely to give high performance.

(ii) One-dimensional, compressible-flow theory allows ideal (frozen-composition, ideal-gas, frictionless) flow in the nozzle to be calculated, and hence the corresponding ideal thrust. This is usually slightly smaller than that which would follow from the assumption that the gases were in thermodynamic equilibrium throughout the nozzle.

If there is a very large pressure decrease in the nozzle, the equilibrium composition may change with distance. The actual composition change will lag somewhat behind the equilibrium change, because chemical rearrangements proceed only at finite speed.

B) CHEMICAL KINETICS

This subject allows one to determine the rate of composition change. A typical reaction is the reassociation of :-

H + OH, to form H_2O,

as the pressure and temperature decrease along the length of the nozzle.

Because of the high pressure and temperature in the rocket combustion chamber, reaction rates are high enough for the gas composition to be nearly in equilibrium at every point. The combustion process is not chemically but physically controlled.

C) PHYSICAL PROCESSES

- The rate-controlling process is often the rate of vaporisation of one or both of the liquid propellants. These vaporise and burn as described in Chapter 7, except that droplets of oxidiser are present as well as those of fuel.
- Mixing in jets and boundary layers are also processes which affect the burning rate to some extent. However, they will be neglected in the present chapter, as being of secondary importance.
- Droplet aerodynamics does have an appreciable role to play because the drag coefficient of a droplet depends on the Reynolds number of the relative motion, and perhaps on other factors, e.g. rate of vaporisation.

8.3 MATHEMATICAL MODEL

A) DESCRIPTION

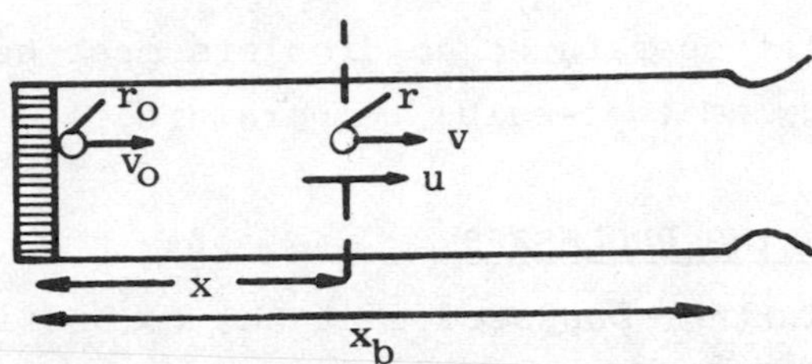

(i) Variables considered

The idealised combustion chamber represented in the sketch is considered, and described in terms of the following variables :

- Independent variables :
  x, the distance along the combustion chamber from the injector face. One thus has a one-dimensional, steady-state model.

- Dependent variables :

droplet radius r (the same for all droplets	at a given x
droplet radius v (the same for all droplets	
gas velocity u (the same for all gas streams	

Note that, for simplicity, the two propellants will be treated as identical in respect of the properties relevant to vaporisation. Of course it is possible to handle the general case also; but to do so would distract from the present purpose.

(ii) Quantities independent of x

It is supposed that the following are constant :

- Gas temperature and gas composition; so the vapour is supposed to burn quickly.
- Gas pressure; so the Mach number and friction factor are taken as low in the combustion chamber.
- Droplet temperature; so droplets must be injected at equilibrium (wet-bulb) temperature.

B) CONTROLLING PROCESSES

It is further supposed that the combustion rate is influenced solely by :

- Vaporisation controlled by conduction and diffusion, as described in Chapter 7.
- Droplet motion, controlled by inertia and Stokes' Law.
- Gas velocity, controlled by the mass-conservation principle.

C) APPROXIMATE ANALYSIS

Before proceeding to derive the complete theory, it is useful to pause to consider what form the result is likely to take. If there were no drag interaction between gas and droplets, the latter would maintain their injection velocity, v_o, throughout. Then, since from Chapter 7 one has :

$$t_b = \frac{r_o^2}{2\Gamma} \rho_{liq} \ln (1 + B) \qquad , (8.3\text{-}1)$$

the length for complete burning x_b will be given by :

$$x_b = v_o t_b$$

i.e.

$$\boxed{x_b = \frac{v_o r_o^2}{2\Gamma} \rho_{liq} \Big/ \ln (1 + B)} \qquad . (8.3\text{-}2)$$

In practice, the droplet velocity will at first diminish as a consequence of the drag exerted by the gas; later it may increase, for the same reason, because the gas velocity increases with x. Therefore it may be expected that equation (8.3-2) will require modification by some function of the ratio of the droplet injection velocity to the final gas velocity. This expectation will be borne out below.

D) EQUATIONS

- Relation between time and space variations.

In place of the dr/dt which appears in the droplet burning theory, one can write :

$$\frac{dr}{dt} = v\,\frac{dr}{dx} \qquad , (8.3\text{-}3)$$

because the increment in time for a droplet, dt, equals dx/v.

- Radius variation.

From the implications of Chapter 7, therefore, one has :

$$-v\,\frac{dr}{dx} = \frac{1}{r}\;\frac{\Gamma}{\rho_{liq}}\;\ln\,(1 + B) \qquad , (8.3\text{-}4)$$

where r now takes the place of r_o (since one is concerned with no other radii but the droplet-surface radius), and B is given an appropriate form. This form will be different according to whether the fuel droplets or the oxidizer droplets are in question; but it will be based on identical principles.

- Newton's Second Law of Motion, for the droplet, implies :

$$v\,\frac{dv}{dx} = \frac{F}{\rho_{liq}\frac{4}{3}\pi r^3} \qquad , (8.3\text{-}5)$$

where F is the force exerted by the gas on the droplet in the x direction.

- Stokes' Law for droplet drag implies :

$$F = 6\,\pi\, r\,(u - v)\mu \qquad , (8.3\text{-}6)$$

where μ is the (suitably averaged) viscosity of the gas. However, the drag of a vaporising droplet may be expected to diminish, as compared with the rigid sphere envisaged by Stokes' Law, just like the heat-transfer coefficient (see sections 4.4 D and 5.2 D).

- Combination of (8.3-5), (8.3-6) and incorporation of the "diminution factor" f given by :

$$f \equiv B^{-1}\,\ell n\,(1 + B) \qquad , (8.3\text{-}7)$$

leads to :

$$v\frac{dv}{dx} = \frac{9}{2}\,\frac{\mu}{\rho_{liq}}\,\frac{(u - v)}{r^2}\,.f \qquad . (8.3\text{-}8)$$

Here it should be noted that the form of f is as given in the following sketch; and that, if the Reynolds number of the relative motion of the droplet and gas exceeds about 10, the right-hand side should be multiplied by a factor in excess of unity.

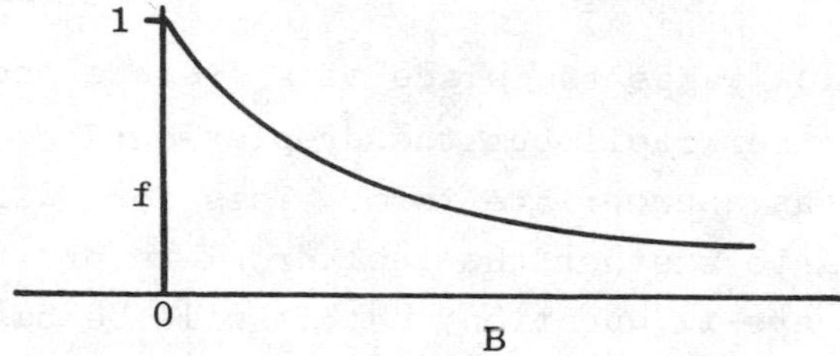

- Mass conservation.

Because the gas is formed from the products of vaporisation and burning of the liquid, the following can be written :

$$\rho u = \left\{ 1 - \left(\frac{r}{r_o}\right)^3 \right\} G \qquad , (8.3\text{-}9)$$

where : $\rho \equiv$ gas density,

$r_o \equiv$ droplet radius at injection,

$G \equiv$ total mass velocity of propellant, whether gas liquid, along the duct.

Here it has been presumed that the droplet density is very much greater than the gas density. The quantity in the curly bracket, of course, equals the proportion of the injected propellant which has already entered the gas phase; the droplet velocity, it is interesting to note, has no influence on this quantity.

E) DIMENSIONLESS FORM OF EQUATIONS

(i) Motive

In order to distinguish essentials from accidentals, the total number of variables will now be reduced. At the same time the labour of writing is diminished by the use of dimensionless variables.

(N.B. There is always a penalty, viz. the need to interpret the new variables in terms of physical ones when thinking about them. Whether the advantages outweigh the disadvantages depends upon the individual and his circumstances.)

(ii) New variables

The following definitions are employed :

$$r/r_o \equiv \zeta, \quad \text{(zeta)} \qquad , \quad (8.3\text{-}10)$$

$$\rho u/G \equiv \omega, \quad \text{(omega)} \qquad , \quad (8.3\text{-}11)$$

$$\rho v/G \equiv \chi, \quad \text{(chi)} \qquad , \quad (8.3\text{-}12)$$

$$\frac{x}{r_o{}^2}\,\frac{\rho\Gamma}{\rho_{liq}}\,\frac{\ln(1+B)}{G} \equiv \xi \ \text{(ksi)} \qquad , \quad (8.3\text{-}13)$$

$$\frac{9}{2}\,\frac{\mu}{\Gamma}\,\frac{f(B)}{\ln(1+B)} \equiv S \qquad . \quad (8.3\text{-}14)$$

Note that S is a propellant property, independent of x, v_o,

r_o or G.

(iii) Resulting equations

Equations (8.3-4), (8.3-8) and (8.3-9) now become, respectively :

Radius variation: $$\chi \zeta \frac{d\zeta}{d\xi} = -1 \qquad (8.3\text{-}15)$$

Newton 2: $$\chi \frac{d\chi}{d\xi} = S \frac{1}{\zeta^2} (1-\zeta^3-\chi) \qquad (8.3\text{-}16)$$

Mass conservation: $$\omega = 1 - \zeta^3 \qquad (8.3\text{-}17)$$

(N.B. (8.3-17) has already been substituted to produce (8.3-16).)

(iv) Boundary conditions

At the injector plane, the droplet radius and velocity have their initial values. In mathematical terms, this is expressed as :

$$\xi = 0 : \zeta = 1 \qquad , (8.3\text{-}18)$$

$$\chi = \chi_o \qquad . (8.3\text{-}19)$$

These two boundary conditions suffice; for there are two first-order differential equations to solve.

(v) Expected form of solution

The function : $\xi_b (S, \chi_o)$;

will be of particular interest, for it allows calculation of the necessary length for complete combustion. ξ_b will be obtained by finding the value of ξ for which ζ falls to zero. Then the length of chamber for complete burning, x_b will follow from :

$$\frac{x_b}{r_o^2} \frac{\rho\Gamma}{\rho_{liq}} \frac{\ln(1+B)}{G} = \xi_b \qquad . (8.3\text{-}20)$$

The variations of ζ, ω and χ with ξ can be expected to take the form shown in the following sketch.

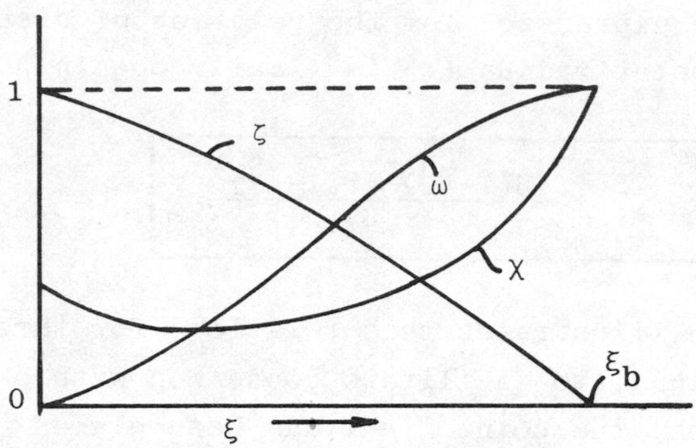

F) SOLUTION OF THE EQUATIONS

(i) Elimination of

- Division of (8.3-16) by (8.3-15) yields :

$$\frac{1}{\zeta}\frac{d\chi}{d\zeta} = -\frac{S}{\zeta^2}(1 - \zeta^3 - \chi),$$

$$\text{i.e.}\quad \frac{d\chi}{d\zeta} - \frac{S}{\zeta}\chi = -\frac{S}{\zeta}(1 - \zeta^3). \tag{8.3-21}$$

This is a linear equation, soluble by multiplication by an integrating factor : ζ^{-S}. Integration leads via :

$$\frac{d\chi}{d\zeta}\zeta^{-S} - S\chi\zeta^{-S-1} = -S(\zeta^{-S-1} - \zeta^{-S+2}) \quad ,$$

$$\text{to:}\quad \chi\zeta^{-S} = -S\left(\frac{\zeta^{-S}}{-S} - \frac{\zeta^{-S+3}}{-S+3}\right) + \text{const.} \tag{8.3-22}$$

- From boundary conditions (8.3-18) and (8.3-19) (i.e. $\chi = \chi_o$ at $\zeta = 1$), there results :

$$\chi_o = +S\left(\frac{1}{+S} + \frac{1}{-S+3}\right) + \text{const.} \quad ,$$

$$= -\frac{3}{S-3} + \text{const.} \quad . \tag{8.3-23}$$

Therefore the integration constant in (8.3-22) equals :

$$\chi_o + 3/(S - 3).$$

Hence, the expression for the relation of droplet velocity (χ) to droplet radius (ζ) is finally obtained, and is :

$$\boxed{\chi = \chi_o\ \zeta^S + \frac{\{S(1-\zeta^3) - 3(1-\zeta^S)\}}{(S - 3)}} \quad . \text{(8.3-24}$$

The sketch illustrates this relation, for three values of S. Note that, to facilitate connexion with physical coordinates, the point $\xi = 1$ has been placed on the left.

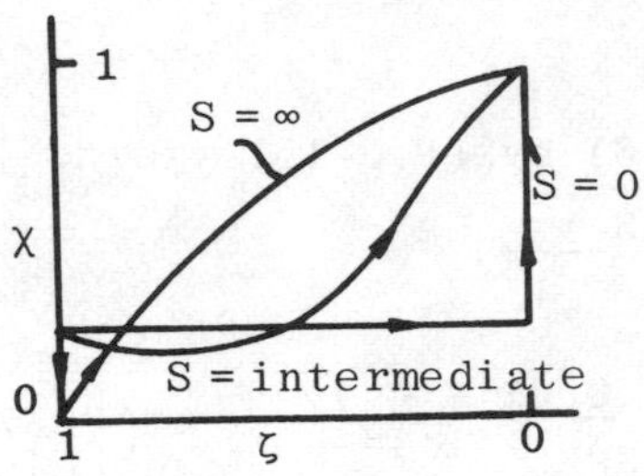

(ii) <u>Solution in terms of ξ</u>

Substitution of (8.3-24) into (8.3-15) yields :

$$\frac{d\xi}{d\zeta} = -\zeta + \frac{S}{(S-3)}\ \zeta^4 - (\chi_o + \frac{3}{S-3})\ \zeta^{S+1} \quad . \text{(8.3-25)}$$

This may be integrated as follows :

$$\xi = -\tfrac{1}{2}\zeta^2 + \frac{1}{5}\frac{S}{(S-3)}\ \zeta^5 - \frac{1}{(S+2)}\ (\chi_o + \frac{3}{S-3})\zeta^{S+2} + \text{const.} \quad . \text{(8.3-26)}$$

• Insertion of boundary condition at $\xi = 0$ leads finally to :

$$\boxed{\xi = \tfrac{1}{2}(1-\zeta^2) - \frac{S}{5(S-3)}(1-\zeta^5) + \frac{\left(\chi_o + \frac{3}{S-3}\right)}{(S+2)}(1-\zeta^{S+2})} \quad . \quad (8.3\text{-}27)$$

This is the required $\zeta \sim \xi$ curve, giving the variation of droplet radius with distance.

• The combustion ceases (i.e. $\zeta = 0$), where :

$$\xi = \boxed{\xi_b = \frac{\chi_o + \frac{3}{10}S}{2 + S}} \quad . \quad (8.3\text{-}28)$$

(iii) Special cases

It is useful to consider two extreme cases, distinguished by the values of the propellant property S. They are :-

• When $S \rightarrow 0$, i.e. when drag is negligible, one has :

$$\xi \rightarrow \tfrac{1}{2}\chi_o(1 - \zeta^2) \quad , \quad (8.3\text{-}29)$$

and:

$$\xi_b \rightarrow \tfrac{1}{2}\chi_o \quad . \quad (8.3\text{-}30)$$

The last equation has an identical meaning to that of (8.3-2), which resulted from the "approximate analysis".

• When $S \rightarrow \infty$, i.e. when the drag is dominant, one deduces :

$$\xi = \frac{3}{10} - \tfrac{1}{2}\zeta^2 + \frac{1}{5}\zeta^5 \quad , \quad (8.3\text{-}31)$$

and:

$$\xi_b = \frac{3}{10} \quad . \quad (8.3\text{-}32)$$

(iv) Graphical representation of the formulae

The following graphs (from Spalding, 1959) represent some of the main features of the solutions.

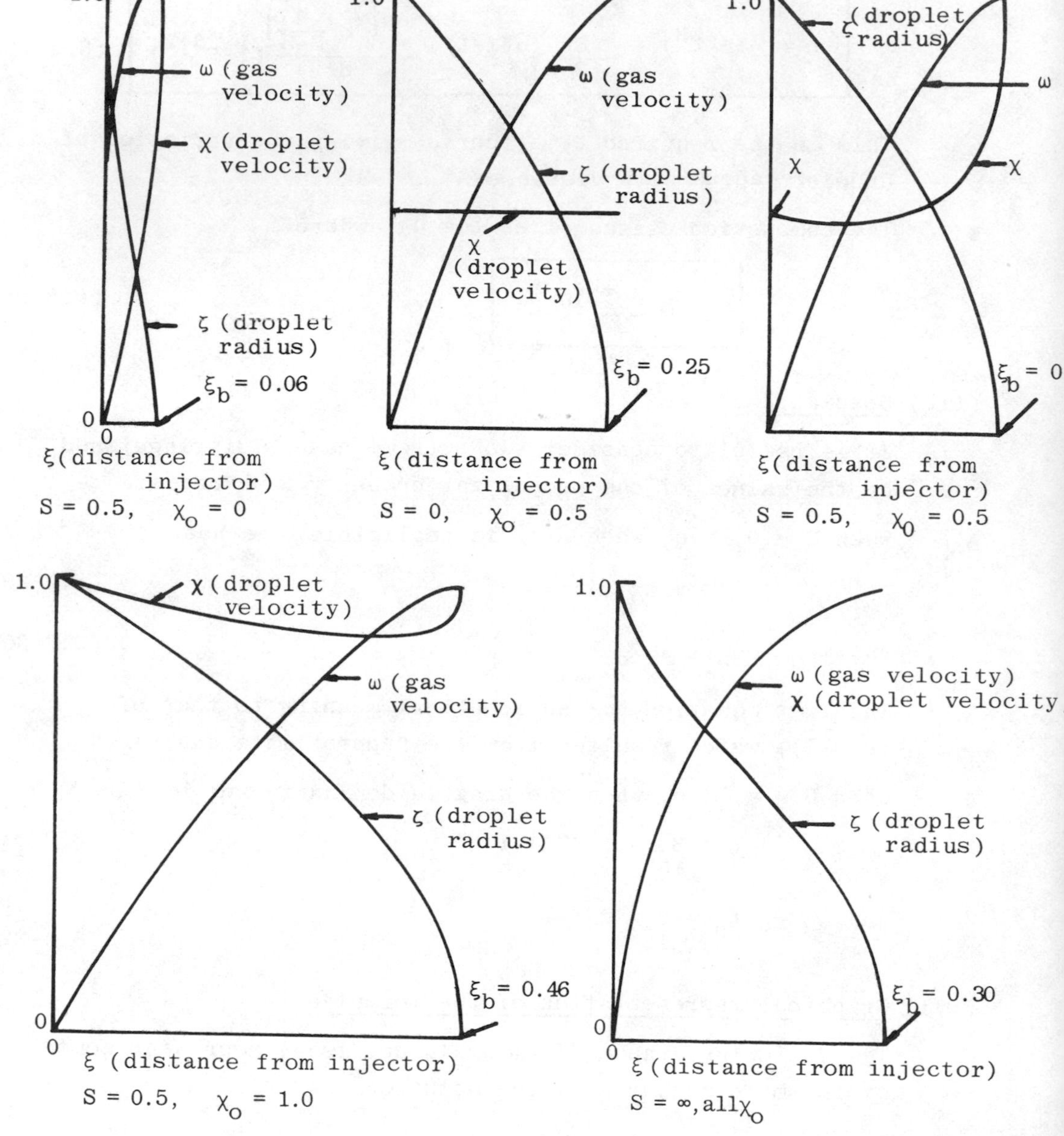

1.0
ω (gas velocity)
χ (droplet velocity)
ζ (droplet radius)
ξ_b = 0.06
0
0
ξ(distance from injector)
S = 0.5, χ_o = 0
1.0
ω (gas velocity)
ζ (droplet radius)
χ (droplet velocity)
ξ_b = 0.25
ξ(distance from injector)
S = 0, χ_o = 0.5
1.0
ζ (droplet radius)
ω
χ
χ
ξ_b = 0
ξ(distance from injector)
S = 0.5, χ_0 = 0.5
1.0
χ (droplet velocity)
ω (gas velocity)
ζ (droplet radius)
ξ_b = 0.46
0
0
ξ (distance from injector)
S = 0.5, χ_o = 1.0
1.0
ω (gas velocity)
χ (droplet velocity
ζ (droplet radius)
ξ_b = 0.30
0
0
ξ (distance from injector)
S = ∞, all χ_o

Notes :

- χ_o = .5 means injection velocity = ½ final gas velocity, by reason of equation (8.3-12); this is higher than is usual in practice.
- S = .5 means $\frac{9}{2}\frac{\mu}{\Gamma}\frac{f(B)}{\ln(1+B)}$ = 0.5; this is of the usual order of magnitude, as can be seen from the following calculation: $S = \frac{9}{2} \times Pr \times \frac{1}{B}$, if f = {ln(1 + B)}/B. B is of around 10, Pr of the order of 1. But of course B can vary from, say, 5 to 20; so S varies also.
- Obviously, it is possible for the relative velocity of droplets and gas to be quite large.
- The influences of the "friction parameter" S and the dimensionless injection velocity χ_o on the minimum chamber length for complete combustion are revealed by the following graph. Obviously, for χ_o in the region of 0.5, S has very little influence on ξ_b; so knowledge of its value is not important.
- Since S is usually of the order of 0.5, the value of χ_o has a large influence on the combustion-chamber length.

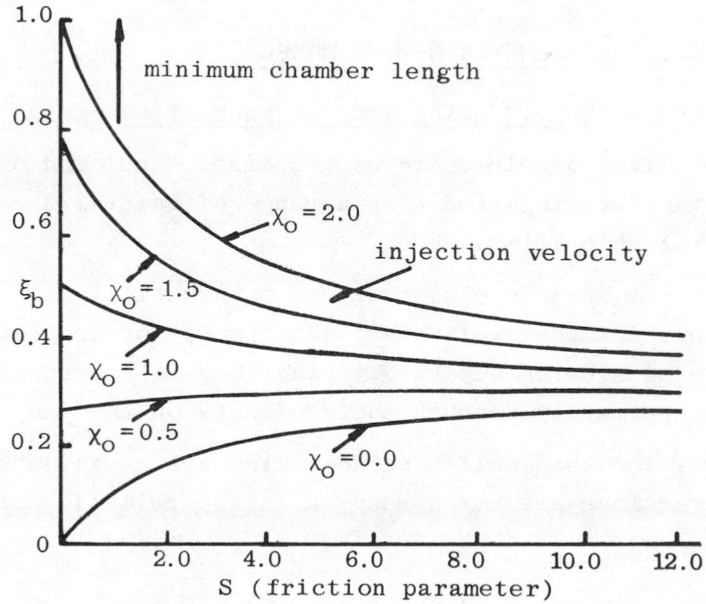

8.4 DISCUSSION

A) PRACTICAL IMPLICATIONS

(i) The importance of droplet size

Whatever the values of S and χ_o, the minimum chamber length x_b is always proportional to the square of the droplet radius at injection, as equation (8.3-20) shows. Therefore, fine atomisation is of very great importance.

(ii) The importance of injection velocity

The volumetric rate of combustion is equal to G/x_b. Rearrangement of (8.3-20) reveals :

$$\frac{G}{x_b} = \frac{1}{\xi_b} \frac{r_o^2 \rho_{liq}}{\rho\Gamma \ell n(1+B)} \quad ; \quad (8.4\text{-}1)$$

which, since it is desirable for the combustion chamber for a given thrust to be as small as possible, shows that ξ_b should also be as small as possible.

It follows that χ_o should be small, i.e. the droplet injection velocity should be small compared with the fluid gas velocity. Therefore long thin combustion chambers are best from this point of view.

B) POSSIBLE REFINEMENTS OF THE MODEL

(i) Allowance for differing droplet properties

In reality, droplets are of two kinds, fuel and oxidizer; and they are injected with a range of initial sizes and initial velocities.

It is not hard to elaborate the mathematical model to account for these effects : each family of droplets requires to be considered separately; and they interact with each other primarily through their effects on the gas velocity.

Since, for the small S values which are usual, these interactions are not strong, a fairly reliable and simple rule is: the largest droplets control the minimum combustion-chamber length; therefore concentration of attention on these droplets will not lead to inappropriate design decisions.

(ii) Two- and three-dimensional effects

If the droplets were extremely finely atomised, the combustion would still not be completed instantaneously; for it is necessary for the unevennesses in fuel-oxidizer ratio across the face of the injector to become evened out. To understand the mixing processes, and their effects on the distance to completion of combustion, it is necessary to study jet theory. This is the subject of the next four chapters. Subsequently it would be possible to combine the droplet and jet aspects into a single mathematical model of rocket combustion. However, this has not yet been done.

(iii) Refined analysis of droplet vaporisation

Since the droplets are not, as a rule, injected at their equilibrium-vaporisation temperatures, and since there may be a significant Reynolds-number effect on the drag and vaporisation rates, refinements are needed in the analysis.

It is not hard to make these; but they necessitate numerical rather than analytical solution techniques for the differential equations. The analysis of Chapter 5 can be used for the droplet-temperature variation; and there is no difficulty about introducing functions of $(\chi - \omega)$ into (8.3-15) and (8.3-16) to allow for the Reynolds-number influences.

(iv) Chemical-kinetic effects

In a few rocket-motor situations, it is possible that the chemical reaction rates cannot keep pace with the imposed rates of mixing. In these circumstances, further consideration must be given to the "chemical loading" of the combustion space. In principle, it is possible to "quench" the flame by injecting a sufficiently large flow of cold propellant into it.

No analysis of this effect will be given here, but similar phenomena will be discussed in Chapters 16 and 17 below.

c) CLOSURE

The rocket-motor designer specifies the number, size and inclination of the injector holes, not the diameter of the droplets. Therefore, in order to complete the mathematical model, a link between injector dimensions and droplet-size distribution is needed. It is still not possible to make this link in any theoretical manner.

In practice, measurements of droplet size are made in physical-model tests, with non-burning liquids. Their relevance to the real situation is tenuous.

8.5 REFERENCES

PRIEM R J (1957)
"Propellant vaporization as a criterion for rocket engine design; calculations of chamber length to vaporize a single n-Heptane drop".
NACA TN 3985.

PRIEM R J (1957)
"Propellant vaporization as a criterion for rocket-engine design; calculations using various log-probability distributions for Heptane drops".
NACA TN 4098.

SPALDING D B (1959)
"Combustion in liquid-fuel rocket motors".
Aero Quarterly, Vol. 10, pp 1 - 27.

SUTTON G P (1956)
"Rocket propulsion elements".
2nd edition, Wiley, New York.

EXERCISES TO FACILITATE ABSORPTION OF MATERIAL OF CHAPTER 8

MULTIPLE-CHOICE PROBLEMS

Select the most appropriate answer (A, B, C, D or E) according to the usual rules.

8.1 The most common type of atomisation system for a large high-performance, liquid-propellant rocket motor is :

A swirl atomiser.

B rotating disc or cup.

C steam atomiser.

D impinging jet.

E target plate.

8.2 All of the following could be used as the propellants of a liquid-fuelled rocket motor, except :

A aniline and nitric acid.

B kerosine and liquid hydrogen.

C hydrogen peroxide and a catalyst.

D hydrazine hydrate and hydrogen peroxide.

E kerosine and liquid oxygen.

8.3 All the following statements are true except :

A the gases flow through a rocket nozzle so rapidly that there is insufficient time for their composition to change.

B at the high temperatures prevailing in rocket motors, oxidation reactions do not go to completion even in equilibrium.

C chemical reactions proceed so rapidly in the combustion chamber, that the gases are nearly in equilibrium.

D some chemical reactions take place in the nozzle.

E the combustion rate is controlled by vaporisation.

8.4 The mathematical model of rocket combustion treated in Chapter 8 is characterised by all the following, except :

A at any x value, all droplets have the same diameter.

B the cross-sectional area of the chamber is independent of x.

C the velocities of gas and liquid are equal at any given x.

D the gas velocity is zero where x equals zero.

E the temperature and composition of the gas are taken as uniform except in the immediate vicinity of droplets.

According to the mathematical model of a liquid-propellant rocket motor, as x increases :

8.5 The droplet velocity,

8.6 The gas velocity,

8.7 The droplet radius,

8.8 The droplet temperature,

A remains constant.

B increases steadily.

C decreases steadily.

D first rises and then falls.

E first falls and then rises.

The equations :

8.9 $\frac{dr}{dt} = v \frac{dr}{dx}$,

8.10 $- v \frac{dr}{dx} = \frac{1}{r} \frac{\Gamma}{\rho_{liq}} \ln (1 + B)$,

8.11 $v \frac{dv}{dx} = \frac{9}{2} \frac{\mu}{\rho_{liq}} \frac{(\mu - v)}{r^2} f (B)$,

8.12 $$\rho u = \left\{1 - \left(\frac{r}{r_0}\right)^3\right\} G ,$$

express :

A the mass-conservation principle.

B the vaporisation-rate law.

C the shift of view-point from the individual droplet to the individual cross-section of the chamber.

D Stokes' Law.

E Stokes' Law combined with Newton's Second Law of Motion.

8.13 The equation :

$$\xi_b = (\chi_0 + \frac{3}{10}S)/(2 + S)$$

implies all the following statements, except :

A if S were infinite, the minimum chamber length would not depend on the injection velocity.

B the size of droplet which is formed by an impinging-jet atomiser is smaller, the higher is the injection velocity.

C when χ_0 equals 0.6, ξ_b does not depend on S.

D the larger the injection velocity, the greater the necessary chamber length.

E a long narrow chamber can have a smaller volume than a short fat one for the same thrust.

8.14 All the following statements about the quantity S are true, except :

A it is dimensionless.

B it is proportional to the Prandtl number of the gas.

C it cannot take negative values.

D it increases in magnitude with increases in the temperature of the gas.

E it is usually around 0.5.

In the sketch, the number(8.15), (8.16), (8.17), (8.18),

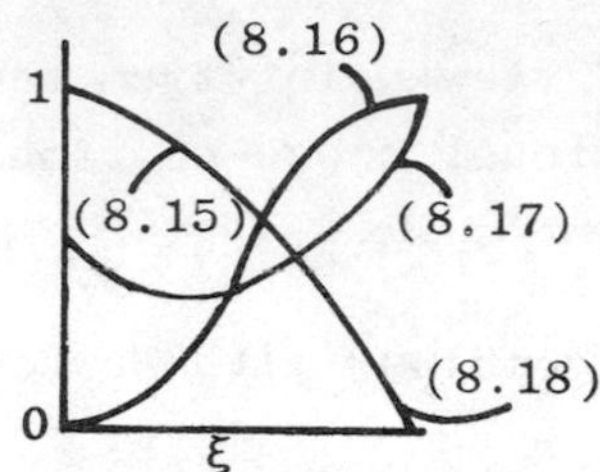

indicates :

A the location where combustion is complete.

B the variation of droplet temperature.

C the variation of gas velocity.

D the variation of droplet velocity.

E the variation of droplet diameter.

8.19 If a propellant injector is re-designed, so as to have the same total orifice area but a larger number of injection holes, the effect will probably be to :

A reduce the thrust of the motor.

B make a longer chamber necessary.

C reduce the necessary length of chamber.

D raise the temperature of the gases in the combustion chamber.

E move the point at which gas and droplets have equal velocity farther from the injector plate.

If the mathematical model were refined to take account of :

8.20 The fact that droplets are injected at temperatures below that of the wet bulb,

8.21 The finite Reynolds number of the relative motion, the length required for complete combustion would :

A certainly increase.

B certainly decrease.

C certainly remain unchanged.

D change in a direction that cannot be predicted.

E no longer be calculable.

8.22 The droplets have their minimum velocity at the section at which this velocity equals the gas velocity
<u>because</u>
the Reynolds number of the relative motion is there very small.

8.23 If an injector atomised most of the propellant very finely, but produced a small proportion of large droplets, the motor length for complete combustion would be greater than if all the droplets had the large diameter
<u>because</u>
the gas velocity, and so the average droplet velocity, would be higher.

8.24 If the droplets are injected at initial temperatures below their wet-bulb temperature, their vaporisation times become shorter
<u>because</u>
the larger temperature difference between gas and liquid will increase the rate of heat transfer.

8.25 It is possible to design a liquid-propellant rocket motor from first principles

<u>because</u>

theories exist which connect droplet size and velocity with injector dimensions, and these provide the required input to the theory of Chapter 8.

8.26 The length of rocket motor needed to burn all the propellant is always smaller than the droplet-vaporisation time divided by the injection velocity

<u>because</u>

the droplets always slow down immediately after injection.

8.27 ξ_b can be interpreted as the minimum length of motor divided by twice the product of vaporisation time and maximum gas velocity

<u>because</u>

the latter is represented by G/ρ and the former by $(\frac{1}{2}r_o^2/\Gamma)\ \ell n\ (1 + B)$.

ANSWERS TO MULTIPLE-CHOICE PROBLEMS

Answer	Problem number (8's omitted)
A	3, 8, 12, 18, 20, 23
B	2, 6, 10, 13, 21, 22
C	4, 7, 9, 16, 19, 27
D	1, 14, 17, 24, 26
E	5, 11, 15, 25

CHAPTER 9

THE LAMINAR JET

9.1 INTRODUCTION

A) PURPOSE

- Many practical combustion systems involve the injection of a jet of fuel into an atmosphere containing oxygen. Examples are :- diesel, gas-turbine and rocket engines; furnaces; domestic heating appliances. Often the fuel is gaseous from the start; or, if it is in liquid form, it is so well atomised that the vaporisation time is short compared with the subsequent mixing time.
- Typically, the flame looks like the one in the following sketch.

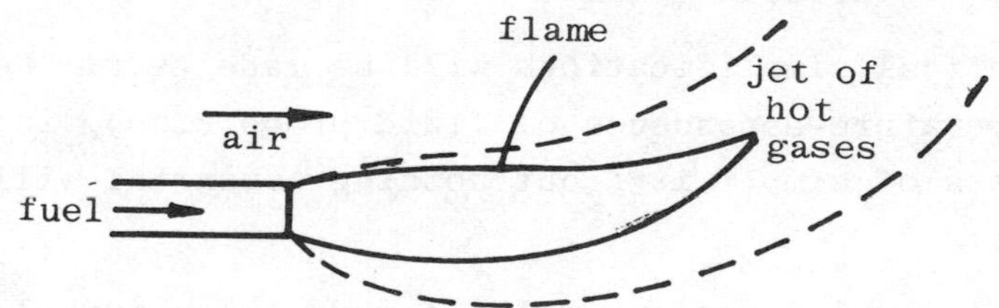

If the buoyancy forces act along the line of the injection, and there are no other features (e.g."cross-wind") promoting three-dimensionality, an axi-symmetrical flame and mixing region are formed, as shown :

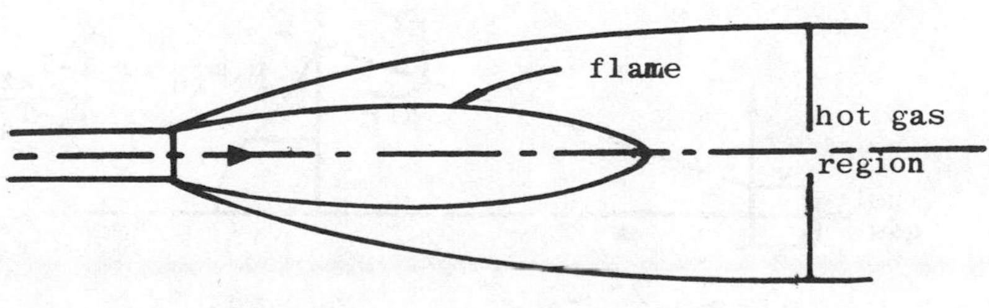

- To design the equipment, it is necessary at least to calculate the length of the flame, and it would be helpful to know other things, e.g. velocity profile, temperature profile, radiant heat flux, etc.
- Often the chemical reaction proceeds as fast as possible, as for liquid-droplet combustion (these are physically-controlled flames). This means the flame shape is limited by diffusional mixing of fuel and oxygen.
- It is therefore necessary to be able to compute diffusional mixing rates. That is the purpose of this and the next three chapters.

B) METHOD

- The order will be :

  1. laminar jet mixing; 2. laminar flame; 3. turbulent jet mixing; 4. turbulent flame.
- Mathematical simplifications will be made (e.g. neglect of temperature-dependence of fluid properties), in the interests of simplicity; but nothing essential will be lost thereby.
- Physical simplifications will be made, when turbulence is in question, because the laws of turbulence are incompletely known.

C) THE LAMINAR-JET PROBLEM

The situation now to be considered is represented by the sketch below, and also by the following :-

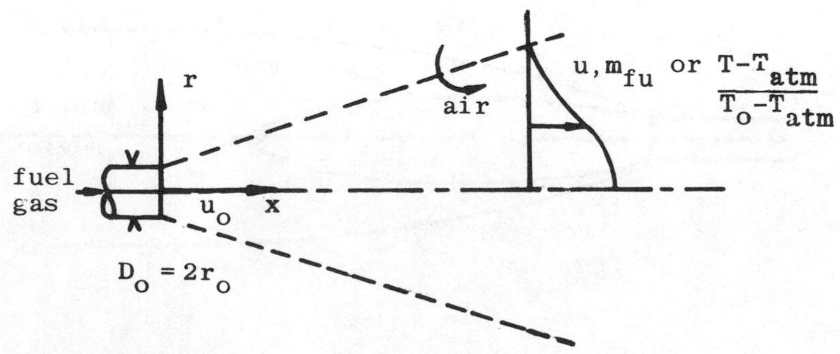

- the surrounding air is at rest;
- chemical reaction is absent;
- the density, viscosity and other properties of the gases are uniform;
- the flow is steady;
- buoyancy is absent;
- axial symmetry prevails;
- the pressure in the fluid is uniform;
- diffusion, heat conduction and viscous action in the axial (x) direction are negligible;
- the flow is laminar;
- the Schmidt number μ/Γ equals unity;
- the Prandtl number μ/Γ_k equals unity.

The complete analysis is lengthy; the following presentation is restricted to the beginning and the end, and a few intermediate stepping stones.

It will be shown that, at large values of x/D_o, the profiles of u, m_{fu} and $(T - T_\infty)/(T_{ax} - T_\infty)$ become "self-similar", and depend upon r/x alone. Especial attention will be devoted to this self-similar behaviour.

9.2 MATHEMATICAL ANALYSIS

A) SYMBOLS USED

u	axial velocity,
v	radial velocity,
m	mass fraction of fuel,
T	gas temperature,
μ	viscosity,
ρ	density,
Γ	exchange coefficient of fuel gas,
f	mixture fraction based on the fuel and air as reference states.

Note that, since the situation being dealt with is a non-reacting one, f and m_{fu} will always be equal; so only the latter need be referred to explicitly. Further, the similarities between heat transfer and mass transfer permit discussion of T also to be foregone.

B) DIFFERENTIAL EQUATIONS

- Mass conservation :

$$\frac{\partial}{\partial x}(\rho u\, r) + \frac{\partial}{\partial r}(\rho v\, r) = 0 \qquad . \quad (9.2\text{-}1)$$

- Momentum conservation in axial direction :

$$\frac{\partial}{\partial x}(\rho ur.u) + \frac{\partial}{\partial r}(\rho vr.u) = \frac{\partial}{\partial r}\left(\mu r\,\frac{\partial u}{\partial r}\right) \qquad . \quad (9.2\text{-}2)$$

- Injected-fluid (fuel) conservation :

$$\frac{\partial}{\partial x}(\rho ur.f) + \frac{\partial}{\partial r}(\rho vr.f) = \frac{\partial}{\partial r}\left(\Gamma r\,\frac{\partial f}{\partial r}\right) \qquad . \quad (9.2\text{-}3)$$

Since ρ, μ and Γ are being taken as uniform, and the Schmidt number as unity, it is possible to define :

$$\nu \equiv \frac{\mu}{\rho} = \frac{\Gamma}{\rho} \qquad , \quad (9.2\text{-}4)$$

and rewrite the equations as :

$$\frac{\partial u}{\partial x} + \frac{\partial v}{\partial r} + \frac{v}{r} = 0 \qquad , \quad (9.2\text{-}5)$$

$$u\,\frac{\partial u}{\partial x} + v\,\frac{\partial u}{\partial r} = \frac{\nu}{r}\,\frac{\partial}{\partial r}\left(r\,\frac{\partial u}{\partial r}\right) \qquad , \quad (9.2\text{-}6)$$

$$u\,\frac{\partial f}{\partial x} + v\,\frac{\partial f}{\partial r} = \frac{\nu}{r}\,\frac{\partial}{\partial r}\left(r\,\frac{\partial f}{\partial r}\right) \qquad . \quad (9.2\text{-}7)$$

C) SOLUTION FOR LARGE VALUES OF x

(i) Definitions of jet invariants

I_u and I_f are defined by :

$$I_u \equiv \nu^{-1} \int_o^\infty u^2 \, r \, dr \quad , \quad (9.2\text{-}8)$$

$$= \nu^{-1}\left(\tfrac{1}{2} \, u_o^2 r_o^2\right) \quad ; \quad (9.2\text{-}9)$$

$$I_f \equiv \nu^{-1} \int_o^\infty uf \, r \, dr \quad , \quad (9.2\text{-}10)$$

$$= \nu^{-1}\left(\tfrac{1}{2} \, u_o r_o^2\right) \quad . \quad (9.2\text{-}11)$$

These quantities are equal to $1/(\pi\mu)$ times the rates of flow of axial momentum and of fuel respectively; and, since there are sources of neither momentum nor fuel along the length of the jet, their values are independent of x, and equal to those at the injection plane, where $x = 0$.

(ii) Boundary conditions

At large radius, one has :

$$r \to \infty : \quad u = 0 \quad , \quad (9.2\text{-}12)$$

$$f = 0 \quad . \quad (9.2\text{-}13)$$

At the entrance plane, one has :

$$\left.\begin{matrix} x = 0 \\ r \leq r_o \end{matrix}\right\} \left\{\begin{matrix} u = u_o & , \quad (9.2\text{-}14) \\ f = 1 & , \quad (9.2\text{-}15) \end{matrix}\right.$$

$$\left.\begin{matrix} x = 0 \\ r > r_o \end{matrix}\right\} \left\{\begin{matrix} u = 0 & , \quad (9.2\text{-}16) \\ f = 0 & , \quad (9.2\text{-}17) \end{matrix}\right.$$

as already implied.

(iii) Solution of the differential equations

It can be verified by substitution that the following relations satisfy the differential equations (9.2-5) (9.2-6) and (9.2-7), and the boundary conditions (9.2-12) and (9.2-13). They do not satisfy the boundary conditions for the injection plane ($x = 0$) in detail; but they agree with these conditions in respect of the integral relations (9.2-8 and 9) and (9.2-10 and 11). They can therefore be accepted as solutions which are valid for large values of x, at which the effects of the details of the profiles of u and f at x = 0 have become insignificant.

The equations are :

$$u = \frac{3}{4}\frac{I_u}{x}(1 + \xi^2/4)^{-2} \quad , \quad (9.2\text{-}18)$$

$$v = \left(\frac{3}{8} I_u \nu\right)^{\frac{1}{2}} \frac{\xi}{x}(1 - \xi^2/4)(1 + \xi^2/4)^{-2} \quad , \quad (9.2\text{-}19)$$

$$f = \frac{3}{4}\frac{I_f}{x}(1 + \xi^2/4)^{-2} \quad , \quad (9.2\text{-}20)$$

where the single non-dimensional space variable ξ is defined by :

$$\xi \equiv \left(\frac{3}{8}\frac{I_u}{\nu}\right)^{\frac{1}{2}} \frac{r}{x} \quad . \quad (9.2\text{-}21)$$

(iv) Discussion

- Since ξ is constant for fixed r/x, the jet can be regarded as radiating from the injection point in a conical manner.
- The values of u and f on the axis ($\xi = 0$) vary reciprocally with longitudinal distance x, in accordance with

$$u_{ax}x = \frac{3}{4} I_u \quad , \quad (9.2\text{-}22)$$

$$f_{ax}x = \frac{3}{4} I_f \quad . \quad (9.2\text{-}23)$$

- Inspection of these formulae confirms that the solution cannot be valid at small x; for u_{ax} cannot exceed the entrance velocity u_o; nor can f_{ax} exceed unity. It follows that the solutions are valid only for the condition :

$$x > \frac{3}{4}\frac{I_u}{u_o} \quad ,$$

$$\text{i.e.} \quad \frac{x}{r_o} > \frac{3}{8}\frac{u_o r_o}{\nu} \quad . \quad (9.2\text{-}24)$$

This formula indicates that the number of nozzle radii downstream at which the injection details are submerged is of the order of the Reynolds number based on the nozzle conditions.

- The radial profiles of u and f, when normalised by their respective axial velocities, are identical, namely :

$$\frac{u}{u_{ax}} = \frac{f}{f_{ax}} = \frac{1}{(1+\xi^2/4)^2} \quad . \quad (9.2\text{-}25)$$

The sketch represents the shape qualitatively. The significance of the value $\xi = 1.287$ is that it is often useful to characterise the width of a jet by the value of $r_{\frac{1}{2}}$, i.e. the radius at which the velocity has one half its axial velocity. The conclusion is :

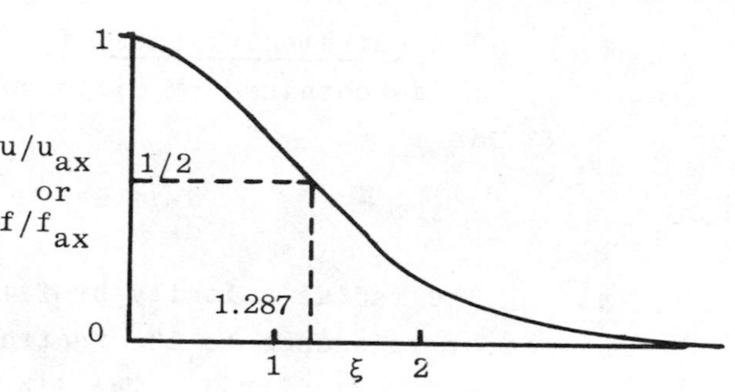

$$\frac{r_{\frac{1}{2}}}{x} = 1.287 \left(\frac{8\nu}{3I_u}\right)^{\frac{1}{2}} \quad , \quad (9.2\text{-}26)$$

$$\text{i.e. } \frac{r_{\frac{1}{2}}}{x} = 1.287 \times \left(\frac{16}{3}\right)^{\frac{1}{2}} \left(\frac{\nu}{u_o r_o}\right)$$

$$= 2.97 \left(\frac{\nu}{u_o r_o}\right) \quad . \quad (9.2\text{-}27)$$

This shows that the angle of the jet is inversely proportional to the Reynolds number.

- The total mass rate of flow in the jet, M, can be obtained by integration, as follows :

By definition :

$$M \equiv \int_o^{\infty} 2\pi r \rho u dr \quad . \quad (9.2\text{-}28)$$

Insertion of (9.2-18) leads to :

$$M = 2\pi\rho x^2 \left(\frac{8}{3}\frac{\nu}{I_n}\right)\left(\frac{3}{4}\frac{I_u}{x}\right)\int_o^\infty \frac{\xi d\xi}{(1+\xi^2/4)^2}$$

i.e. $\boxed{M = 8\pi\mu x}$. (9.2-29)

This remarkable result shows that the mass rate of flow increases with x, but is actually independent of the velocity of the injected fluid or the radius of the nozzle. Of course, this is true only for the condition of (9.2-24).

- The entrainment rate, i.e. the rate of increase of M with x, is obtained by differentiation of (9.2-29); the result is :

$$\frac{dM}{dx} = 8\pi\mu \qquad . (9.2\text{-}30)$$

- The radial velocity profile is most interestingly normalised by reference to the "entrainment velocity", v_{ent}, i.e. $-(dM/dx)/(2\pi r\rho)$. The finding, from (9.2-19) and (9.2-30), is

$$\frac{v}{v_{ent}} = \left(\frac{\xi^2}{4}\right)\left(1 - \frac{\xi^2}{4}\right)\Big/\left(1 + \frac{\xi^2}{4}\right)^2 \qquad . (9.2\text{-}31)$$

At large radii ($\xi \to \infty$) this tends to -1, as expected; at small radii v/v_{ent} tends to $\xi^2/4$.

Obviously v changes from positive to negative as ξ increases beyond 2.

9.3 MIXING PATTERNS

A) CONTOURS OF u AND f

The equations describing lines of constant u or f can be found by re-arrangement of (9.2-18) or (9.2-20) to yield :

$$r = \frac{x}{R}\frac{8}{\sqrt{3}}\left[\left(\frac{3}{8}R\frac{r_o}{x}\cdot\frac{u_o}{u}\right)^{\frac{1}{2}} - 1\right]^{\frac{1}{2}} \qquad , (9.3\text{-}1)$$

or an equivalent expression with f^{-1} in the place of u_o/u. Here the symbol R has been introduced for the Reynolds number; thus :

$$R \equiv \frac{r_o u_o}{\nu} \qquad . \quad (9.3\text{-}2)$$

Curves obeying (9.3-1), for fixed values of u/u_o (or f), have the form shown in the following sketch.

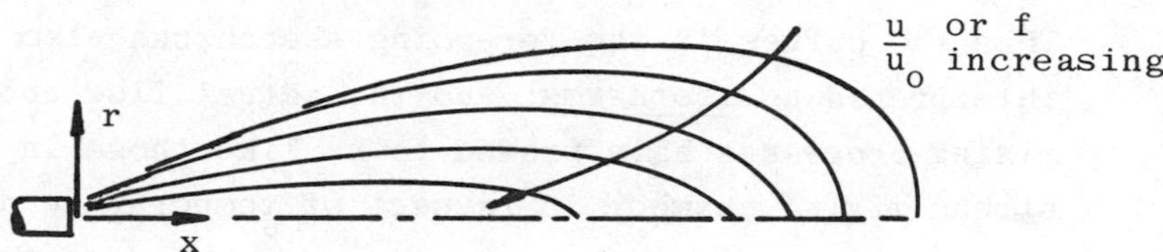

B) INTERPRETATION

Along one of the envelopes indicated in the sketch, f is constant. The material to be found there, although it has arrived at its location and state by a complex set of flow and molecular-mixing processes, can be regarded as consisting of material which could have been formed in a steady-flow mixing chamber of the kind described in section 6.2 C. Moreover this is true of momentum per unit mass as well as of the composition; for the relation :

$$f = \frac{u}{u_o} \qquad , \quad (9.3\text{-}3)$$

which is implied in (9.2-18) and (9.2-20) is also what would be derived in the two-stream mixing process in which the F stream shared its momentum (u_o per unit mass) with the A stream (which has zero momentum). In the present case, there is an exact similarity between the processes of mass transfer and momentum transfer.

c) TEMPERATURE DISTRIBUTIONS

It is easy to show, by writing down and inspecting the appropriate differential equation, and postulating equality of λ/c and Γ, that the temperature field is related to the f field by :

$$\frac{T - T_\infty}{T_o - T_\infty} = f \qquad . (9.3\text{-}4)$$

Thus the curves in the foregoing sketch can also be interpreted as isotherms; and the actual flow and molecular-mixing processes have proved to be like those in the two-stream mixing chamber in respect of temperature also.

The derivation of this is left to the reader.

9.4 THE EFFECTS OF DEPARTURES FROM THE POSTULATED CONDITIONS

A) IF THE SCHMIDT NUMBER μ/Γ DID NOT EQUAL UNITY

If μ/Γ equals a constant different from unity, it is possible to show that the u distribution is unchanged, and that :

$$\frac{f}{f_{ax}} = \left(\frac{u}{u_{ax}}\right)^{\mu/\Gamma} \qquad . (9.4\text{-}1)$$

Thus, if μ/Γ is less than unity, as is common for gases, the concentration profile is broader than that of velocity; for μ/Γ less than unity, as is true of liquids, the reverse relation holds.

Of course, f_{ax} no longer obeys equation (9.2-23); for $\mu/\Gamma < 1$, f_{ax} has smaller values; for $\mu/\Gamma > 1$, f_{ax} has larger values.

B) WHEN THE VALUES OF μ, ρ AND Γ VARY, FOR EXAMPLE WITH f OR T

When the fluid properties vary, analytical solution of the equations ceases to be possible, in general. Then numerical means must be adopted for solving the differential equations. Such means are available.

No qualitative changes in the behaviour of the jet result from the variability of the properties.

C) WHEN BUOYANCY FORCES ARE PRESENT

If the jet is injected vertically upward, and the density of the injected fluid is slightly lower than that of the surroundings, buoyancy effects cause the momentum flux in the jet to increase with x, without however departures from axial symmetry being caused thereby. This process is easily calculated numerically.

When the buoyancy forces have a component at right angles to the jet axis, departures from axial symmetry result. Such effects are harder to calculate.

D) WHEN THE SURROUNDING FLUID IS ALSO IN MOTION

In most jet-mixing situations of practical engineering, the surrounding fluid is not at rest, as has been postulated above, but is in motion with a velocity which may vary with x. This motion considerably influences the mixing pattern. Numerical methods of computation are normally required for the quantitative prediction of these effects.

9.5 REFERENCES

- H SCHLICHTING (1968)
  "Boundary-layer theory".
  McGraw-Hill, New York, 6th Ed.

- SPALDING D B (1978)
  "GENMIX - A General Computer Program for Two-Dimensional Parabolic Phenomena"
  HMT Series Number 1, Pergamon Press, Oxford

EXERCISES TO FACILITATE ABSORPTION OF MATERIAL OF CHAPTER 9

ANALYTICAL PROBLEMS

9.1 Demonstrate by reference to the appropriate differential equation that, in a laminar axi-symmetrical jet for which $\lambda/c = \Gamma = \mu$, the temperature profiles and the concentration profiles are identical in shape. List all the further assumptions that it is necessary to make in order to obtain the result.

9.2 If $\bar{f}$ stands for the average f value of the material flowing in a laminar axi-symmetrical jet, show that :

$$\bar{f}/f_{ax} = 1/3$$

9.3 Show that, for a laminar axi-symmetrical jet, the sectional Reynolds number defined as $r_{\frac{1}{2}}\, u_{ax}/\nu$ in the fully-developed region is 1.114 times $r_o u_o/\nu$.

MULTIPLE-CHOICE PROBLEMS

9.4 The pressure may reasonably be taken as uniform in the laminar-jet problem
because
the viscosity of a gas is almost independent of pressure.

9.5 The velocity distribution predicted for an axi-symmetrical uniform-property laminar jet is likely to be found in the flame of a Bunsen burner
because
such flames are usually laminar and axi-symmetrical.

9.6 The profiles of u/u_{ax} and f/f_{ax} are identical for the model of the lecture
because
the Schmidt number is supposed to have a constant value throughout the flow.

9.7 The far-downstream solution for u and f in the laminar jet cannot hold in the vicinity of the nozzle
<u>because</u>
neither u nor f can anywhere have values in excess of the injected fluid.

9.8 If the jet were directed vertically upward and the injected fluid were slightly lighter than the surrounding fluid, I_u would decrease with x
<u>because</u>
gravity would cause the surrounding fluid to sink.

9.9 The mass rate of flow in a laminar axi-symmetrical jet is independent of the injection rate
<u>because</u>
neither the width of the jet nor the velocity in it, at a given x, are altered by a change in u_o.

9.10 Near the axis of the jet the radial-outflow velocity is positive
<u>because</u>
$\partial u/\partial x$ is negative, the continuity equation implies

$$\frac{\partial u}{\partial x} + \frac{1}{r}\frac{\partial (vr)}{\partial r} = 0$$

and v equals zero on the axis.

9.11 The shear stress in the fluid is finite on the axis
<u>because</u>
$\partial u/\partial x$ is finite there.

9.12 At large values of r/x, the product vr tends to a finite positive value
<u>because</u>
the jet spreads in the shape of a cone.

9.13 The value of x at which a given constant-f contour attains its maximum radius can be found from the relation :

$$\frac{\partial}{\partial x}\left\{ x\left[\left(\frac{3}{8}\, R\, \frac{r_o}{x}\cdot\frac{1}{f}\right)^{\frac{1}{2}} -1\right]^{\frac{1}{2}}\right\} = 0$$

<u>because</u>

the expression in the curly bracket is equal to the radius of the contour as a function of x.

9.14 The rate of entrainment of fluid into an axi-symmetrical laminar jet, per unit length, depends only on the viscosity of the fluid

<u>because</u>

an increase in jet velocity is accompanied by a decrease in jet radius.

When the injection conditions of a laminar jet are changed by :

9.15 doubling the injection velocity;

9.16 doubling the nozzle radius;

9.17 doubling the kinematic viscosity;

9.18 doubling the mass flow without change of velocity,

the value of f on the axis at a fixed distance from the nozzle is likely to be multiplied by a factor of :

A 2

B 4

C 8

D ½

E ¼

For a laminar axi-symmetrical jet, the expression :

9.19 $(1 + \xi^2/4)^{-2}$;

9.20 $2.97/R$;

9.21 $8\pi\mu x$;

9.22 $(\xi^2/4)(1 - \xi^2/4)(1 + \xi^2/4)^{-2}$;

9.23 $\frac{3}{8} R$;

represents

A the lower limit of x/r_o at which the self-similar solutions can prevail.

B the radial profile of f or u.

C the mass flow rate in the jet.

D the value of $r_{\frac{1}{2}}/x$.

E the variation of the radial velocity with radius.

9.24 All the following laminar jet flows necessitate three-dimensional numerical treatments for their prediction except that for which :

A the density varies and the jet is inclined to the direction of gravity.

B the surrounding atmosphere is in motion along a line oblique to the direction of injection.

C the jet impinges on a plane surface inclined to its axis.

D the nozzle cross-section is elliptical rather than circular.

E the injected gas reacts exothermically with the atmosphere.

For a laminar axi-symmetrical jet, the graphical representation of the variation of x of :

9.25 the mass flow rate,

9.26 the velocity on the axis,

9.27 the entrainment rate,

9.28 the radius where $u = \frac{1}{2}u_{ax}$,

9.29 the radius where $f = \frac{1}{2}f_{ax}$,

9.30 the radius where $f = \frac{1}{2}$,

9.31 the radial velocity at infinity,

9.32 the radial velocity on the axis,

9.33 the product $u_{ax}x$ could be :

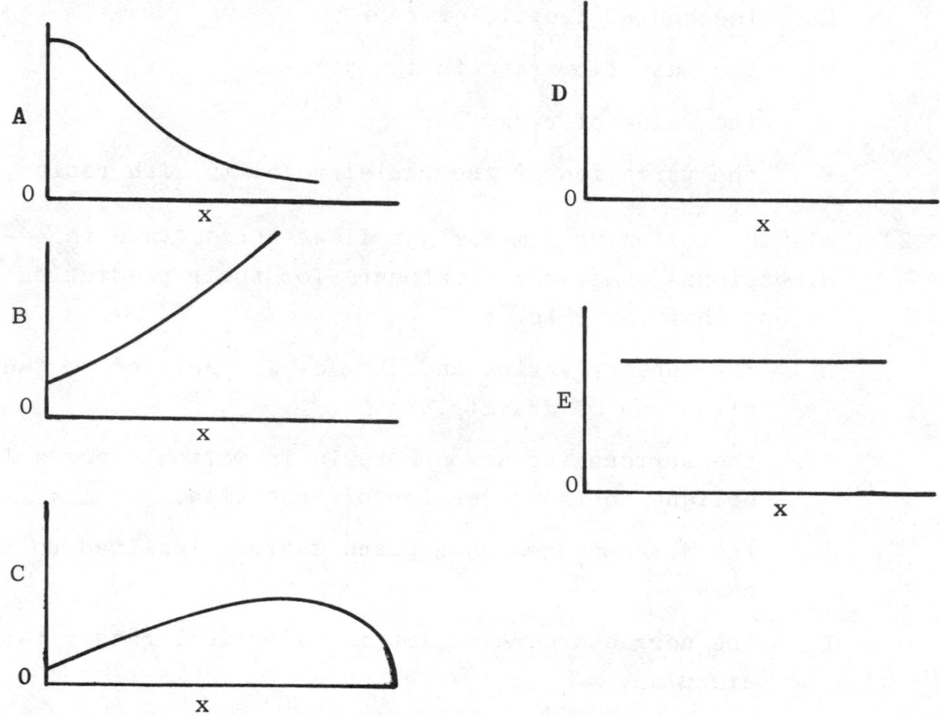

ANSWERS TO MULTIPLE-CHOICE PROBLEMS

Answer	Problem number (9's omitted)						
A	7,	10,	14,	15,	18,	23,	26
B	4,	6,	16,	19,	25,	28,	29
C	9,	13,	21,	30			
D	5,	11,	12,	20,	31,	32	
E	8,	17,	22,	24,	27,	33	

CHAPTER 10

THE LAMINAR DIFFUSION FLAME

10.1 INTRODUCTION

A) PURPOSE

The theory of the laminar jet will be used as the basis for a theory of the laminar diffusion flame. The task is to predict the length, width, shape and structure of the flame formed when a fuel gas is injected from an orifice into a still atmosphere.

Because the laminar-jet theory was for uniform properties whereas substantial temperature (and therefore property) variations are present in flames, the theory will be only approximate. However, the qualitative predictions are correct; and the quantitative predictions can be made so by the use of appropriate average property values.

B) EXPERIMENTAL FACTS

The sketch illustrates the type of flame in question. Little oxygen is detectable within the reaction-zone envelope; and little fuel is detectable outside it. The reaction zone is a thin envelope from which blue light is emitted.

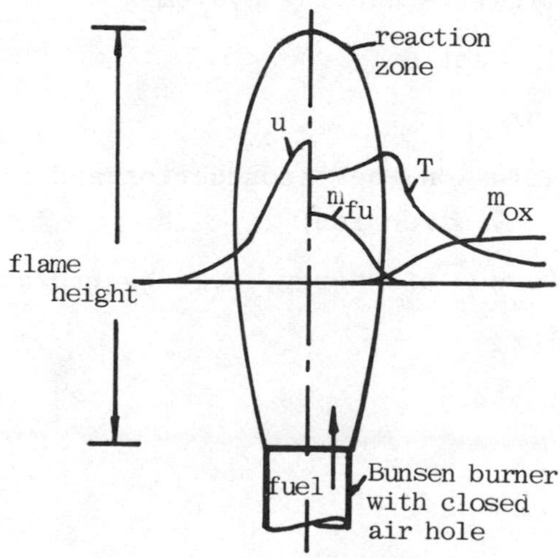

The height of the flame increases with increase in nozzle velocity, as indicated on the right. Eventually turbulence ensues, which leads to a diminution of flame height, followed by a tendency for the height to remain constant.

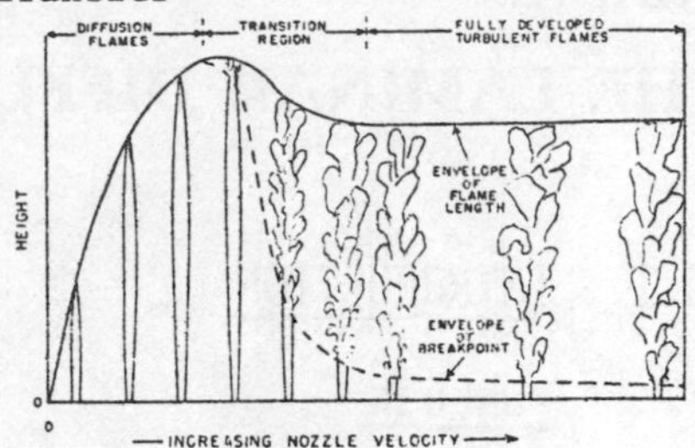

Progressive change in flame type with increase in nozzle velocity.

An increase of diameter also increases flame height, at fixed velocity. However, in the laminar region, the flame height correlates best with volumetric injection rate, i.e. with the combination $u_o r_o^2$.

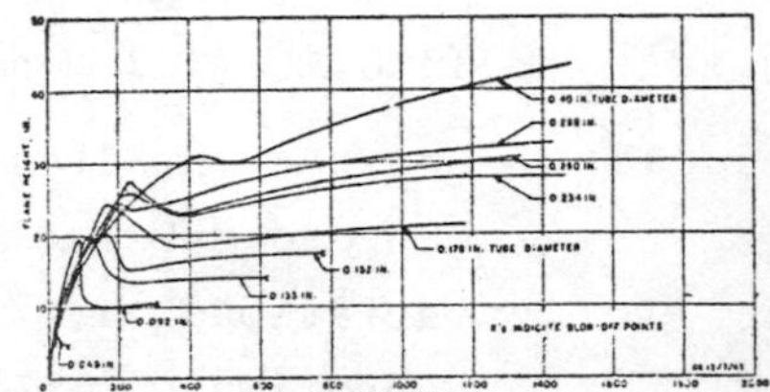
City gas flames in free air. Effect of volume flow rate and tube diameter on flame height, 100 percent gas.

10.2 MATHEMATICAL MODEL

A) ASSUMPTIONS

It is presumed that the phenomenon is characterised by the following features :

- axial symmetry;
- steady, laminar flow;
- a Simple Chemically-Reacting System;
- fast chemical reaction;
- uniform pressure;
- negligible diffusion, heat conduction and viscous action in the axial (x) direction;
- negligible sources of energy, e.g. by radiation;
- low Mach number;
- negligible buoyancy;

- equality of μ, Γ's and λ/c;
- uniform density and viscosity.

It is the last three assumptions which are most at variance with reality.

B) DIFFERENTIAL EQUATIONS

It will be useful to present the differential equations in three different forms : the first is fairly general; the second incorporates the restrictions of the SCRS; and the third incorporates the further restrictions of uniformity of properties.

(i) First form

- Continuity (i.e. mass conservation) :

$$\frac{\partial}{\partial x}(\rho ur) + \frac{\partial}{\partial r}(\rho vr) = 0 \qquad ; \quad (10.2\text{-}1)$$

- Axial-momentum conservation :

$$\frac{\partial}{\partial x}(\rho ur.u) + \frac{\partial}{\partial r}(\rho vr.u) = \frac{\partial}{\partial r}\left(\mu r\frac{\partial u}{\partial r}\right) \qquad ; \quad (10.2\text{-}2)$$

- Fuel conservation :

$$\frac{\partial}{\partial x}(\rho ur.m_{fu}) + \frac{\partial}{\partial r}(\rho vr.m_{fu}) = \frac{\partial}{\partial r}\left(\Gamma_{fu}r\frac{\partial m_{fu}}{\partial r}\right) + rR_{fu} \quad ; \quad (10.2\text{-}3)$$

- Oxygen conservation :

$$\frac{\partial}{\partial x}(\rho ur.m_{ox}) + \frac{\partial}{\partial r}(\rho vr.m_{ox}) = \frac{\partial}{\partial r}\left(\Gamma_{ox}r\frac{\partial m_{ox}}{\partial r}\right) + rR_{ox} \quad ; \quad (10.2\text{-}4)$$

- Energy conservation (with $h \equiv cT + H\, m_{fu}$) :

$$\frac{\partial}{\partial x}(\rho ur.h) + \frac{\partial}{\partial r}(\rho vr.h) = \frac{\partial}{\partial r}\left(\lambda r\frac{\partial T}{\partial r}\right) + H\frac{\partial}{\partial r}\left(\Gamma_{fu}r\frac{\partial m_{fu}}{\partial r}\right) \quad . \quad (10.2\text{-}5)$$

(ii) Introduction of the SCRS simplifications

- Because the fuel and oxygen are presumed to form a Simple Chemically-Reacting System, one can employ :

$$R_{fu} = R_{ox}/s \quad , \quad (10.2\text{-}6)$$

and :

$$\mu = \Gamma_{fu} = \Gamma_{ox} = \lambda/c \quad . \quad (10.2\text{-}7)$$

- Then equations (10.2-3) and (10.2-4) can be combined to form :

$$\frac{\partial}{\partial x}(\rho ur.\phi) + \frac{\partial}{\partial r}(\rho vr.\phi) = \frac{\partial}{\partial r}\left(\mu r\frac{\partial \phi}{\partial r}\right) \quad , \quad (10.2\text{-}8)$$

where :

$$\phi \equiv m_{fu} - m_{ox}/s \quad ; \quad (10.2\text{-}9)$$

Further, since the mixture fraction f may be defined by the linear equation of section 6.2 C :

$$f \equiv \frac{(m_{fu} - m_{ox}/s) - (m_{fu} - m_{ox}/s)_A}{(m_{fu} - m_{ox}/s)_F - (m_{fu} - m_{ox}/s)_A} \quad , \quad (10.2\text{-}10)$$

it may be proved, by substitution, that f obeys a similar equation, namely :

$$\frac{\partial}{\partial x}(\rho ur.f) + \frac{\partial}{\partial r}(\rho vr.f) = \frac{\partial}{\partial r}\left(\mu r\frac{\partial f}{\partial r}\right) \quad . \quad (10.2\text{-}11)$$

- Introduction of the equality represented by (10.2-7) into the energy equation (10.2-5), permits this to be reduced to :

$$\frac{\partial}{\partial x}(\rho ur.h) + \frac{\partial}{\partial r}(\rho vr.h) = \frac{\partial}{\partial r}\left(\mu r\frac{\partial h}{\partial r}\right) \quad . \quad (10.2\text{-}12)$$

- The similarity of the equations for u, f and h deserves note; it is this which permits a single form of solution to serve for all three equations.

(iii) Uniform-property form

Insertion of the assumptions that μ and ρ are uniform, introduction of the symbol ν for μ/ρ, and combination with the continuity equation, lead finally to the same set of equation as was studied, in Chapter 9, for the laminar jet without chemical reaction, namely :

$$\frac{\partial u}{\partial x} + \frac{\partial v}{\partial r} + \frac{v}{r} = 0 \quad , \quad (10.2\text{-}13)$$

$$u\frac{\partial u}{\partial x} + v\frac{\partial u}{\partial r} = \frac{\nu}{r}\frac{\partial}{\partial r}\left(\frac{r\partial u}{\partial r}\right) \quad , \quad (10.2\text{-}14)$$

$$u\frac{\partial f}{\partial x} + v\frac{\partial f}{\partial r} = \frac{\nu}{r}\frac{\partial}{\partial r}\left(r\frac{\partial f}{\partial r}\right) \quad , \quad (10.2\text{-}15)$$

$$u\frac{\partial h}{\partial x} + v\frac{\partial h}{\partial r} = \frac{\nu}{r}\frac{\partial}{\partial r}\left(r\frac{\partial h}{\partial r}\right) \quad . \quad (10.2\text{-}16)$$

c) SOLUTIONS

(i) Boundary conditions

- At large radius one has :

$$r \to \infty : \quad u = 0 \quad , \quad (10.2\text{-}17)$$

$$f = 0 \quad , \quad (10.2\text{-}18)$$

$$h = h_\infty \quad . \quad (10.2\text{-}19)$$

- At the entrance plane, the variables have the values u_o, 1 and h_o; however, as in Chapter 9, these will be satisfied only in an integral manner, by the relations :

$$x = 0: \begin{cases} I_u \equiv \nu^{-1}\int_o^\infty u^2 r\,dr \quad , \\ \quad = \nu^{-1} u_o^2 r_o^2/2 \quad ; \quad (10.2\text{-}20) \\ I_f \equiv \nu^{-1}\int_o^\infty ufr\,dr \quad , \\ \quad = \nu^{-1} u_o r_o^2/2 \quad ; \quad (10.2\text{-}21) \\ I_h \equiv \nu^{-1}\int_o^\infty u\,(h-h_\infty) r\,dr \quad , \\ \quad = \nu^{-1} u_o(h_o-h_\infty) r_o^2/2 \quad . \quad (10.2\text{-}22) \end{cases}$$

Of course, the integrals I_u, I_f and I_h are independent of longitudinal distance x, because there are no source terms in the differential equations.

(ii) Solutions

It may be proved by substitution that the solutions of the equations for u, f and h are, as is obvious from the analysis in Chapter 9 :

$$\frac{ux}{I_u} = \frac{fx}{I_f} = \frac{(h - h_\infty)x}{I_h} = \frac{3}{4}(1 + \xi^2/4)^{-2} \quad , (10.2\text{-}23)$$

where ξ is defined as before by :

$$\xi = \left(\frac{3}{8}\frac{I_u}{\nu}\right)^{\frac{1}{2}} \frac{r}{x} \quad . (10.2\text{-}24)$$

In terms of the Reynolds number of the nozzle flow R ($\equiv u_o r_o/\nu$), these relations become :

$$\frac{u}{u_o} . \frac{x}{r_o} . \frac{1}{R} = f . \frac{x}{r_o} . \frac{1}{R} = \frac{(h - h_\infty)}{(h_o - h_\infty)} . \frac{x}{r_o} . \frac{1}{R} = \frac{3}{8}\left(1 + \frac{3}{16} R \frac{r^2}{x^2}\right)^{-2} \quad . (10.2\text{-}25)$$

10.3 DISCUSSION

A) GENERAL FEATURES

The most noticeable feature of the solutions represented by (10.2-25) is that they are identical to those of Chapter 9; the fact that combustion has occurred appears to have made no difference. There is indeed no primary effect of combustion on the distribution of velocity and of conserved properties; and the secondary effects, which manifest themselves in practice through the influences of temperature on density and transport properties, have been deliberately ignored.

It can be seen that, once again, the axial values of velocity, mixture fraction and enthalpy difference fall off as the reciprocal of x; and the radial profiles fall asymptotically to zero as r/x increases.

Equation (10.2-25) represents the far-downstream ($x/r \gg \frac{3}{8}R$) solution; close to the nozzle, the equation can fit reality only in an integral sense. It is, however, quite adequate for the purpose of this book.

B) FLAME SHAPE

- If the reaction rates are infinitely fast, fuel and oxygen cannot be simultaneously present in finite concentration; then the reaction zone is a surface at which both fuel and oxygen have negligible concentrations. When the reaction rates are finite but still large, as is usually the case in practice, the region of overlapping m_{fu} and m_{ox} is still very small; for practical purposes therefore the flame can be regarded as being a surface located where :

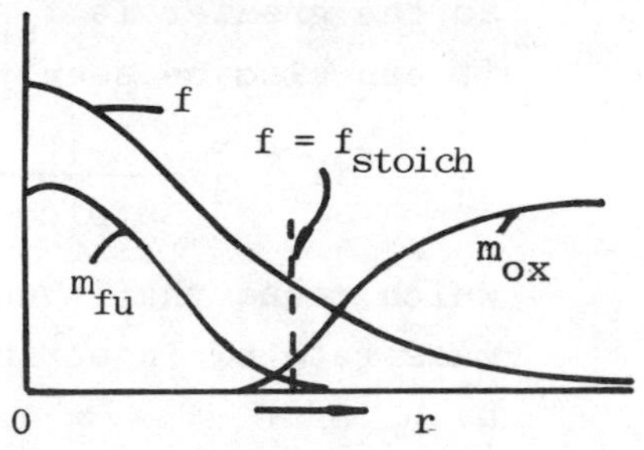

$$m_{fu} - m_{ox}/s = 0 \; ; \qquad (10.3\text{-}1)$$

$$f = f_{stoich} \; ,$$

$$= \frac{m_{ox,\infty}/s}{1+m_{ox,\infty}/s} \; . \qquad (10.3\text{-}2)$$

The last result follows from inserting (10.3-1) and boundary-condition information in (10.2-30); and it has already been derived, in a different context, in Chapter 6.

- The radius of the flame at any particular x is found by inserting f_{stoich}, calculated from (10.3-2), into (10.2-25), and rearranging; the result is :

$$\frac{r_{fl}}{x} = \left[\frac{16}{3R} \left\{ \left(\frac{3}{8} R \frac{r_o}{x} \cdot \frac{1}{f_{stoich}} \right)^{\frac{1}{2}} - 1 \right\} \right]^{\frac{1}{2}} \; . \qquad (10.3\text{-}3)$$

- When this is expressed graphically, it gives a flame shape as indicated in the following sketch.

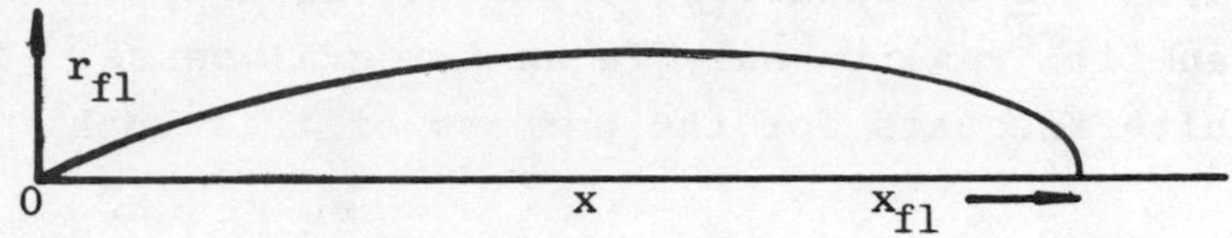

- The flame length, x_{fl}, is given by equating r_{fl} to zero in (10.3-3). The conclusion is :

$$\frac{x_{fl}}{r_o} = \frac{3}{8}\,\frac{R}{f_{stoich}} \qquad ; \ (10.3\text{-}4)$$

so the greater is f_{stoich}, the longer is the flame. It can also be seen that :

$$x_{fl} = \frac{3}{4}\,\frac{1}{f_{stoich}\,\nu}\left(\frac{\pi u_o r_o^{\,2}}{2}\right) \qquad ; \ (10.3\text{-}5)$$

which means that the flame length is proportional to the mass rate of injection of fuel, and not otherwise affected by u_o or r_o.

- The maximum width of the flame can be found by differentiatio of r_{fl} with respect to x. The conclusion is that it occurs where $x = 9x_{fl}/16$, and that the maximum radius is given by :

$$\frac{r_{fl,max}}{x_{fl}} = \frac{3}{4}\cdot\frac{(3)^{\frac{1}{4}}}{R^{\frac{1}{2}}} \qquad . \ (10.3\text{-}6)$$

Of course, since x_{fl} itself increases with the Reynolds number, R, the maximum flame width varies as $R^{\frac{1}{2}}$.

10.4 PRACTICAL IMPLICATIONS

A) INFLUENCE OF THE AIR REQUIREMENT OF THE FUEL ON FLAME LENGTH

Equations (10.3-4) and (10.3-2), when combined, result in :

$$\frac{x_{fl}}{r_o} = \frac{3}{8} R \left(1 + \frac{s}{m_{ox,\infty}}\right) \qquad (10.4\text{-}1)$$

This means that the flame length increases with the stoichiometric constant of the reaction, s; and it increases also as the oxygen content of the atmosphere is diminished.

To fix this in mind, one might say that the fuel "has to go farther" to find its oxygen, the more oxygen it wants and the less oxygen there is in the air.

B) HOW TO DESIGN FOR A SHORT FLAME

Often it is desirable that the combustion of the fuel should be completed in a short length from the point of injection. Since equation (10.3-5) has shown that the flame length is proportional to the rate of flow of fuel, this can be achieved only by reducing the rate of fuel flow per nozzle.

Short combustion chambers can be designed, it appears, only if the fuel is injected through a large number of nozzles, each contributing a small fraction of the total flow.

This qualitative conclusion has a much wider validity than the present laminar-diffusion-flame theory. This is why fuel is injected through many small orifices in rocket motors, and through several in gas turbines. In furnaces, on the other hand, long flames are often desired for uniformity of heat transfer over the furnace length; and so the fuel is often supplied through a single nozzle.

C) THE RELIABILITY OF THE PRESENT THEORY FOR DESIGN PURPOSES

Because the important effects of buoyancy and variations of ρ, μ, Γ etc. have been neglected, the quantitative predictions of flame length and shape will be erroneous unless suitable average values of ρ etc. are inserted; and one cannot tell what values are suitable without solving the equations exactly.

Fortunately, this is easy if a standard computer program is used, for example GENMIX (Spalding, 1978).

The influence of buoyancy is to shorten the flame because it increases the velocity of flow of gas in the jet, which means that the jet radius is smaller and gradients correspondingly steeper; the steeper gradients increase the rates of inter-diffusion.

The increases of properties μ, Γ and λ with temperature also tend to increase the rate of entrainment of surrounding air into the jet; but the increased specific volume widens the jet and so reduces the gradients. It is not possible to predict which effect on flame length is the greater without performing detailed calculations.

10.5 REFERENCES

- BURKE S P & SCHUMANN T E W (1928)

  Ind. Eng. Chem., Vol. 20, pp 998 - 1004.

  These authors developed an early theory of diffusion flames confined in a uniform-sectioned pipe.

- WOHL K, GAZLEY C, & KAPP N (1949)
  "Diffusion flames."
  Third Symposium on Combustion, Williams and Wilkins, Baltimore, pp 288 - 300.

  These authors, and those of the following paper, presented experimental data on laminar diffusion flames burning in air, together with a simple theory.

- HOTTEL H C & HAWTHORNE W R (1949)
  "Diffusion in laminar flame jets."
  Third Symposium on Combustion, Williams and Wilkins, Baltimore, pp 254 - 266.

SPALDING D B (1978)
"GENMIX - A General Computer Program for Two-Dimensional Parabolic Phenomena"

HMT Series, Number 1, Pergamon Press, Oxford

The computer program in this book permits the computation, by way of numerical solution of the parabolic partial differential equations, of many "boundary-layer" phenomena, of which the laminar diffusion flame is one. Variable fluid properties, and buoyancy, are as easy to include as to ignore.

EXERCISES TO FACILITATE ABSORPTION OF MATERIAL OF CHAPTER 10

ANALYTICAL PROBLEMS

10.1 What will be the ratio of the lengths of two diffusion flames, one of hydrogen and the other of methane, burning in atmospheric air, if the nozzle diameters and injection velocities are the same in both cases?

Answer : $x_{fl,H_2}/x_{fl,CH_4} = 1.95$

10.2 Prove that the maximum diffusion-flame width occurs at a distance from the nozzle equal to 9/16 of the distance to the tip of the flame.

MULTIPLE-CHOICE PROBLEMS

In the laminar diffusion flame indicated in the sketch, the oxygen-concentration along the radii 10.3, 10.4 etc., could be represented by curves of the form :

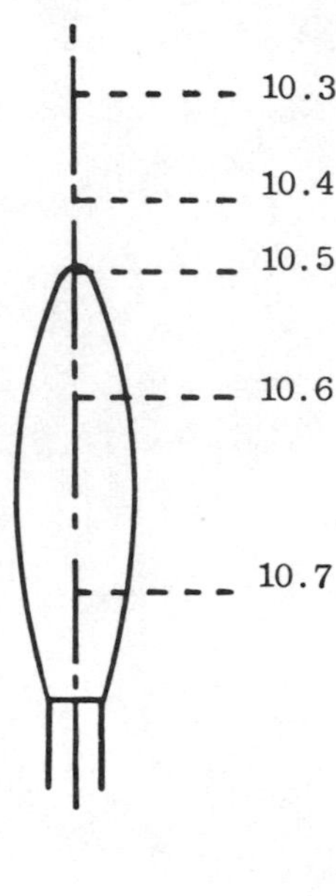

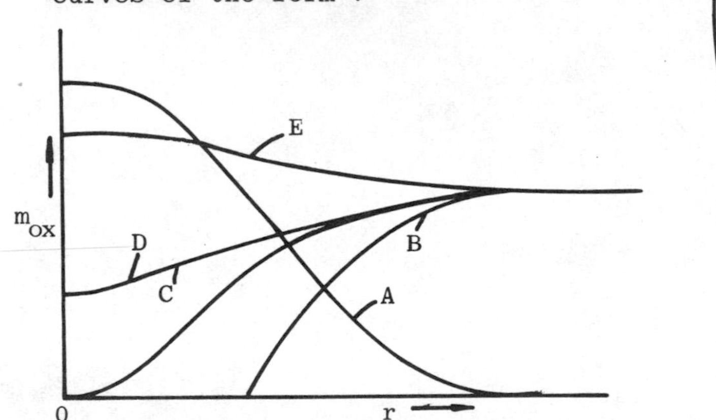

In the same laminar diffusion flame, the variation along the axis of :

10.8 m_{fu}

10.9 f

10.10 m_{ox}

10.11 m_{prod}

10.12 m_{dil}

could be represented by :

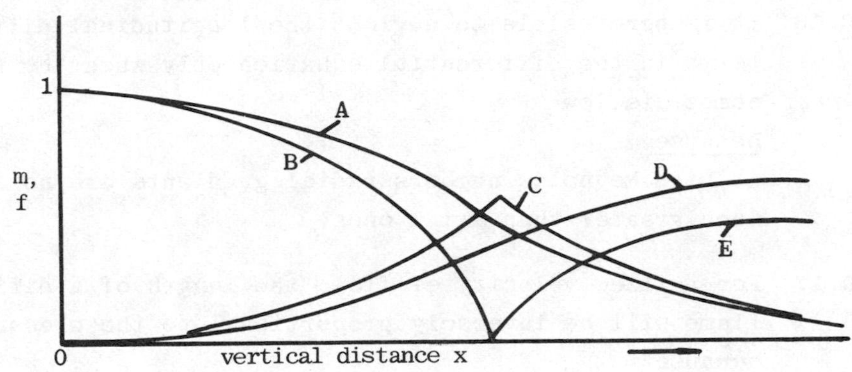

Mark the following statements in the usual way.

10.13 Little free oxygen is present within the diffusion-flame envelope

because

the diffusion coefficient of oxygen is much lower than that of fuel.

10.14 In a laminar diffusion flame, the distribution of nitrogen concentration obeys the differential equation

$$\rho u \frac{\partial m_{N_2}}{\partial x} + \rho v \frac{\partial m_{N_2}}{\partial r} = \frac{1}{r}\frac{\partial}{\partial r}\left(r\Gamma_{N_2}\frac{\partial m_{N_2}}{\partial r}\right)$$

because

N_2 is an inert gas and its exchange coefficient is nearly equal to that of O_2.

10.15 In the SCRS the quantity $\phi \equiv m_{fu} + m_{prod}/(1 + s)$ obeys the differential equation

$$\rho u \frac{\partial \phi}{\partial x} + \rho v \frac{\partial \phi}{\partial r} = \frac{1}{r} \frac{\partial}{\partial r} \left(r \Gamma_\phi \frac{\partial \phi}{\partial r} \right)$$

<u>because</u>

the rate of generation of combustion products equals (1+s) times the rate of generation of fuel.

10.16 It is permissible to neglect the longitudinal diffusion terms in the differential equation only when the Reynolds number is low

<u>because</u>

at high Reynolds numbers radial gradients are no longer much greater than axial ones.

10.17 For a fixed velocity of flow, the length of a diffusion flame will be inversely proportional to the pressure,

<u>because</u>

f_{stoich} is directly proportional to pressure.

10.18 In any real laminar diffusion flame, there is always a region in which both fuel and oxidant have non-zero concentrations

<u>because</u>

the volumetric rate of chemical reaction must vanish when either reactant has zero concentration.

ANSWERS TO MULTIPLE-CHOICE PROBLEMS

Answer	Problem number (10's omitted)
A	9, 18
B	6, 7, 8, 14
C	5, 11, 13, 15, 17
D	3, 4, 12
E	10, 16

CHAPTER 11

THE TURBULENT JET

11.1 INTRODUCTION

A) PURPOSE

When the Reynolds number of a jet ($\equiv u_o R_o/\nu$) exceeds about 15,000, the flow becomes turbulent. In engineering and the environment, high Reynolds numbers are common; so turbulent jets occur more frequently than laminar ones. This is just as true of flames as of non-reacting flows.

Turbulent flows exhibit special features, such as independence of further increase of Reynolds number, which render them simpler, in some respects, than laminar ones. The present purpose is, by concentrating attention on the turbulent jet in stagnant surroundings, to expose some of the major properties of turbulent flows in general.

B) SOME CHARACTERISTICS OF TURBULENT FLOWS

(i) Random fluctuations

Measurements made in a turbulent flow reveal that, even though their time-average values at a particular point stay constant, random rapid fluctuations occur.

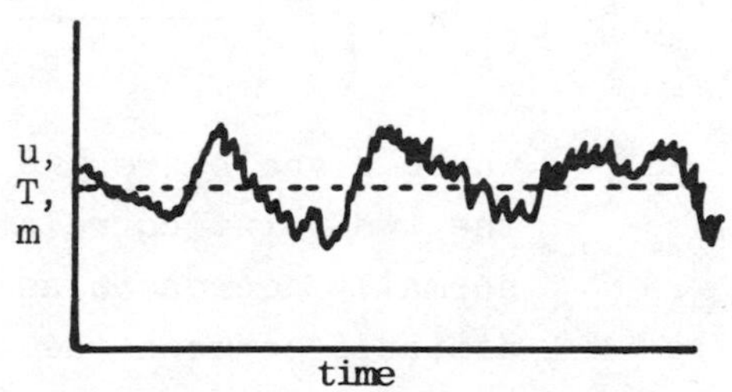

These can be described in part by :

- time-average values, e.g.

$$\bar{u} \equiv \left[\left(\int u dt \right) / t \right]_{t \to \infty} \qquad . \quad (11.1\text{-}1)$$

- root-mean-square fluctuations, e.g.

$$u' = \left[\left(\int (u - \bar{u})^2 \, dt \right) / t \right]^{\frac{1}{2}}_{t \to \infty} \quad . \quad (11.1\text{-}2)$$

(ii) Correlations

- The fluctuations of two different variables at the same point may be "correlated"; for example, even where $\bar{u}$ and $\bar{v}$ are zero, the quantity defined by :

$$\overline{(u - \bar{u})\,(v - \bar{v})} \equiv \left[\left(\int (u - \bar{u})\,(v - \bar{v}) dt \right) / t \right]_{t \to \infty} \quad .(11.1\text{-}3)$$

may not be.

- Many of the correlations have physical significance, for example :-

$$-\rho \left\{ \overline{(u - \bar{u})\,(v - \bar{v})} \right\} \equiv \text{"turbulent shear stress"};$$

$$-\rho \left\{ \overline{(m - \bar{m})\,(v - \bar{v})} \right\} \equiv \text{"turbulent diffusion flux"};$$

$$-c\rho \left\{ \overline{(T - \bar{T})\,(v - \bar{v})} \right\} \equiv \text{"turbulent heat flux"};$$

$$\tfrac{1}{2} \left\{ \overline{(u - \bar{u})^2 + (v - \bar{v})^2 + (w - \bar{w})^2} \right\} \equiv \text{"turbulent kinetic energy"}.$$

- In addition to the above "one-point" correlations, "two-point" correlations are also of interest, for example :

$$C_{ab} \equiv \left[\left(\int \overline{(u - \bar{u})_a \,(u - \bar{u})_b} \, dt \right) \Big/ t \right]_{t \to \infty} \quad , \quad (11.1\text{-}4)$$

where a and b are two distinct points. The magnitude of the two-point correlation normally decreases as the distance between the two points, r_{ab}, increases. The width of the $C_{ab} \sim r_{ab}$ profile is a measure of the "turbulence scale".

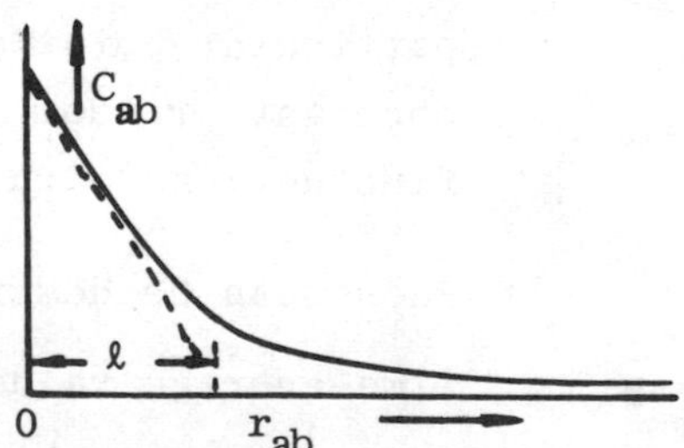

(iii) "Effective" transport properties

- An "effective viscosity" (also called "turbulent" or "eddy" viscosity) can be defined by :

$$-\rho\,\overline{(u-\overline{u})(v-\overline{v})} \equiv \mu_{eff}\frac{\partial\overline{u}}{\partial y} \qquad . \quad (11.1\text{-}5)$$

This is much larger than the laminar viscosity in a turbulent flow; and it varies from place to place.

- An "effective exchange coefficient" for mass transfer can be defined by :

$$-\rho\,\overline{(m-\overline{m})(v-\overline{v})} \equiv \Gamma_{eff}\frac{\partial\overline{m}}{\partial y} \qquad . \quad (11.1\text{-}6)$$

This is usually of the same order of magnitude as μ_{eff}, but often between 30% and 100% higher.

- An "effective conductivity" for heat transfer can be defined by :

$$-c\rho\,\overline{(T-\overline{T})(v-\overline{v})} \equiv \lambda_{eff}\frac{\partial\overline{T}}{\partial y} \qquad . \quad (11.1\text{-}7)$$

This is usually quite closely equal to the "effective exchange coefficient" for mass transfer.

- The last two statements can be summed up by defining "effective" Schmidt and Prandtl numbers by :

$$\sigma_{eff,m} \equiv \frac{\mu_{eff}}{\Gamma_{eff}} \qquad , \quad (11.1\text{-}8)$$

and :

$$\sigma_{eff,T} = \frac{c\mu_{eff}}{\lambda_{eff}} \qquad , \quad (11.1\text{-}9)$$

coupled with the statement :

$$\sigma_{eff,m} = \sigma_{eff,T} \approx 0.5 \sim 1.0 \quad . \quad (11.1\text{-}10)$$

- μ_{eff}, etc. are not necessarily "isotropic"; this means, for example, that the heat may be transferred more easily in one direction than another. Little is known about this

(iv) Similarities of turbulent and laminar flows

Because the effective transport properties, defined as above, are normally finite and positive, and since they do not vary by large factors in a single flow, turbulent flows appear much like laminar ones with augmented values of the transport properties.

This will be apparent in respect of the turbulent jet, discussed below; thus, the material of Chapter 9, which concerned the laminar jet, will prove to be directly applicable.

c) SOME THEORETICAL IDEAS

(i) The influence of the local mean-flow field

- If it is supposed that μ_{eff} is mainly influenced by near-by features of the mean flow field, dimensional analysis leads to the conclusion :

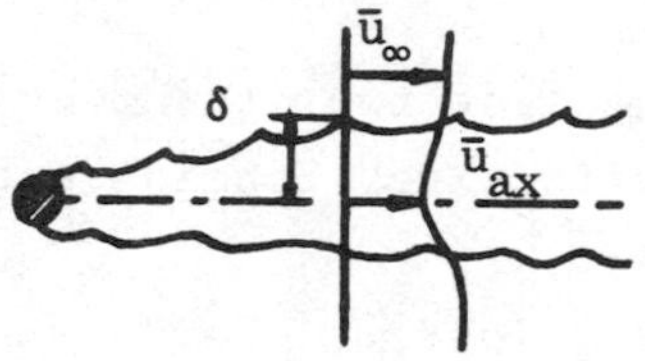

$\bar{u}_\infty$ δ $\bar{u}_{ax}$

$$\mu_{eff} = \text{constant} \quad \rho|\bar{u}_\infty - \bar{u}_{ax}|\delta \quad . \quad (11.1\text{-}11)$$

Experience shows the constant typically to be of the order of 0.01; but its actual value varies from one type of flow to the next.

- If a Reynolds number is formed with μ_{eff} and local features, thus :

$$Re_{turb} \equiv \frac{\rho |\bar{u}_\infty - \bar{u}_{ax}| \delta}{\mu_{eff}} \qquad , (11.1\text{-}12)$$

this Re_{turb} of course turns out to be a constant, of the order of 100.

(ii) Smaller-scale features

- If μ_{eff} is to be connected with strictly local quantities, the "turbulent kinetic energy" k and the length scale ℓ (see above) suggest themselves. Then dimensional analysis implies :

$$\mu_{eff} = \text{constant } \rho k^{\frac{1}{2}} \ell \qquad , (11.1\text{-}13)$$

wherein the constant depends upon just how the length-scale is defined.

- If ℓ is defined by reference to the local dissipation rate of energy, $\varepsilon\rho$, by :

$$\ell \equiv k^{3/2}/\varepsilon \qquad , (11.1\text{-}14)$$

the constant is of the order of 0.1. Variations from one flow to another are not large.

(iii) Prandtl's mixing-length model

- If $k^{\frac{1}{2}}$ is taken as equal to $\ell |\partial\bar{u}/\partial y|$, and the constant in (11.1-13) is defined to equal unity, ℓ is called the "mixing length", ℓ_m; thus :

$$\mu_{eff} \equiv \rho \ell_m{}^2 \left|\frac{\partial \bar{u}}{\partial y}\right| \qquad . (11.1\text{-}15)$$

- Typically, ℓ_m is found to be proportional to, and of the order one-tenth of, the width of the mean-flow region. The constant varies from one flow to another.

D) SOME VALUES

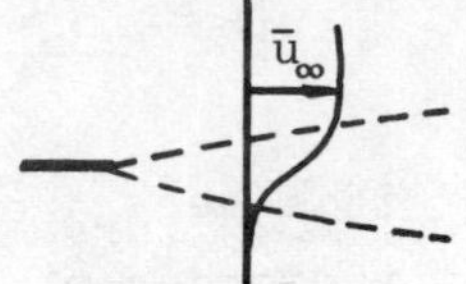

(i) Constants in μ_{eff} formulae

- Plane mixing layer :
  $\mu_{eff} \approx 0.0043\ \rho\bar{u}_\infty\delta$.(11.1-16)

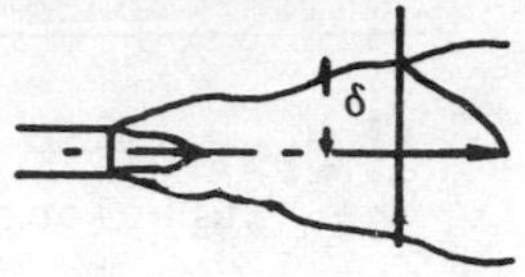

- Plane turbulent jet :
  $\mu_{eff} \approx 0.0148\ \rho\bar{u}_{ax}\delta$.(11.1-17)

- Plane wake :
  $\mu_{eff} \approx 0.0376\ \rho\left|\bar{u}_\infty-\bar{u}_{ax}\right|\delta$.(11.1-18)

- Axi-symmetrical jet :
  $\mu_{eff} \approx 0.0102\ \rho\bar{u}_{ax}\delta$.(11.1-19)

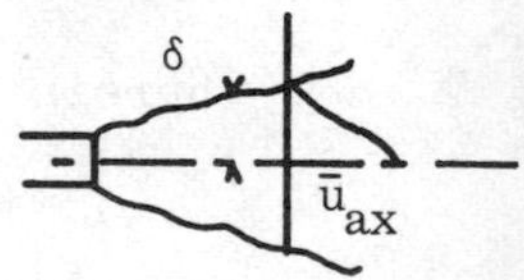

(ii) Mixing-length constants

- Plane mixing layer :

$$\ell_m \approx 0.07\delta \quad .\ (11.1\text{-}20)$$

- Plane turbulent jet :

$$\ell_m \approx 0.09\delta \quad .\ (11.1\text{-}21)$$

- Plane wake :

$$\ell_m \approx 0.16\delta \quad .\ (11.1\text{-}22)$$

- Axi-symmetrical jet :

$$\ell_m \approx 0.075\delta \quad .\ (11.1\text{-}23)$$

(iii) The determination of $\boldsymbol{\delta}$

The "thickness" of a mixing layer cannot be determined unequivocally. By contrast, for a jet, $r_{\frac{1}{2}}$ can be so determined.

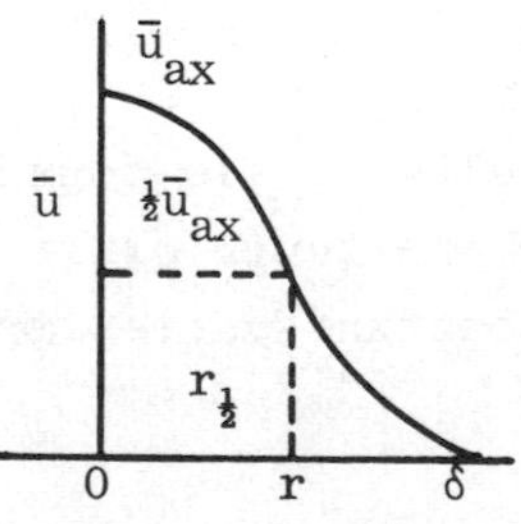

Profiles of velocity in turbulent jets and wakes are fairly uniform; and they imply that $\delta/r_{\frac{1}{2}} \approx 2.5$. This ratio has been used in the above μ_{eff} and ℓ_m/δ specification. For example, the formulae for the axi-symmetrical jet can be re-written in terms of $r_{\frac{1}{2}}$ as :

$$\mu_{eff} \approx 0.0256 \quad \rho\overline{u}_{ax} \, r_{\frac{1}{2}} \qquad ; \; (11.1\text{-}24)$$

$$\ell_m \approx 0.1875 \, r_{\frac{1}{2}} \qquad . \; (11.1\text{-}25)$$

11.2 MATHEMATICAL MODEL OF THE AXI-SYMMETRICAL JET IN STAGNANT SURROUNDINGS

A) THE SITUATION CONSIDERED

The situation is precisely as for the laminar jet, section 9.1C, except that the effective viscosity from equation (11.1-24) replaces the laminar viscosity.

B) THE AXIAL VELOCITY

From equations (9.2-17) and (11.1-24), we have :

$$\frac{r_{\frac{1}{2}}}{x} = 2.97 \left(0.0256 \, \frac{u_{ax} r_{\frac{1}{2}}}{u_o r_o} \right) \qquad . \; (11.2\text{-}1)$$

Hence

$$\frac{u_{ax}}{u_o} = 13.15 \, \frac{r_o}{x} \qquad , \; (11.2\text{-}2)$$

i.e.

$$\boxed{\frac{u_{ax}}{u_o} = 6.57 \, \frac{D_o}{x}} \qquad . \; (11.2\text{-}3)$$

(Note that the overbars have been omitted, as no longer necessary).

(ii) The jet angle

From equations (9.2-9) and (11.1-24) , there results:

$$I_u = \frac{1}{2} \frac{u_o^2 r_o^2}{0.0256\, u_{ax} r_{\frac{1}{2}}} \qquad . \quad (11.2\text{-}4)$$

Combination with (9.2-22) yields :

$$u_{ax}\, x = \frac{3}{4} \cdot \frac{1}{2} \frac{u_o^2\, r_o^2}{0.0256\, u_{ax} r_{\frac{1}{2}}} \qquad ,$$

i.e.
$$\frac{u_{ax}^2}{u_o^2} = 14.6 \frac{r_o^2}{r_{\frac{1}{2}} x} \qquad . \quad (11.2\text{-}5)$$

Finally, elimination of r_o between this equation and (11.2-2) leads to the desired result :

$$\frac{r_{\frac{1}{2}}}{x} = \frac{14.6}{(13.15)^2} \qquad ,$$

i.e.
$$\boxed{\frac{r_{\frac{1}{2}}}{x} = 0.0845} \qquad . \quad (11.2\text{-}6)$$

(iii) μ_{eff} related to the momentum flux

Combination of equations (11.1-24) (11.2-2) and (11.1-6) yields :

$$\mu_{eff} = 0.0256 \times 13.15 \times 0.0845\, \rho u_o r_o \qquad ,$$

$$= .02845\, \rho u_o r_o \qquad . \quad (11.2\text{-}7)$$

Let the injected momentum flux be ascribed the symbol F; thus :

$$F \equiv \rho u_o^2 \pi r_o^2 \qquad . \quad (11.2\text{-}8)$$

Then μ_{eff} is related to F by :

$$\boxed{\mu_{eff} = 0.01605\, \rho^{\frac{1}{2}}\, F^{\frac{1}{2}}} \qquad . \quad (11.2\text{-}9)$$

(iv) The entrainment rate

From equation (9.2-30), the rate of entrainment is proportional to the viscosity. Application of this formula to the turbulent jet, with the viscosity calculated from (11.2-8), leads to:

$$\frac{dM}{dx} = 0.404\,\rho^{\frac{1}{2}}F^{\frac{1}{2}} \qquad . \quad (11.2\text{-}10)$$

This formula is useful, even for jets of non-uniform density and irregular nozzle profile, because F is constant in a jet; so the jet retains its "memory" of this quantity, far downstream, where the details of injection conditions have been "forgotten". In that case, ρ must be taken as the density of the surrounding fluid.

(v) Profiles of velocity and concentration

- The longitudinal velocity profile can be represented, from (9.2-25), as :

$$\frac{u}{u_{ax}} = \frac{1}{\left\{1 + 0.414(r/r_{\frac{1}{2}})^2\right\}^2} \qquad . \quad (11.2\text{-}11)$$

Obviously, this sets u/u_{ax} equal to ½ when r equals $r_{\frac{1}{2}}$.

- The radial velocity can be similarly deduced from (9.2-31) as :

$$\frac{2\pi r\rho v}{0.404\rho^{\frac{1}{2}}F^{\frac{1}{2}}} = 0.414\,\frac{\left(\frac{r}{r_{\frac{1}{2}}}\right)^2\left\{1 - 0.414\left(\frac{r}{r_{\frac{1}{2}}}\right)^2\right\}}{\left\{1 + 0.414\left(\frac{r}{r_{\frac{1}{2}}}\right)^2\right\}^2} \qquad . \quad (11.2\text{-}12)$$

- The profile of mixture fraction f can be taken as identical with that of u/u_o, if Γ_{eff} is taken as equal to μ_{eff}. Otherwise equation (9.4-1) must be invoked.

11.3 EXPERIMENTAL INFORMATION ON THE SPREAD OF A TURBULENT JET

(i) The axial velocity

Measurements agree well with prediction of equation (11.2-3) for $x > 6.5\ D_o$.

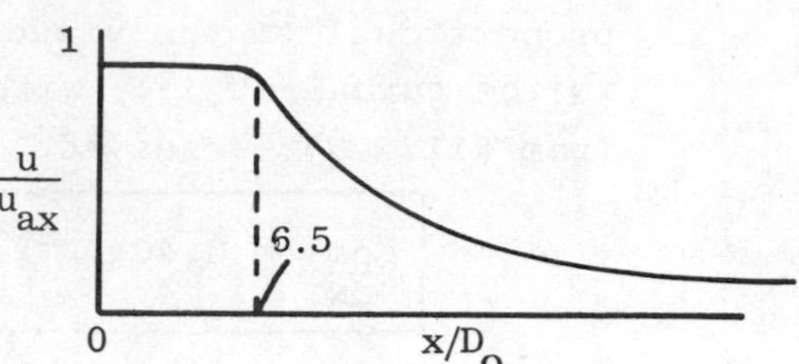

(ii) The jet angle

The value of $r_{\frac{1}{2}}/x$ is found to equal about 0.085, after the jet has become fully developed, in accordance with the predictions.

(iii) The entrainment rate

- The entrainment rate is measured directly by supplying air to the chamber until the pressure within it exactly equals that in the external atmosphere (Ricou and Spalding, 1961).

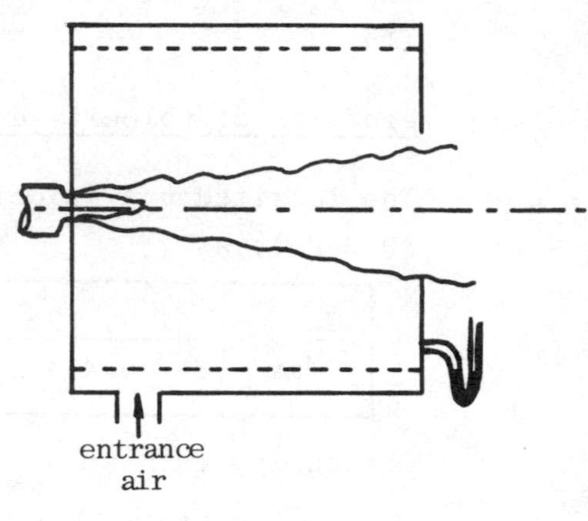

- The result is :

$$\frac{dM}{dx} \approx 0.28\ \rho^{\frac{1}{2}}\ F^{\frac{1}{2}} \quad , \quad (11.3\text{-}1)$$

i.e. considerably less than predicted by equation (11.2-9).

- The explanation is that the uniform-μ_{eff} model is unrealistic in the outer region, where μ_{eff} falls to the laminar value. Consequently, the actual

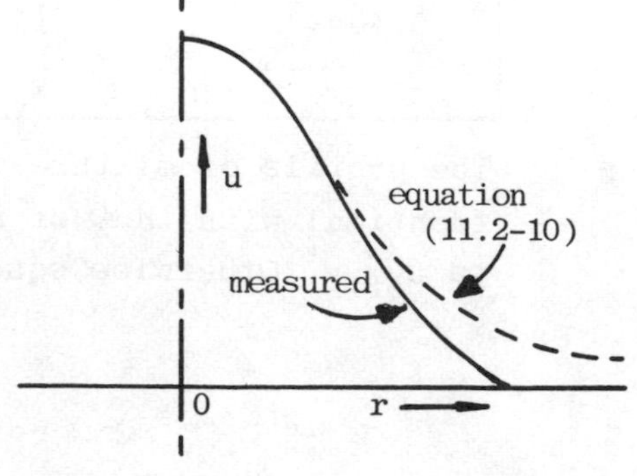

velocity profile differs from the predicted one, as sketched. The total mass flow rate is appreciably smaller than predicted because the outer region bulks large in the integral :

$$M = 2\pi\rho \int_0^\infty u\, r\, dr \qquad . \quad (11.3\text{-}2)$$

(iv) The f-field

- The axial profile obeys the equation :

$$f_{ax} \approx 5\, D_o/x \quad , \ (11.3\text{-}3),$$

for x greater than $5\, D_o$. This means that the composition difference decays more rapidly than the velocity difference; it corresponds to σ_{eff} being in excess of unity.

- The radial profile corresponds, as indicated. If equation (9.4-1) is employed, it can be deduced that $\sigma_{eff} \approx 0.7$.

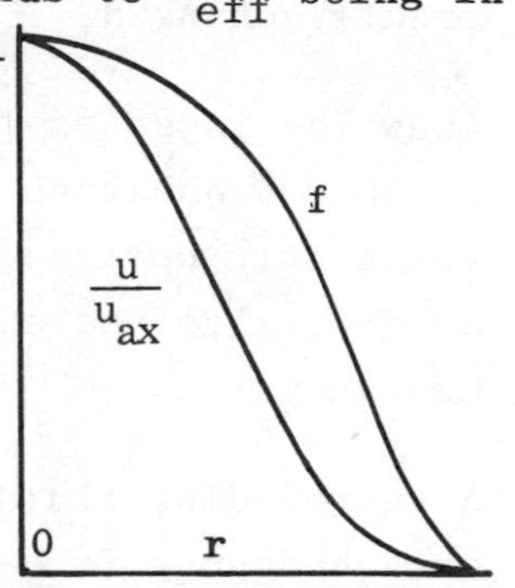

11.4 DISCUSSION

A) TYPICAL FEATURES OF TURBULENT FLOWS ILLUSTRATED BY THE JET IN STAGNANT SURROUNDINGS

- The time-mean flow is similar to a laminar one, but with considerably augmented transport properties ($\mu_{eff} >> \mu$ at high Re).

  The flow pattern (angle of spread, dimensionless velocity profiles, concentration fields) is independent of Reynolds number, once this is high enough for the flow to be fully turbulent.

B) SOME EFFECTS NOT ACCOUNTED FOR IN THE MODEL

- If the surrounding fluid is in axi-symmetrical motion, the rate of spread is diminished; so is the rate of entrainment which is proportional to $(u_{ax}-u_\infty)$ rather than to u_{ax}.

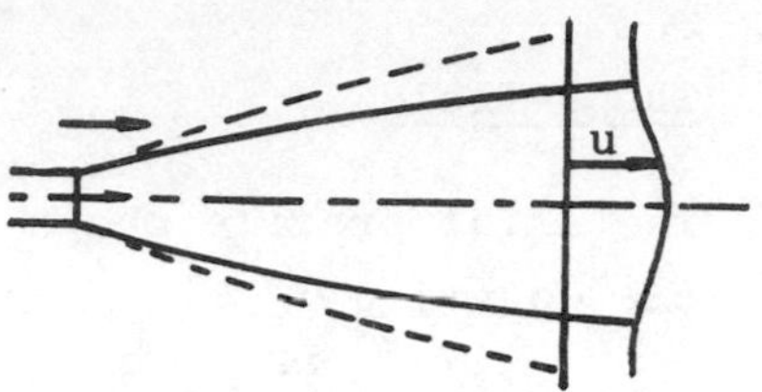

However, the shape of the transverse velocity profile (i.e. $\frac{u-u_\infty}{u_{ax}-u_\infty}$ versus $\frac{r}{r_{\frac{1}{2}}}$) is not greatly changed.

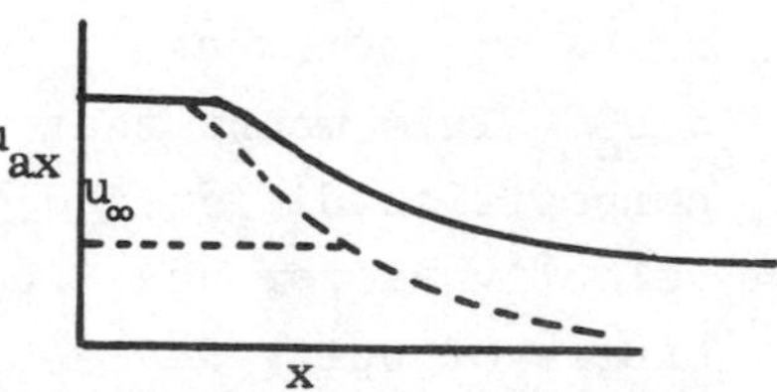

- When the injected fluid is less dense than the surroundings, and it is injected vertically upwards, the momentum increases with distance, as does also the entrainment rate, as a consequence of buoyancy.

- A surrounding fluid flow oblique to the angle of the jet, or a buoyancy force which is not aligned with the injection, produce three-dimensionality.

- Strong density variations, such as those caused by chemical reaction, also affect the flow, both by changing local effective transport properties (since ρ appears in equation 11.1-24) and by augmenting buoyancy effects.

- All these influences can be accounted for by numerical methods.

11.5 REFERENCES

- ABRAMOVICH G N (1963)
  "The Theory of Turbulent Jets"
  MIT Press, Cambridge, Massachusetts, USA.

- LAUNDER B E & SPALDING D B (1972)
  "Mathematical Models of Turbulence"
  Academic Press, London and New York.

- RAJARATNAM N (1976)
  "Turbulent Jets"
  Elsevier Scientific Publishing Company, Amsterdam.

- RICOU F P & SPALDING D B (1961)
  "Measurement of Engrainment by Axi-Symmetrical Turbulent Jets"
  J Fluid Mechanics, vol 11, pp 21-32.

- SCHLICHTING H (1960)
  "Boundary-Layer Theory"
  McGraw-Hill, New York, 4th edition.

- SPALDING D B
  "Turbulence Models (3rd Issue)"
  Imperial College, London, Heat Transfer Section Report, Number HTS/76/17

EXERCISES TO FACILITATE ABSORPTION OF MATERIAL OF CHAPTER 11

ANALYTICAL PROBLEMS

11.1 If measurements of the instantaneous velocities at one point in a turbulent flow show u to equal $\left[\bar{u} + (u^*/\sqrt{2})\sin \omega t\right]$ and v to equal $\left[\bar{v} + (v^*/\sqrt{2})\sin (\omega t + \alpha)\right]$, show that the turbulent shear stress equals $\rho u^* v^*/4$.

11.2 Show that the effective Reynolds number of a turbulent jet, defined as $u_{ax} r_{\frac{1}{2}} \rho/\mu_{eff}$ equals 39.0.

11.3 Show that the experimental relation for the entrainment rate into a turbulent jet implies that the average f value of the material flowing in the jet is :

$$\left[1 + \frac{1}{3.17}\frac{x}{D_o} \cdot \left(\frac{\rho_\infty}{\rho_o}\right)^{\frac{1}{2}}\right]^{-1}$$

MULTIPLE-CHOICE PROBLEMS

11.3 The effective viscosity of a turbulent fluid can be related to the kinetic energy of the fluctuating motion k, and to the eddy size ℓ, by :

A $\mu_{eff} = k^{\frac{1}{2}}\ell$

B $\mu_{eff} = \rho k^{\frac{1}{2}}\ell$

C $\mu_{eff} = \rho k \ell^{\frac{1}{2}}$

D $\mu_{eff} = \rho k^{\frac{1}{2}}/\ell$

E $\mu_{eff} = k^{\frac{1}{2}}\ell/\rho$

11.4 In a turbulent jet of atmospheric air, in stagnant surroundings, at a distance 5m from the nozzle the axial velocity is 10m/s. Since the laminar kinematic viscosity of air is of the order of 10^{-5} m^2/s, the effective viscosity is likely to exceed the laminar viscosity by a factor of :

A 10

B 10^2

C 10^3

D 10^4

E 10^5

11.5 In a turbulent jet of circular cross-section in uniform surroundings at rest, all the following quantities are likely to be independent of longitudinal distance except :

A $u_{ax}^2 r_{\frac{1}{2}}^2$

B $u_{ax} r_{\frac{1}{2}}^2 (T_{ax} - T_\infty)$

C $(T_{ax} - T_\infty)/f_{ax}$

D $u_{ax} r_{\frac{1}{2}}^2$

E $u_{ax}/(T_{ax} - T_\infty)$

For a turbulent jet of circular cross-section in uniform surroundings at rest, the quantity :

11.6 $r_{\frac{1}{2}}$

11.7 u_{ax}

11.8 M

11.9 f_{ax}

11.10 F

varies in proportion to x raised to the power :

A -1

B 0

C 2

D 1

E $\frac{1}{2}$

At a fixed large axial distance from the orifice, in a round-sectioned turbulent jet, the effect of doubling :

11.11 the velocity of flow through the orifice,

11.12 the diameter of the orifice,

11.13 the densities of both the injected and surrounding fluids,

11.14 the laminar viscosities of both fluids,

11.15 the laminar viscosity of the injected fluid only,

is likely to be :

A no change.

B a doubling of the width of the jet.

C a doubling of the momentum flow rate.

D a doubling of the velocity on the axis.

E a doubling of the density on the axis.

In a turbulent jet in stagnant surroundings

11.16 the mixing length,

11.17 the effective viscosity,

11.18 the turbulence energy,

11.19 the maximum shear stress in the section,

is likely to vary with distance from the orifice as :

A x

B 0

C x^{-1}

D x^{-2}

E x^{-3}

11.20 The constant-μ_{eff} theory of turbulent jets over-estimates the mass flow rate at any section, if the constant is such as to provide the correct rate of spread,
because
the experimental velocities exceed the predicted ones at the larger radii.

11.21 A light jet injected vertically upward into a denser medium entrains more fluid in a given length than one of equal density with the medium
because
buoyancy causes the momentum flux in the jet to increase with height.

11.22 The entrainment-chamber method of measuring entrainment rates is not applicable to the case in which the injected-fluid density differs from the density of the entrained fluid
because
zero pressure difference across the exit orifice would no longer imply that the correct amount of fluid was being supplied.

11.23 Turbulent flows often exhibit a pattern which is independent of the Reynolds number
because
doubling the velocity differences halves the effective viscosities.

11.24 The value of the mixing length ℓ_m divided by the width, δ, of a turbulent mixing region, is of the order of :

A 1

B .1

C .01

D .001

E .0001

11.25 The constant in the formula μ_{eff} = constant × $\rho\delta$ × velocity difference, for a turbulent mixing region, is of the order :

A 1

B 0.1

C 0.01

D 0.001

E 0.0001

If the mixing length is supposed to be independent of radius in a turbulent jet, knowledge of the velocity-profile shape allows us to conclude that

11.26 the effective viscosity,

11.27 the shear stress,

A is independent of radius.

B is always finite and positive except at the axis.

C is always finite and positive except at the axis and the outer edge of the layer.

D is finite at all radii.

E exhibits both positive and negative values.

Of the following curves, the one which could represent, for a jet injected into a stream of fluid flowing in the same direction but at a lower uniform velocity,

11.28 the $r_{\frac{1}{2}} \sim x$ curve,

11.29 the f_{ax} curve,

11.30 the u_{ax} curve,

11.31 the $u_{ax}x$ curve,

11.32 the flux of excess momentum ($\equiv 2\pi\int_0^\infty \rho u\,(u - u_\infty)rdr$),

is :

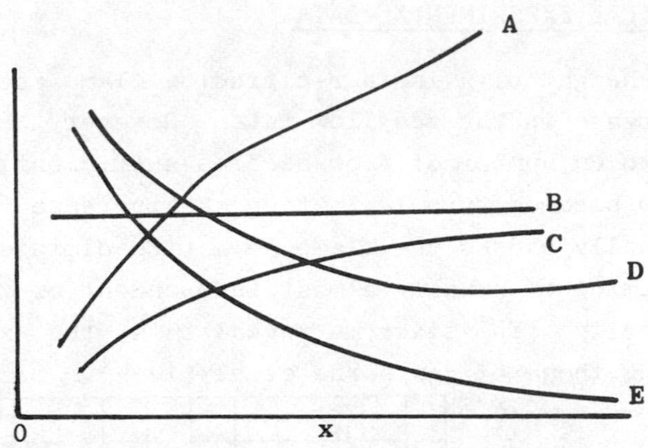

ANSWERS TO MULTIPLE-CHOICE QUESTIONS

Answer	Problem number (11's omitted)
A	7, 9, 14, 15, 16, 21, 31
B	3, 10, 17, 24, 32
C	20, 23, 25, 26, 28
D	4, 5, 6, 8, 11, 12, 18, 19, 30
E	13, 22, 27, 29

CHAPTER 12

THE TURBULENT DIFFUSION FLAME

12.1 INTRODUCTION

A) TYPICAL EXPERIMENTAL DATA

The height of a laminar diffusion flame increases with increase in the gas flow rate. However, when the Reynolds number of flow becomes sufficiently great, the flow becomes turbulent; then the increased mixing rate actually causes the flame height to diminish at first, becoming eventually almost independent of the injection velocity. The diagrams reveal this, the experiments being those of Hawthorne et al (1949).

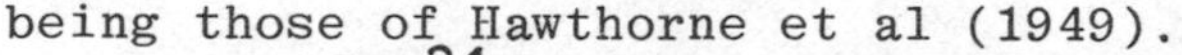

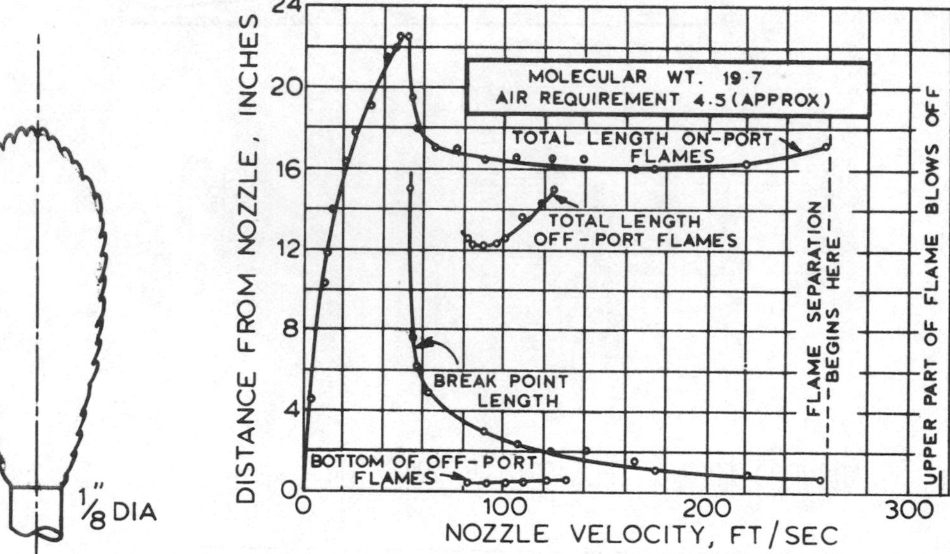

In the flow-rate region for which x_{fl} is independent of u_o, the lengths of flame divided by diameter of orifice are given by the following table. In every case, the fuel gas before injection, and the surrounding air, are at room temperature and pressure.

Fuel	Formula	x_{fl}/D
Carbon monoxide	CO	45
Hydrogen	H_2	130
Acetylene	C_2H_2	175
Propane	C_3H_8	295

Experiments with nozzles of various diameters show that x_{fl} increases linearly with diameter. From the table it is obvious that different fuels have significantly different non-dimensional heights. One purpose of the present chapter is to explain how this comes about.

B) SPECIAL FEATURES OF TURBULENT DIFFUSION FLAMES

- The measurement of flame height is not easy; for the turbulence of the flame renders the flame unsteady and its instantaneous shape "tattered". "Fragments" of burning gas detach themselves from the main body, and fly upwards, diminishing in size. Most flame height measurements in the literature were made "by eye"; precision is not to be expected.

- The turbulence affects not only x_{fl} but the whole reaction zone; in comparison with that of a laminar diffusion flame, this is thick, and in violent motion.

- A consequence is that, if the time-average values of m_{fu} and m_{ox} are measured along a radius, their curves overlap significantly.

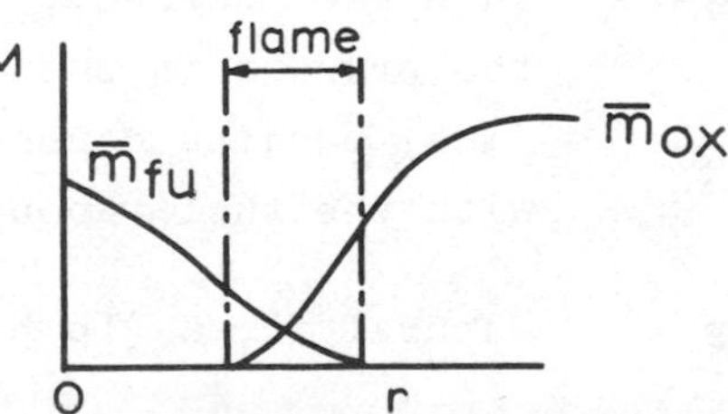

- Although at first sight this suggests a chemical-kinetic

influence, the truth is different : the over-lapping does not imply that fuel and oxygen are present at the same place and the same time; but the fluctuations result in one reactant being in large excess at one moment, and the other one at the next moment.

c) PRACTICAL RELEVANCE OF TURBULENT DIFFUSION FLAMES

Although the ideal turbulent diffusion flame which is to be described is rare in practice, many practical flames have features in common. Examples include :-

- The industrial furnace, in which a jet of steam-atomised oil is injected steadily into a combustion space. Radiation from the flame heats the roof and the material (e.g. molten iron) beneath it.

Air
Fuel

- In the liquid-propellant rocket motor, the stream of droplets resulting from liquid-jet impingement may, if they vaporise quickly,behave more like a jet of injected vapour than like a cloud of droplets. Thus the turbulent diffusion flame is an alternative model for rocket combustion.

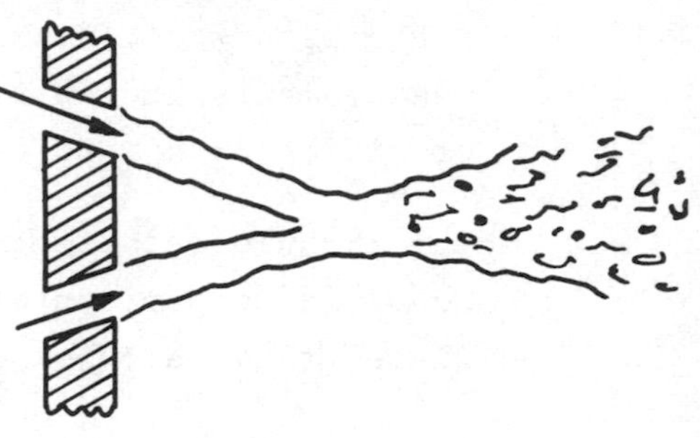

- In diesel engines, the situation is similar, except that the process is unsteady. Even when atomisation is very fine, burning cannot be instantaneous; the mixing of air with the fuel vapours is still needed.

- Natural fires (in houses, forests, prairies, etc.) are

often of the turbulent diffusion flame kind; however, usually the rate of supply of fuel to the flame is not then given externally but is controlled by heat transfer from the flame.

12.2 THE IDEAL TURBULENT DIFFUSION FLAME

A) MOTIVATION

The axi-symmetrical turbulent diffusion flame will be studied as an ideal case, to illuminate the general features of turbulent diffusion-flame phenomena, while still preserving analytical ease; understanding is the main objective; but practical conclusions will be drawn.

All simplifications which can be made without loss of essential features will be made.

B) ASSUMPTIONS DEFINING THE MODEL

The model will be similar to that for laminar flames, except that the effective viscosity will replace the laminar viscosity, as in Chapter 11. All effects of density variation will be ignored, except that the momentum flux will be calculated using the actual density of the injected fluid, i.e. the fuel.

Allowance will be made for the presence of fluctuations of f, and of quantities linked with f (m_{fu}, m_{ox}, T, etc.).

Γ_{eff}'s, and λ_{eff}/c, which are taken as equal according to the SCRS assumptions, will also be taken as equal to the effective viscosity; for, since the latter is being presumed uniform, a convenient further approximation is worth making.

c) MATHEMATICAL DESCRIPTION OF THE IDEAL TURBULENT DIFFUSION FLAME

(i) Time-average values of u and f

The equations of Chapter 11 can be taken as valid, because the presumed uniformity of properties permits no influence of the combustion on the flow pattern. From equation (11.2-3) and (11.2-11) one can write :

$$\frac{\bar{u}}{u_o} = \bar{f} = 6.57 \frac{D_o}{x}\left(\frac{\rho_o}{\rho_\infty}\right)^{\frac{1}{2}} \left[1 + 58\left(\frac{r}{x}\right)^2\right]^{-2} \qquad . \quad (12.2\text{-}1)$$

(ii) Fluctuations

- If it is supposed that, at any point, f fluctuates between a low value f_- and a high value f_+, and if the proportions of time spent at intermediate values are known, the time-average values of m_{fu}, m_{ox}, etc., can be determined.

- The simplest presumption will be made here, namely that the fluid spends equal times at the limits and none in between.

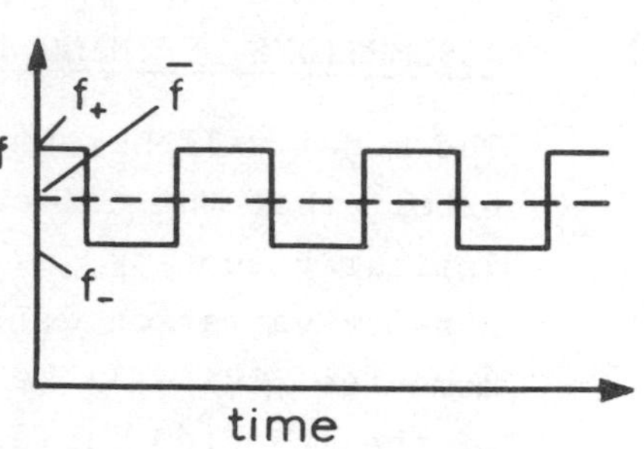

Then :

$$\bar{f} = \frac{1}{2}(f_+ + f_-) \qquad , \quad (12.2\text{-}2)$$

and

$$f' = \frac{1}{2}(f_+ - f_-) \qquad . \quad (12.2\text{-}3)$$

- The value of f' depends upon many factors, but an approximate formula, giving the correct order of magnitude is :

$$f' = \ell_m \left|\frac{\partial \bar{f}}{\partial y}\right| \qquad , \quad (12.2\text{-}4)$$

where y is the direction of maximum variation of $\bar{f}$.

(iii) Time-average values of m_{fu}, m_{ox}, T.

The diagram reveals the following features :-

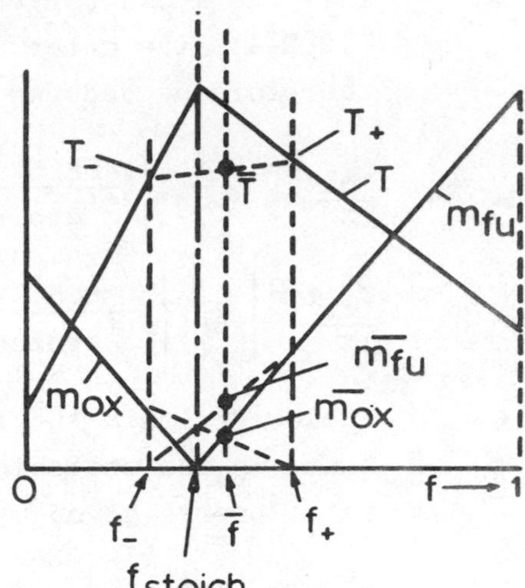

- $\bar{T} = \frac{1}{2}(T_+ + T_-)$; and this is lower than the T corresponding to $\bar{f}$.
- $\bar{m}_{fu} = \frac{1}{2}(m_{fu+} + m_{fu-})$; and this is greater than the m_{fu} corresponding to $\bar{f}$.
- $\bar{m}_{ox} = \frac{1}{2}(m_{ox+} + m_{ox-})$; and this exceeds the m_{ox} corresponding to $\bar{f}$.
- If f_{stoich} did not lie between f_+ and f_- however, as it does in the sketched example, $\bar{T}$, $\bar{m}_{fu}$ and $\bar{m}_{ox}$ would all correspond to $\bar{f}$.

The diagram thus explains why it is possible to find finite amounts of unburned fuel and oxygen even where $\bar{f}$ equals f_{stoich}.

(iv) Flame shape

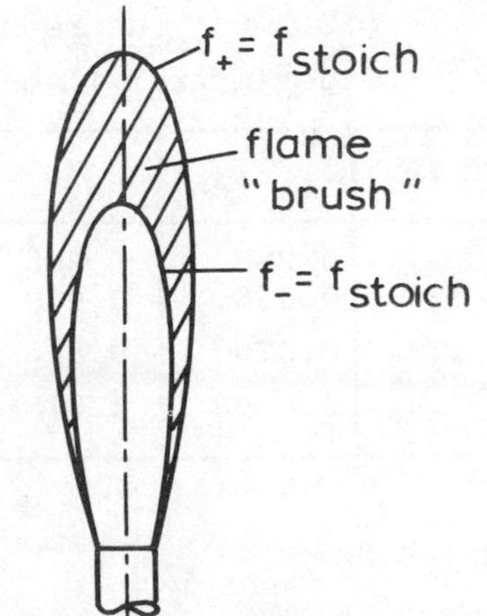

- The reaction zone will be a finite volume, enclosed by surfaces on which f_- and f_+ equal f_{stoich}.
- The outer envelope has f_+ equal to f_{stoich} because there the gas is mainly fuel-lean; but reaction just

occurs at the instants when the mixture ratio has its maximum fuel content.

- If f' is known (or can be estimated, e.g. from equation (12.2-4), the outer and inner radii of the flame can therefore be deduced from equation (12.2-1). Thus:

$$\frac{r_{out}}{x} = \left[\frac{1}{58}\left\{\frac{6.57}{(f_{stoich}-f')}\frac{D_o}{x}\left(\frac{\rho_o}{\rho_\infty}\right)^{\frac{1}{2}} -1\right\}^{\frac{1}{2}}\right]^{\frac{1}{2}} \quad ; \quad (12.2\text{-}5)$$

$$\frac{r_{in}}{x} = \left[\frac{1}{58}\left\{\frac{6.57}{(f_{stoich}+f')}\frac{D_o}{x}\left(\frac{\rho_o}{\rho_\infty}\right)^{\frac{1}{2}} -1\right\}^{\frac{1}{2}}\right]^{\frac{1}{2}} \quad . \quad (12.2\text{-}6)$$

- Enclosed within the inner and outer envelopes of the flame is the stoichiometric surface, of which the radius is of course given by :

$$\frac{r_{stoich}}{x} = \left[\frac{1}{58}\left\{\frac{6.57}{f_{stoich}}\frac{D_o}{x}\left(\frac{\rho_o}{\rho_\infty}\right)^{\frac{1}{2}} -1\right\}^{\frac{1}{2}}\right]^{\frac{1}{2}} \quad . \quad (12.2\text{-}7)$$

- The lengths of the inner, outer and stoichiometric envelopes are given by setting the radii equal to zero. For example:

$$\frac{x_{stoich}}{D_o} = \frac{6.57}{f_{stoich}}\left(\frac{\rho_o}{\rho_\infty}\right)^{\frac{1}{2}} \quad . \quad (12.2\text{-}8)$$

12.3 COMPARISON WITH EXPERIMENT

A) FLAME LENGTH

- For the four gaseous fuels of section 12.1A, the length of the stoichiometric contour is easily calculated from (12.2-8); the results, and their comparisons with the experimental flame lengths, appear in the following table.

Fuel	f_{stoich}	ρ_o/ρ_∞	x_{stoich}/D_o	x_{fl}/D_o	x_{stoich}/x_{fl}
CO	0.288	0.965	21.7 ,	45	0.483
H_2	0.0281	0.069	61.6	130	0.474
C_2H_2	0.070	0.895	89.0	175	0.508
C_3H_8	0.0598	1.515	135.8	295	0.461

Here f_{stoich} has been calculated from equation (6.2-19), with the assumption that the mass fraction of oxygen in the atmosphere is 0.232. The density ratios are equal to the molecular-weight ratios.

- Inspection of the last column shows that x_{stoich} is predicted to be about half the experimental flame length x_{fl}. This is mainly explicable on the grounds that the average density in the flame is much lower than the atmospheric density; thus, to put 0.25 ρ_∞ in place of ρ_∞ would bring x_{stoich} close to x_{fl}. However, the fluctuations must also play a part; for the length of the outer envelope x_{out} exceeds that of the stoichiometric one, according to :

$$\frac{x_{out}}{x_{stoich}} = \frac{f_{stoich}}{f_{stoich} - f'} \quad . \quad (12.2\text{-}9)$$

B) OTHER FEATURES

Qualitative agreement is found between the predictions and the measurements in respect of profiles of velocity, time-average concentration, time-average and fluctuating temperatures, etc.; however, the neglect of the density variations leads to quantitative differences. Partly because of the density reduction, the flame tends to have a larger diameter than the foregoing equations would indicate.

12.4 DISCUSSION

A) MORE REFINED MODELS OF THE AXI-SYMMETRICAL TURBULENT DIFFUSION FLAME

Provided numerical methods of analysis are employed, e.g. the GENMIX computer program (Spalding, 1978), the following additional features can be taken into account:-

- density variations;

- buoyancy effects;
- generation, dissipation, convection and diffusion of turbulence energy, length scale and concentration fluctuations;
- radiative heat loss.

This does not mean that the additional refinements procure complete agreement with experimental information; but this is chiefly because no comprehensive attempt to refine both experiments and mathematical models has yet been made.

B) COMBINATION OF DROPLET-BURNING AND TURBULENT-JET MODELS

- The time taken for an injected fuel particle to travel along the axis to the stoichiometric contour can be calculated as follows. Let the distance to the start of the hyperbolic axial-velocity decay be x_1, defined by:

$$x_1 = 6.57\,(\rho_o/\rho_\infty)^{\frac{1}{2}}\,D_o \qquad . \quad (12.4\text{-}1)$$

Then the time to reach the tip of the stoichiometric contour, t_{stoich}, is given by:

$$t_{stoich} = \int_o^{x_{stoich}} \bar{u}_{ax}^{-1}\,dx$$

$$= \frac{x_1}{u_o} + \frac{1}{x_1 u_o}\int_{x_1}^{x_{stoich}} x\,dx \qquad ;$$

i.e.

$$\boxed{t_{stoich} = \frac{1}{2}\frac{x_1}{u_o}\left[1 + \left(\frac{x_{stoich}}{x_1}\right)\right]^2} \qquad . \quad (12.4\text{-}2)$$

If the liquid fuel is injected in the form of droplets, as in a diesel engine, perhaps mixed with steam as in a furnace, the time for vaporisation can be calculated as described in earlier chapters and compared with t_{stoich}

from equation (12.4-2)

If the droplet-vaporisation time exceeds the jet-mixing time, the droplet model is the one to use in design; this is usually the situation in rocket motors. At the other extreme, the jet-mixing model is the one to use; this is true of flames in large furnaces. The intermediate situation, which probably characterises diesel engines, has not been studied in any detail.

12.5 REFERENCES

1. BARON T (1954)
"Reactions in turbulent free jets; the turbulent diffusion flame".
Chem.Eng.Progr., Vol. 50, p 73.

2. BECKER H A, HOTTEL H C & WILLIAMS G C (1967)
"The nozzle-fluid concentration field of the round turbulent free jet".
J. Fluid Mech., Vol. 30, p 285.

3. HAWTHORNE W R, WEDDELL D S & HOTTEL H C (1949)
"Mixing and diffusion in turbulent gas jets".
Third Symposium on Combustion.
Williams and Wilkins, Baltimore, Maryland, p 266.

4. KREMER H (1964)
Zur Ausbreitung inhomogener turbulenter Freistrahlen und turbulenter Diffusionsflammen, Dissertation Techn.
Hochschule Karlsruhe. Also VDI Berichte No. 95, p 55 (1966); Int. Z. Gaswaerme, Vol. 15, p 39 (1966).

5. KREMER H (1967)
"Mixing in a plane free-turbulent-jet diffusion flame".
Eleventh Symposium (International on Combustion).
The Combustion Institute, pages 799 to 806.

6. SPALDING D B (1970)
"Mathematische Modelle turbulenter Flammen".
Vorträge der VDI-Tagung Karlsurhe Verbrennung und Feuerungen". VDI-BERICHTE nr 146 Düsseldorf: VDI; Verlag, pp 25-30.

7. SPALDING D B (1971a)

"Mixing and chemical reaction in steady confined turbulent flames".

Thirteenth Symposium on Combustion, The Combustion Institute, Pittsburgh, p 649.

8. SPALDING D B (1971b)

"Concentration fluctuations in a round turbulent free jet".

Chemical Engineering Science, Vol. 26, pp 95-107.

9. SPALDING D B (1978)

"GENMIX - A General Computer Program for Two-Dimensional Parabolic Phenomena"

HTS Series, Number 1, Pergamon Press, Oxford.

10. WOHL K, GAZLEY C & KAPP N (1949)

"Diffusion flames".

Third Symposium on Combustion Flame and Explosion Phenomena, Williams and Wilkins, Baltimore, pp 288-299.

EXERCISES TO FACILITATE ABSORPTION OF THE MATERIAL OF CHAPTER 12

ANALYTICAL PROBLEMS

12.1 A furnace has been heated by town's gas, injected into nearly stagnant air through a nozzle of 1 cm diameter at a velocity of 10 m/s.

It is desired to replace the town's gas by methane. If the length of the flame, and the heat input to the furnace, are to be the same, what must be the nozzle diameter and the injection velocity for the methane?

Data: Fuel	Town's gas	Methane	Units
Calorific value	540	1067	Btu/ft^3
Air requirement	4.71	9.52	vol.air / vol.gas
Density	.43	.552	density gas / density air

Answer: $D_0 = 578$ cm, $u_0 = 15.1$ m/s

12.2 Calculate the values of f_{stoich} for pure CO, H_2, C_2H_2 and C_3H_8, burning in an atmosphere of air ($m_{ox} = 0.232$).

Answers: See text

12.3 What will be the ratio of the turbulent-diffusion-flame lengths of H_2 and CH_4, burning in atmospheric air?

Answer: $H_2:CH_4 = .68$

12.4 Prove that the maximum radius of the stoichiometric envelope of a turbulent diffusion flame, according to the theory given in the text, is: $0.227\, D_o(\rho_o/\rho_\infty)^{\frac{1}{2}}(1+s/m_{ox,\infty})$.

12.5 All the following statements are true, except:

A The length of flame produced by a jet of steam and oil injected into a stream of air is almost independent of the fineness of the atomisation.

B The flame mentioned in A is a region of turbulent burning gas into which air is entrained from the surrounding stream.

C A jet of warm air, injected into cooler air at rest, spreads out to form a turbulent jet of conical shape.

D The angle of the jet mentioned in C is independent of the Reynolds number, if this is high enough.

E The flame length in a rocket motor is always controlled by droplet vaporisation, no matter how small are the droplets.

12.6 When 1 kg of fuel is mixed, in a steady-flow combustion process, with 20 kg of air, and burns completely, the value of f for the combustion products is:

A 1/20.

B 0.

C 20/21.

D 1/(0.232 × 20).

E 1/21.

For equilibrium mixtures of a pure fuel with air, the curve on the sketch marked: 12.7, 12.8, 12.9, 12.10, could represent:

A specific enthalpy.

B mass fraction of combustion products.

C mass fraction of oxygen.

D mass fraction of unburned fuel.

E mass fraction of nitrogen.

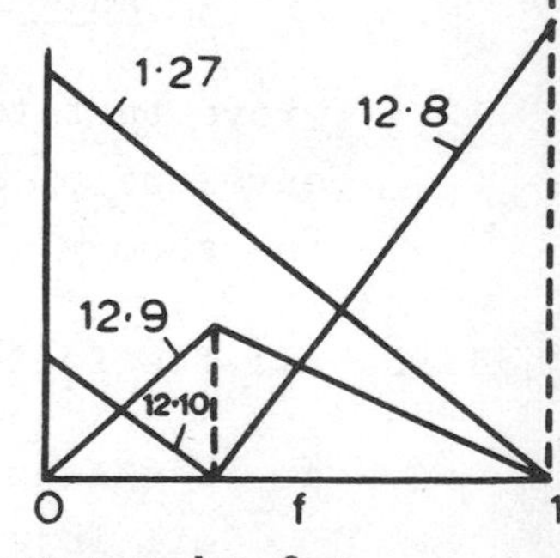

A hydrocarbon gas requires 3.5 kg of oxygen per kg for

complete combustion. It is mixed with air (containing .232 kg O_2 per kg), and comes to equilibrium with it. The value(s) of f which correspond (s) to:

12.11 pure fuel,

12.12 pure air,

12.13 $m_{fu} = 0$,

12.14 $m_{ox} = 0$,

12.15 $m_{fu} = m_{O_2}$,

is(are) :

A 0.

B 1.

C 0.0622.

D from 0 to 0.0622.

E 0.0622 to 1.

The value of f_{stoich} for:

12.16 hydrogen burning with pure oxygen,

12.17 methane burning with air,

12.18 a liquid hydrocarbon of formula $(CH_2)_n$, burning with air,

12.19 solid carbon burning with pure oxygen,

is:

A 0.273.

B 0.08.

C 0.111.

D 0.0548.

E 0.0634.

Samples of gas are extracted from a turbulent diffusion flame, formed by hydrogen burning in air. Although the gases may not be in equilibrium when they enter the sampling probe, they are brought to equilibrium by passage through a catalyst vessel before entering a gas analysis device. When the meter indicates:

12.20 $m_{H_2} = 0$,

12.21 $m_{O_2} = 0$,

12.22 $m_{O_2} = m_{H_2} = 0$,

12.23 $m_{H_2} = 0.1$,

12.24 $m_{O_2} = 0.1$,

it can be concluded that the value of the mixing fraction f in the furnace at the point of sampling is :

A 0.0281.

B 0.0281.

C 0.016.

D 0.0281.

E 0.125.

In the sketch, which represents conditions along a radius in a turbulent diffusion flame, the curve marked: 12.25, 12.26, 12.27, 12.28, could represent:

A the concentration of unburned fuel.

B the concentration of unburned oxygen.

C the concentration of combustion products.

D the mixture fraction f.

E the temperature.

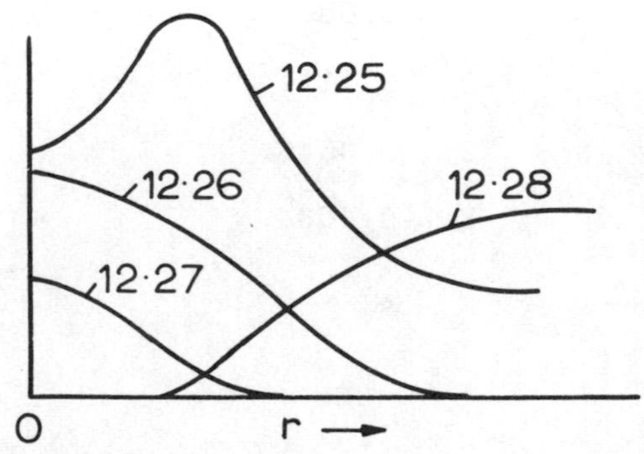

The length of a turbulent diffusion flame, of which the conditions are changed by:

12.29 increasing the nozzle diameter,

12.30 increasing the fuel flow rate,

12.31 diluting the fuel with steam,

12.32 diluting both air and fuel with steam,

12.33 greatly diminishing the fuel flow rate,

is:

A probably decreased.

B probably left unchanged.

C probably increased.

D probably altered, but in a direction that is impossible to determine from the data.

E rendered more easy to determine visually.

12.34 Coal gas, used for domestic supplies, has a density of 0.48 kg/m^3 at atmospheric temperature; it requires 12.25 kg of air per kg for complete combustion. The length of the turbulent diffusion flame which it produces, when burning in air from a nozzle of fixed size, is likely to:

A be less than that for CO.

B lie between that for CO and that for H_2.

C lie between that for H_2 and that for C_2H_2.

D lie between that for C_2H_2 and that for C_3H_8.

E exceed that of C_3H_8.

12.35 An industrial gas burner designed for use with coal gas produces a turbulent diffusion flame. The fuel is changed to methane, but the flame length is to remain unchanged. The nozzle diameter must therefore be increased by a factor of:

A 1.617.

B 0.726.

C 1.375.

D 0.618.

E 0.852.

12.36 Knowing that high temperature accelerates chemical reactions, a chemist (C) proposed to shorten the length of a turbulent diffusion flame by pre-heating the gaseous fuel. An engineer (E), who knew that the flame length is physically rather than chemically controlled, argued that the proposal was unsound, and that the flame length would remain unaltered.

Experiment showed:

A that E's prediction was right, but for the wrong reason.

B that E's prediction was right, and his argument was sound.

C that C's prediction was right, but for the wrong reason.

D that C's prediction was right, and his argument was sound.

E that both E and C made incorrect predictions.

ANSWERS TO MULTIPLE-CHOICE PROBLEMS

Answer	Problem number (12's omitted)
A	12, 19, 22, 27, 31, 33
B	9, 11, 20, 28, 30, 34
C	10, 15, 16, 24, 29, 36
D	8, 13, 17, 21, 26, 32, 35
E	5, 6, 7, 14, 18, 23, 25

CHAPTER 13

SURVEY OF KINETICALLY-INFLUENCED PHENOMENA

13.1 INTRODUCTION

A) "PHYSICALLY-CONTROLLED" COMBUSTION PHENOMENA

The processes considered in earlier lectures proceed at rates which can be calculated without precise knowledge of chemical-kinetic constants; it has been enough to presume that the reaction has proceeded to completion.

Combustion processes of this kind are said to be "physically-controlled". Examples are:- droplet burning, combustion in rockets, laminar and turbulent diffusion flames. They are very common in engineering.

B) "KINETICALLY-INFLUENCED" COMBUSTION PHENOMENA

Many other practical combustion processes are influenced by chemical-kinetic constants. They include combustion in:

- gasoline (spark-ignition) engines;
- diesel (compression-ignition) engines;
- domestic gas burners (flame stability);
- aircraft gas turbines (at high altitudes);
- ram-jet engines;
- house and forest fires.

Kinetically-influenced phenomena are those in which perhaps the very existence of the flame, and certainly the rate of combustion within it, are sensitive to the "ease" with which the redistribution of atoms between molecular species can be effected. This "ease" is characterised as follows:-

- it increases with rise of temperature;
- often the rate of reaction can be represented as:

$$\text{rate} = \text{constant} \times (\text{function of concentrations}) \times \exp(-\text{constant}/T), \qquad (13.1\text{-}1)$$

where $T \equiv$ absolute temperature;

- the constants can vary greatly with small changes in molecular or crystalline structure.

The dependence on temperature is such that a rise from 20^0C to 30^0C will typically double the reaction rate. The dependence on molecular structure results in two hydrocarbons, with the same number of carbon and hydrogen atoms in their molecules, having quite different reactivities.

c) TYPICAL KINETICALLY-INFLUENCED PHENOMENA

(i) Ignition

Flames are started by ignition, which usually involves raising the temperature of the reactants by external means, for example:-

- the passage of an electrical spark (gasoline engine, hazardous electrical equipment in a coal mine);
- the ignition of dry grass by wind-borne fragments of burning material;
- compression of combustible gases, as in the diesel engine.

(ii) Stabilisation

Once formed, a flame may adopt a steady condition; inflowing fresh gas is continuously ignited by contact with hot partially-burned gas. This is how the flame is held at the lip of a Bunsen burner.

(iii) Propagation

There are two types of propagation: steady, and unsteady.

The first is exemplified by the inner cone of the Bunsen burner, spreading in from the burner lip; the second is exemplified by the flame which propagates from the spark in a gasoline engine.

(iv) Extinction

Flames may be extinguished by either excessive supply of reactants, or by excessive heat loss. The first is exemplified by "blowing out" a candle flame; the second by the "dying out" of a coke fire when the air supply is restricted.

These four types of kinetically-influenced phenomenon will be examined qualitatively in the remainder of the lecture.

13.2 IGNITION PHENOMENA

A) SPARK IGNITION

(i) Practical relevance

Spark ignition is desired in: gasoline engines, gas turbines, domestic oil burners, etc.

It is desired in: coal mines, oil refineries, etc.

(ii) Nature

A high voltage, imposed between two electrodes separated by a small gap, causes the intervening gas to conduct electricity; the gas is thereby heated to several thousand degrees Celsius.

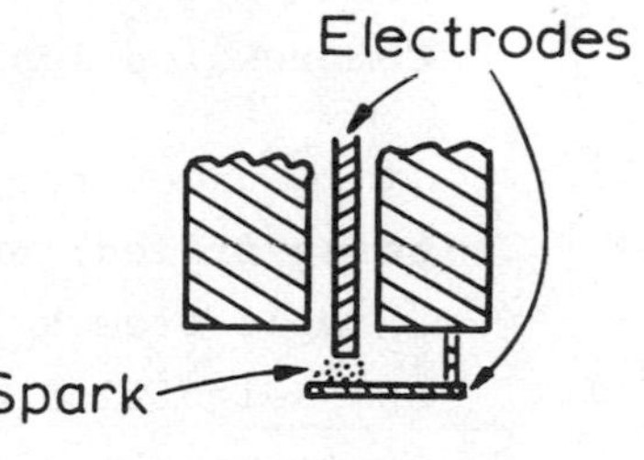

If the gas contains a fuel and oxidant, this heating of the gas may cause a flame to form around the spark gap and spread throughout the mixture; but it may not.

(iii) Interrelationships

Factors favouring successful propagation are:-

High spark energy, high pressure and initial temperature, a mixture ratio in the neighbourhood of stoichiometric.

Factors tending to reduce the probability of flame propagation are:- a large dilution of the reactant gases by inerts, a small distance between the electrodes, a large relative velocity between the gas and the electrodes.

(iv) Questions to be answered in later lectures

Can a mathematical model be devised which satisfactorily explains the above findings?

If the performance of a spark plug under one set of conditions is known, can one deduce what its performance will be under a different set of conditions?

How should one ensure that ignition will take place reliably over the whole operating range of an engine?

B) COMPRESSION IGNITION

(i) Practical relevance

Compression ignition is desired in diesel engines.

It is undesired in, say, oxygen cylinders which, while being filled, are contaminated by oil vapours, for example from a lubricant of a valve. "Knock" in a spark-ignition engine is another undesired compression-ignition phenomenon.

(ii) Types of apparatus for the study of compression ignition

The following kinds may be used:-

- The sudden-compression machine, in which a considerable

mixture is suddenly compressed by a piston-cylinder combination, and the subsequent combustion-induced pressure rise is measured.

- The slow-compression machine, in which there is no attempt to separate the compression and combustion phases, but the course of combustion is deduced from the difference of the actual pressure-volume curve from that for pure compression.

- The shock tube, in which a diaphragm at first separates high-pressure, incombustible gas from low-pressure combustible gas. The diaphragm is ruptured: then a pressure wave runs into the low-pressure gas, causing it to ignite.

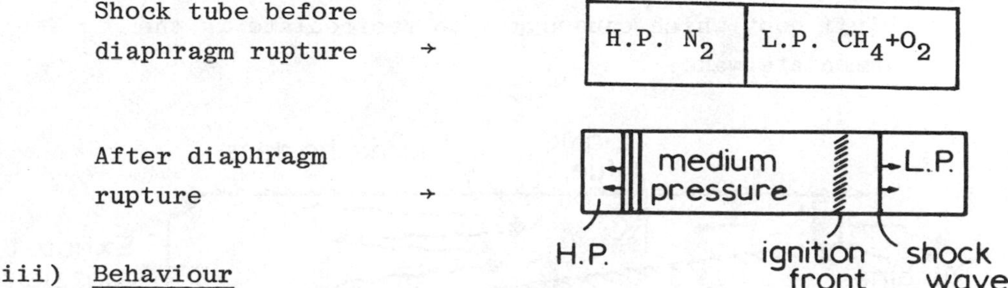

(iii) Behaviour

Most compression-ignition phenomena can be expressed in terms of an "ignition delay", i.e. the time elapsing between arrival at a prescribed state of pressure and temperature and the onset of a significant amount of reaction.

The ignition delay normally increases with: reduction in initial temperature (steeply) and pressure (less steeply), and with the ability of gas to lose heat by conduction to walls.

It varies greatly from one fuel to another at fixed

pressure, temperature and fuel-air ratio; but the fuel-air ratio itself is not very influential.

(iv) Some questions

How can "knock" in spark-ignition engines be eliminated, by preventing compression ignition?

How can "knock" in diesel engines be eliminated, by making compression ignition extremely easy and rapid?

13.3 FLAME STABILISATION BY A BLUFF BODY

A) PRACTICAL RELEVANCE

In steady-flow air-breathing engines, i.e. gas turbines, and ram-jets, the average gas velocity is several tens or even hundreds of metres per second. The usual way of stabilising a flame in such a stream, i.e. of ensuring that it is not blown away, is to provide in the stream a bluff body which causes gas to recirculate in the immediate wake.

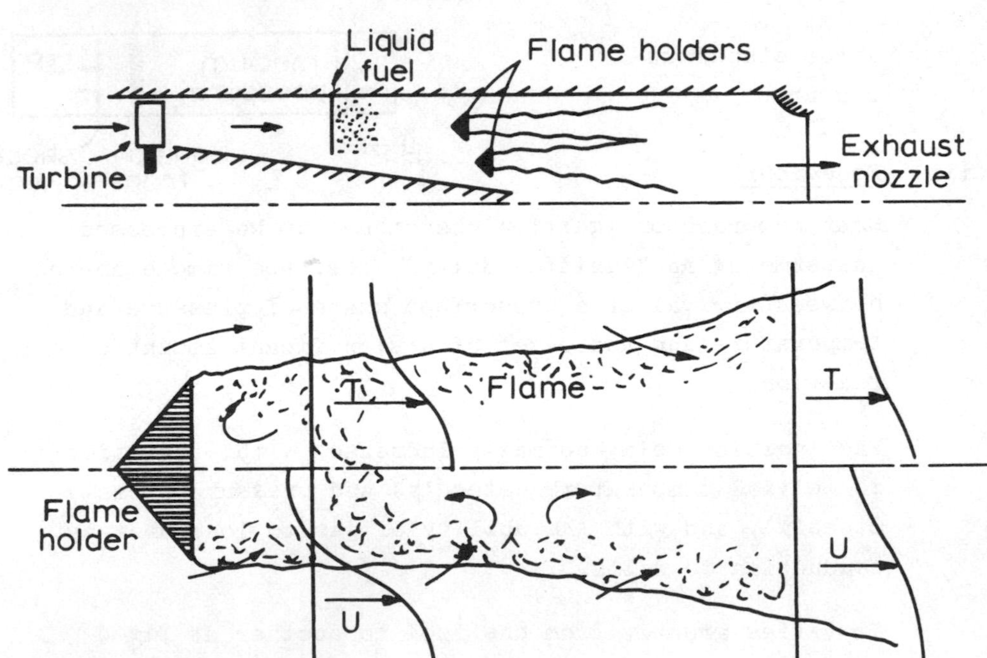

B) BEHAVIOUR

Once lighted, the flame is held by the "bluff-body" flame-stabiliser because recirculation of hot gases ignites the fresh mixture.

It is possible for stabilisation to occur without propagation; the flame is then confined to a small region just downstream of the flame-holder.

(iii) Interrelationships

Flame stabilisation is made more likely by:

- high pressure and temperature of the stream;
- large dimensions of the flame-holder;
- low gas velocity;
- fuel-air ratio near stoichiometric.

(iv) Some questions

How can it be ensured that the pressure drop caused by the bluff body is a minimum, while still retaining reliable flame-holding properties?

Is there any means which will predict whether a newly-designed flame-holder system will work, without actually building and testing it? Can purely theoretical means be used? Can one make and test a small-scale model, and work fom the results of that?

13.4 PROPAGATION

A) PRACTICAL RELEVANCE

Flame propagates from a bluff-body stabiliser, as just described.

It also propagates across the combustion space of a gasoline engine after the spark has successfully ignited a part of the gas.

When the gas taps of a gas cooker are turned on, ignition is effected by propagation along an igniter tube communicating with a pilot flame.

B) SIGNIFICANT PROPERTIES OF PROPAGATING FLAMES

From the practical point of view, knowledge of the speed of propagation (or angle in the gas-turbine situation) is required. This speed is affected, as a rule, by:

- the fuel-air ratio;
- the turbulence level;
- the initial temperature level;
- the pressure.

Flames propagating in ducts can proceed either at about the same speed as if they were unconfined, or at speeds in excess of that of sound. The first type are called deflagrations, the second detonations. The latter exhibit very large pressure increases, and may be highly destructive; incidentally, the detonation speed is not influenced by the chemical-kinetic constants.

C) SOME QUESTIONS

Can the speed of propagation of flame in a gasoline engine be modified by adding small quantities of "dope" to the fuel?

Can measurements of flame speed (which are fairly easy to make) be used to provide quantitative information about the chemical-kinetics constants of the constituents?

Can the occurrence of detonation be predicted? How can its occurrence be avoided?

13.5 EXTINCTION

A) PRACTICAL RELEVANCE

Sometimes extinction is desired, as in all fire-fighting problems. Sometimes it is undesired, as in aircraft gas turbines at high altitudes. Even under normal operating conditions, it would be desirable to operate gas turbines over a wider range of fuel-air ratios; however, the flame becomes extinguished as the fuel flow rate is reduced below a critical value. Prevention of extinction dominates the design of gas-turbine combustors.

An example of desired extinction is the suppression of combustion in the exhaust plume of a rocket motor; this is required so as to make the rocket less visible, and so as not to interfere with guidance systems.

B) BEHAVIOUR

Extinction can be promoted by chemical means, e.g. the addition of small quantities of suppressants. It is also favoured by diminution of the supply of reactants, e.g. replacing O_2 by CO_2, and by cooling, e.g. spraying water into the flame.

"Quenching" (extinction by cooling) can be effected by the proximity of cold solid walls. Thus "flame traps" can be provided by gauzes or metal "honeycombs" having small passages through which gases cannot pass without being severely chilled.

C) SOME QUESTIONS

What are the quantitative relationships between reaction-kinetic properties, transport properties, and solid-wall properties which determine whether quenching will occur?

What governs the fuel-air ratio at which a gas-turbine

combustor will "blow out"? How will this depend upon pressure, approach velocity, linear size of combustor, etc.

Note that, since extinction is in some senses the opposite of stabilisation, the two topics will be dealt with simultaneously in the later lectures.

EXERCISES TO FACILITATE ABSORPTION OF MATERIAL OF CHAPTER 13

MULTIPLE-CHOICE PROBLEMS

13.1 Kinetically-influenced phenomena are those in which:

A No chemical reaction takes place, and the phenomenon depends on physical properties only.

B The rate of chemical reaction is sensitive to the ease with which atoms re-arrange themselves between molecules.

C The rate of reaction is independent of the physical properties of the materials present.

D The rate of chemical reaction is practically independent of the chemical-kinetic constants.

E The reactants are fully mixed before chemical reaction is initiated, so that physical processes exert no decelerating influence on the progress of the reaction.

13.2 Physically-controlled phenomena are those in which:

A

B

C as for 13.1 above.

D

E

13.3 All the following phenomena can be classified as physically-controlled phenomena except:

A the burning of a liquid-fuel droplet injected into a rocket combustion chamber at its equilibrium temperature.

B the burning of a liquid-fuel droplet injected into a rocket combustion chamber at an initial temperature below the equilibrium temperature.

C a turbulent diffusion flame in an industrial furnace.

D the outer flame of an aerated Bunsen burner.

E the inner flame of an aerated Bunsen burner.

13.4 All the following phenomena can be classified as kinetically-influenced except:

A compression ignition of fuel and air in a diesel-engine cylinder.

B the "knock" process in a gasoline-engine cylinder.

C the spread of a turbulent flame through the cylinder of a gasoline engine after ignition and before the onset of "knock".

D the burning of the diesel fuel which is injected into the cylinder after spontaneous ignition has occurred.

E the "pre-ignition" process which sometimes occurs in a gasoline engine as a result of the presence of glowing deposits on the walls.

The phenomenon of:

13.5 spark ignition,

13.6 propagation of flame through a uniform laminar combustible mixture,

13.7 stabilisation of flame by means of a bluff body,

13.8 spontaneous ignition as a consequence of compression,

13.9 extinction as a consequence of excessive heat loss, often takes place in:

A the after-burner system of a jet engine.

B the inner cone of a Bunsen burner.

C the gasoline engine.

D a coke-burning stove, to which the air supply is restricted for many hours.

E a turbulent gaseous diffusion flame.

13.10 All the following tend to diminish the probability that a spark will be successful in igniting a flame except:

A the fuel-air ratio is very much richer than stoichiometric.

B the distance between the electrodes is large.

C there is a significant "wind" of gas past the spark gap.

D the pressure of the gas is low.

E the mixture has been diluted with combustion products from a previous cycle of operation.

13.11 Compression ignition can be said to occur in all of the following except:

A the high-pressure-ratio aircraft gas turbine.

B the gasoline engine under "knocking" conditions.

C the diesel engine.

D the shock tube, used for chemical-kinetic investigations.

E a detonation wave, proceeding through a pre-mixed gas in a long pipe.

13.12 For a shock tube employed for chemical-kinetic investigations, just after the rupture of the diaphragm, all the following statements are true except:

A a rarefaction wave moves into the driver gas.

B a shock wave moves into the reactive gas.

C part of the driver gas and part of the reactive gas are at substantially the same pressure.

D no significant amount of reaction takes place until the shock wave has passed through the gas in question.

E in the vicinity of the original location of the diaphragm, driver gas is flowing in the same direction as the rarefaction wave.

13.13 All the following combustion phenomena take placc with significantly more ease or speed, if the fuel-air ratio is near stoichiometric, except:

A ignition as a result of compression by a piston moving in a cylinder containing the gas.

B ignition by a spark.

C propagation of a laminar flame in a tube.

D stabilisation of a flame by a bluff body.

E stabilisation by a pilot flame.

13.14 All the following combustion phenomena tend to take place more easily or swiftly as a result of an increase in the initial (or upstream) temperature of the gases, except:

A compression ignition in a shock tube.

B ignition by contact with a heated surface.

C propagation of a plane laminar flame.

D extinction through contact with cooled walls.

E stabilisation by means of a bluff body.

13.15 The ignition delay of a suddenly-created mixture of fuel and air is likely to be diminished by all of the following except:

A a high initial temperature.

B a high initial concentration of inert diluent.

C a high initial pressure.

D a low value of the constant in exp (-constant/T).

E a large linear dimension of the containing vessel.

13.16 All of the following properties of flames tend to be augmented by an increase in the general pressure level of the gas, except:

A the ease of stabilisation by a pilot flame.

B the speed of ignition after compression by a piston movement.

C the certainty of ignition by a spark.

D the product of density and speed of propagation of a laminar flame.

E the thickness of a laminar flame.

ANSWERS TO MULTIPLE-CHOICE QUESTIONS

Answer	Problem number (13's omitted)
A	7, 11, 13
B	1, 6, 10, 15
C	5, 8
D	2, 4, 9, 14
E	3, 12, 16

CHAPTER 14

INTRODUCTION TO CHEMICAL KINETICS

14.1 TYPES OF REACTION

A) "OVERALL" REACTIONS

The net effect of a reaction process can be written as:

fuel	+	oxygen	→	products,
e.g. C_2H_6	+	$3\frac{1}{2}\ O_2$	→	$2CO_2 + 3\ H_2O$.
(ethane)		(oxygen)		(carbon dioxide) (steam)

Such a statement describes only the beginning and end of the process, not its path. No encounter between 1 ethane molecule and $3\frac{1}{2}$ oxygen molecules ever occurs. Then how does the reaction proceed?

B) INTERMEDIATES

Each atom from a reactant molecule enters into one or more intermediate compounds, as a rule, before taking its final place in a product molecule. The intermediate compounds are of two types, viz:-

(i) Stable: E.g., in the above, process, there may be found in the flame region: CH_4, C_2H_2, CO, H_2, etc.

(ii) Unstable (radicals): E.g. there may also be found: H, OH, O, C, C_2, CH, etc.

(Un)stable means (in)capable of existing in equilibrium at normal temperatures in more than very low concentrations (e.g. mass fraction $< 10^{-6}$).

The great majority of molecular encounters which result in reaction (called "reaction steps"), involve at least one radical, e.g.:

$O_2 + H \rightarrow OH + O$,

$OH + H_2 \rightarrow H_2O + H$,

$O + H_2 \rightarrow 2OH$,

etc.

So a complete reaction is usually effected by way of a sequence of reaction steps. The term "chain reaction" is used to describe the whole process.

The large molecules, such as ethane, may react with radicals directly; or they may break into smaller molecules as a result of heating.

Sequences of reactions believed to be important in methane oxidation and in the formation of nitrogen oxides have been given in section 6.2 above.

c) THREE TYPES OF REACTION STEP

(i) Collision-controlled bi-molecular

In this type of reaction, which is prominent in the present lectures:

- either molecules (or radicals) A and B collide and, if the energy and orientation are right, react immediately:

$$A + B \rightarrow C + D \quad ;$$

- or molecule (or radical) A collides with inert molecule M and, if the energy and orientation are right, decomposes immediately;

$$A + M \rightarrow C + D + M \quad .$$

"Immediately" means "in a time which is much shorter than the average time between collisions".

(ii) Decomposition-controlled

These reactions are similar to those of (i); but the reaction or decomposition times are long, because the energy of collision has to re-distribute itself among the "degrees of freedom" of the molecule.

Such reactions can be written:

- $A + B \rightarrow (AB)^* \rightarrow C + D$,
- or $A + M \rightarrow A^* + M \rightarrow C + D + M$,

where $(AB)^*$ and A^* are relatively long-lived "activated complexes", i.e. molecules with unusually high internal energies.

(iii) Tri-molecular reactions

In tri-molecular reactions, the "activated complex" $(AB)^*$ is formed as above, but decomposes again ("shakes itself to pieces") unless an inert "third body" M is close at hand to absorb the excess energy, e.g.:

- $H + H + M \rightarrow H_2 + M$,
- or $H + OH + M \rightarrow H_2O + M$.

Since three molecules meet less frequently than two, such reactions proceed only at low rates; but they are still of practical importance, because there is no other way for radicals to combine together to make stable molecules.

D) PURPOSE OF THE PRESENT LECTURE

(i) Reaction-rate formula

A quantitative relation will be derived, from considerations involving the kinetic theory of gases, for the rate of the collision-controlled bi-molecular reaction:

$$A + B \rightarrow C + D \quad ,$$

connecting this rate with:

- the concentrations of reactants;
- their pressure and temperature;
- molecular properties.

(ii) Interactions with material and heat balances

It will then be shown that, when other factors are taken into account, the reaction rate can often be related to temperature in the manner indicated by the following sketch. The form of this curve contains the explanation of a great many combustion phenomena.

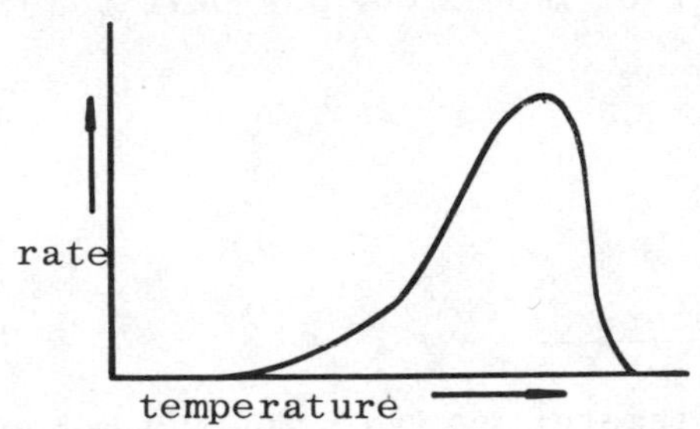

14.2 ELEMENTARY THEORY OF COLLISION-CONTROLLED, BI-MOLECULAR CHEMICAL REACTIONS IN THE GAS PHASE.

A) SOME RESULTS FROM THE KINETIC THEORY OF GASES

(N.B. These will be simply stated here. Sub-section (B) will be devoted to making them seem plausible; but no derivation will be provided. Derivations may be found in Guggenheim, 1960.)

(i) Numerical data

- Avogadro's number N ≡ number of molecules in a kg mole, $= 6.024 \times 10^{26}\ kg^{-1}$.

Therefore the mass of the H atom (atomic wt. = 1.008)

equals 1.008 ÷ (6.024×10^{26}) = 1.673×10^{-27} kg.

Universal Gas Constant $\mathcal{R} \equiv pv\,M/T$, where $p \equiv$ pressure, $v \equiv$ specific volume, $M \equiv$ molecular weight.

$\mathcal{R} = 8.3144 \times 10^3$ joule/kg mole ^{0}K,

= 1.98 cal/g mole ^{0}K .

Boltzmann's Constant, k.

$k \equiv \mathcal{R}/N = 1.3803 \times 10^{-23}$ joule/^{0}K per molecule.

A consequence is that one can write the equation of state of an ideal gas as:

$$p = n\,k\,T \qquad , \; (14.2\text{-}1)$$

where $n \equiv$ number of molecules per unit volume. Note that:

$$n = \frac{N\rho}{M} \qquad . \; (14.2\text{-}2)$$

Molecular diameter, d.

The molecular diameter varies with the molecule, but not much. For most molecules it is about 3.5×10^{-10} m. Molecules do not strictly possess sharp boundaries. However, in some respects they behave like perfectly elastic spheres with definite diameters; d is thus the "equivalent hard-sphere" diameter. For a given molecule, d reduces somewhat with increase in temperature, because the increase in relative velocity at impact causes the molecule to "give".

(ii) Some formulae

The mean kinetic energy per unit mass of gas molecules in thermodynamic equilibrium $\overline{u_{mol}^2}/2$ is given by:

$$\frac{\overline{u^2_{mol}}}{2} = \frac{3}{2}\frac{\mathcal{R}}{M}\,T \qquad ; \ (14.2\text{-}3)$$

the $\mathcal{R}$MS molecular velocity is therefore given by:

$$(\overline{u^2_{mol}})^{\frac{1}{2}} = \sqrt{3\mathcal{R}T/M} \qquad . \ (14.2\text{-}4)$$

Note that this is somewhat greater than the speed of sound ($\sqrt{\gamma\mathcal{R}T/M}$ where γ is of the order of 1.4).

- The number of collisions, in unit volume and unit time, between molecules of types A and B, Z, is given by:

$$Z = \frac{n_A n_B}{s_{AB}}\, d^2_{AB} \left(\frac{\pi\mathcal{R}T}{M^*}\right)^{\frac{1}{2}} \qquad , \ (14.2\text{-}5)$$

where: d_{AB} ≡ mean diameter of molecules,

= $\frac{1}{2}(d_A + d_B)$,

M ≡ mean molecular weight of molecules,

= $M_A\, M_B/(M_A + M_B)$.

s_{AB} ≡ symmetry number,

= 2 when A and B are identical,

= 1 when they are not.

- The viscosity μ of a single-component gas is given by:

$$\mu = \frac{5}{16}\,\frac{(M\mathcal{R}T/\pi)^{\frac{1}{2}}}{N^2_d} \qquad . \ (14.2\text{-}6)$$

This permits d to be deduced from knowledge of the gas viscosity.

It is known that the viscosity of a gas increases with temperature more rapidly than $T^{\frac{1}{2}}$. The reason is, as mentioned above, that d diminishes with increase in T and the corresponding increases of collision energy.

- The energy of collision between molecules is not the same for all collisions, because of the variations in relative velocity and orientation at the moment of impact. The fraction of all collisions having an energy per mole exceeding E, f, is given by:

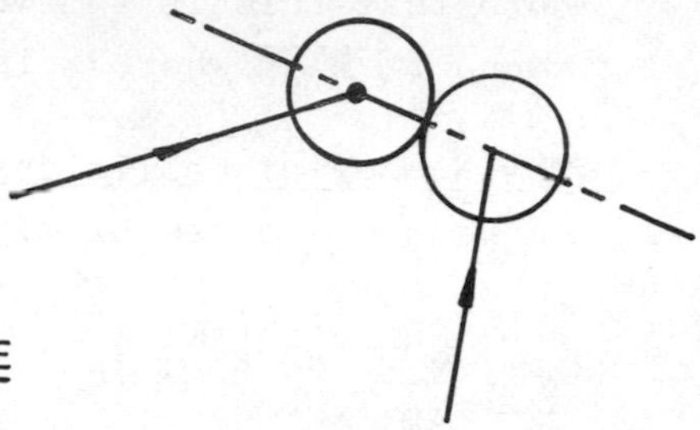

$$f = \exp\{-E/(\mathcal{R}T)\}. \quad (14.2\text{-}7)$$

B) MAKING THE FORMULAE PLAUSIBLE

(i) The collision rate

- The average ($\mathcal{R}$MS) molecular velocity is $\sqrt{3RT/M}$, from equation (14.2-3).

- Therefore one molecule "sweeps out" a volume equal to $\pi d^2\sqrt{3\mathcal{R}T/M}$ in unit time.

  NOTE: d^2, rather than $d^2/4$ because all molecules having centres less than d from the trajectory of the centre of the molecule under observation will be contacted.

- Hence the number of B molecules contacted per unit time by each A molecule is: $n_B\ \pi d^2\sqrt{3\mathcal{R}T/M}$.

- Hence the number of AB collisions per unit volume and time is: $n_A n_B \pi d^2\sqrt{3\mathcal{R}T/M}$.

- Apart from the numerical multiplier, this is the same as the formula for Z, equation (14.2-5). The difference results from the facts that the molecular velocities are not all equal and that, in any case, both sets of molecules are in motion.

(ii) <u>The fraction of sufficiently energetic collisions</u>

- When E is very small, or T is very large, $\exp\{-E/(\mathcal{R}T)\}$ tends to unity; so all the collisions have energy in excess of the critical value.

- When E is very large, or the temperature is small, $\exp\{-E/\mathcal{R}T\}$ is very small indeed; so practically none of the collisions have the critical energy.

- The fraction having energy in excess of the mean kinetic energy of motion is obtained by inserting, from (14.2-3):

$$\frac{E}{M} = \frac{3}{2}\frac{\mathcal{R}}{M}T \quad . \quad (14.2\text{-}8)$$

  Hence the fraction is seen to be: exp (-3/2). It may be concluded that the majority of collisions are less intense than would correspond to head-on impact of an average molecule with a solid wall.

- The variation of $\exp\{-E/(\mathcal{R}T)\}$ with T is represented in the following sketch.

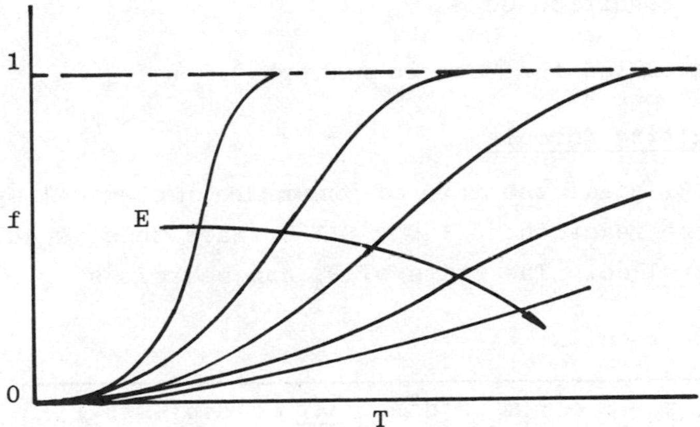

c) CONSEQUENTIAL REACTION-RATE FORMULA

(i) The activation-energy postulate

- Suppose that those collisions between A and B can lead to rearrangement of atoms between molecules which possess an energy of collision along the normal in excess of an "activation energy" E (joules/kg mole). The number of these collisions per unit time and volume is:

$$n_A n_B d^2 \left(\frac{\pi \mathcal{R} T}{M^*}\right)^{\frac{1}{2}} \exp\{-E/(\mathcal{R}T)\} .$$

- In this context, $\exp\{-E/(\mathcal{R}T)\}$ is known as the "Arrhenius term", after the name of the man who noted, from empirical evidence, that it describes the dependence of many reactions on temperature.

(ii) The orientation postulate

- Suppose further that there is yet another condition for successful reaction, namely that the molecules must hit each other at suitably sensitive spots. Let the proportion of sufficiently energetic collisions satisfying this condition be S.

  S is called the "steric factor".

(iii) Resulting formula

- Let R_A equal the rate of consumption of chemical species A in the reaction: $A + B \rightarrow C + D$, when this is collision controlled. The units of R_A can be kg/m^3s.

- There results:

$$\boxed{R_A = - S\, \rho^2 m_A m_B\, N^2 d^2 M_A \left(\frac{\pi \mathcal{R} T}{M^*}\right)^{\frac{1}{2}} \exp\{-E/(\mathcal{R}T)\}} \quad , \ (14.2\text{-}9)$$

where m_A and m_B are the mass fractions of A and B in the gaseous mixture, and ρ is the gas density.

- This is often called the "Arrhenius reaction-rate formula".

14.3 REACTION-RATE FORMULA FOR PRACTICAL USE

A) SIMPLIFICATIONS AND GENERALISATIONS OF THE ARRHENIUS FORMULA

(i) Simplifications

- The influences of pressure and temperature on the reaction rate can be brought to clearer expression by introducing the gas law:

$$\rho = pM/(\mathcal{R}T) \qquad . \quad (14.3\text{-}1)$$

There results:

$$R_A = K\, p^2\, m_A m_B\, T^{-3/2} \exp\{-E/(\mathcal{R}T)\} \qquad , \quad (14.3\text{-}2)$$

where:

$$K \equiv SN^2 d^2 M_A \left(\frac{\mathcal{R}}{M*}\right)^{\frac{1}{2}} \left(\frac{\overline{M}}{\mathcal{R}}\right)^2 \qquad , \quad (14.3\text{-}3)$$

and $\overline{M}$ is the average molecular weight of the mixture, which of course depends upon m_A, m_B and indeed on the concentrations and molecular weights of all the species in the mixture.

- Since S and d can be regarded as empirical constants, so can K.

- The influence of T in the $T^{-3/2}$ term can be approximately expressed by modifying the values of K and E appropriately. The resulting formula, which is convenient when K and E are in any case to be obtained empirically, is:

$$R_A = K'\, p^2 m_A m_B \exp\{-E'/(\mathcal{R}T)\} \qquad , \quad (14.3\text{-}4)$$

where K' and E' are the modified values.

(ii) Generalisation

- Since the collision-controlled reaction is only an idealisation of what actually occurs, the departures from ideality are sometimes represented by expressing the reaction rates in terms of "generalised Arrhenius expressions" such as:

$$R_A = C_1 \, p^{C_2} \, m_A^{C_3} \, m_B^{C_4} \exp(-C_5/T) \qquad . \ (14.3\text{-}5)$$

Then all the C's are regarded as constants, the values of which are to be deduced from experiment.

- Sometimes it is convenient to replace the exponential term by a power of the absolute temperature T. Then one writes:

$$\exp(-C_5/T) \approx (T/T_*)^{C_6} \exp(-C_5/T_*) \qquad , \ (14.3\text{-}6)$$

choosing T_* as a temperature in the range of interest, and C_6 so that both sides exhibit the same dependence on temperature. Thus, by differentiation after taking logarithms:

$$\frac{C_5}{T_*^2} \approx \frac{C_6}{T_*} \qquad ; \ (14.3\text{-}7)$$

Hence:

$$C_6 = C_5/T_*$$

- For most combustion reactions, C_5 is of the order of 2×10^4 ^{0}K. Hence, if T_* is taken as 2×10^3 ^{0}K, C_6 is around 10. Thus, the dependence of reaction rate on temperature is very steep.

B) THE REACTION-RATE BEHAVIOUR OF THE SIMPLE CHEMICALLY REACTING SYSTEM

(i) The situation considered

- Suppose that A and B are fuel and oxidant respectively and that they react in a steady-flow process of uniform enthalpy. Then the balance equations imply:

$$m_{fu} - m_{ox}/s = (m_{fu} - m_{ox}/s)_o \quad , (14.3\text{-}9)$$

and:

$$cT + H\, m_{fu} = (cT + H\, m_{fu})_o \quad , (14.3\text{-}10)$$

where subscript o represents the initial (entry) conditions.

- Let the all-burned condition be represented by subscript 1, and let "reactedness" τ be defined by:

$$\tau \equiv \frac{T - T_o}{T_1 - T_o} \quad . (14.3\text{-}11)$$

Then the above equations also dictate:

$$\tau = \frac{m_{fu} - m_{fu,o}}{m_{fu,1} - m_{fu,o}} \quad , (14.3\text{-}12)$$

$$= \frac{m_{ox} - m_{ox,o}}{m_{ox,1} - m_{ox,o}} \quad . (14.3\text{-}13)$$

These relations are represented by the sketches

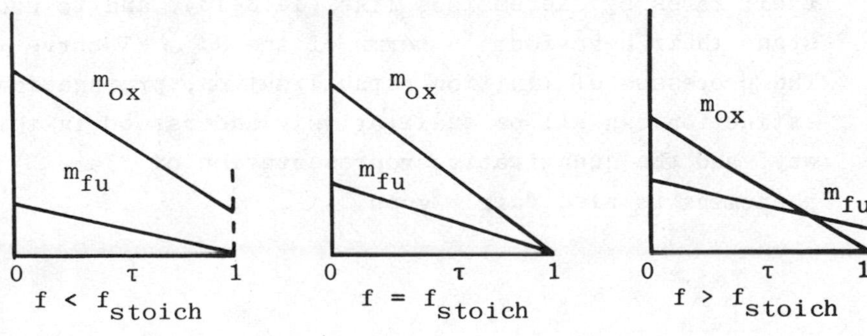

(ii) Dependence of reaction rate on reactedness

- The reaction-rate expression can be the generalised Arrhenius form of equation (14.3-5):

$$R_{fu} = C_1 \, p^{C_2} \, m_{fu}^{C_3} \, m_{ox}^{C_4} \, \exp(-C_5/T) \qquad . \; (14.3\text{-}14)$$

This implies that R_{fu} increases with m_{fu}, m_{ox} and T.

- As reaction proceeds, although T rises, m_{fu} and m_{ox} fall. The consequence is that $-R_{fu}$ falls to zero as τ tends to unity. The sketch illustrates this tendency.

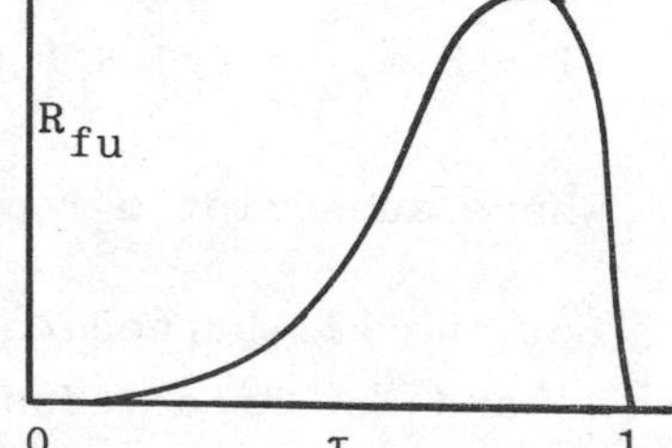

- The concave part of the curve for low τ results from the strong influence of the exponential term.

- R_{fu} is not strictly quite equal to zero at τ = 0; however it is very often negligible there. This depends, of course, on the initial temperature level.

- The shape of the curve is very important for the understanding of combustion phenomena, and should be memorised.

14.4 DISCUSSION

A) PRACTICAL RELEVANCE

Although real reactions proceed by the complex paths indicated earlier, it is often possible to represent their rates by expressions like (14.3-15), and to understand their behaviour in terms of the $R_{fu} \sim \tau$ curve. The processes of ignition, stabilisation, propagation and extinction can all be qualitatively understood in this way; and the quantitative representation of flame phenomena is also fairly good.

B) EXTENSION

Nevertheless, as knowledge of reaction-rate constants increases and the demands for accuracy intensify, it becomes both possible and necessary to represent combustion reactions more precisely. The following aspects can be taken account of:-

- Multiple reactions, for example those described in section 6.2, each with its own set of constants.
- Reverse reactions, i.e. the fact that, as well as $A + B \rightarrow C + D$ we have $C + D \rightarrow A + B$. Their <u>relative</u> rates can be computed from <u>thermodynamic</u> data concerned with equilibrium; so no new chemical-kinetic measurements need be made for the reverse reactions.
- Temperature-dependent specific heats.
- The fact that reactions take place as a consequence of gas∿solid contact.

A further aspect of importance is the interaction between turbulence and chemical reaction; for, though

$$R_{fu} = R_{fu} (m_{fu}, m_{ox}, T) \qquad (14.4\text{-}1)$$

may describe the instantaneous reaction rate, it is <u>not true</u> that the time-average values (denoted by over-bars) are linked by:

$$\overline{R_{fu}} = R_{fu} (\overline{m_{fu}}, \overline{m_{ox}}, \overline{T}) \qquad ; (14.4\text{-}2)$$

for the average value of $m_{fu}\, m_{ox}$, for example, is quite different from $\overline{m_{fu}}\ \overline{m_{ox}}$, as is evident from the discussion of turbulent diffusion flames in Chapter 12.

In the following chapters, attention will be confined to the simple model of reaction described above; however, some further sources of information can be found in the references.

14.5 REFERENCES

1. BAULCH, D L , DRYSDALE, D D , HORNE, D G , & LLOYD, A C (1970).
"High-temperature reaction-rate data".
Dept. Phys. Chem. Repts. No. 1 through 5, Leeds, England.

2. BENSON, S W (1968).
"Thermochemical kinetics".
Wiley, New York.

3. GUGGENHEIM, E A (1960).
"Elements of the kinetic theory of gases".
Pergamon, London.

EXERCISES TO FACILITATE ABSORPTION OF MATERIAL OF CHAPTER 14

ANALYTICAL PROBLEMS

14.1 Show that a bi-molecular reaction in a stoichiometric mixture of fixed pressure and enthalpy can be expected to proceed at a rate given by:

$$\text{Rate} = \frac{a(T - b)^2}{T^{3/2}} \exp(-E/RT) ,$$

and derive expressions for the constants a and b.

14.2 Show that the rate of the above reaction first increases and then decreases with increasing temperature, and derive an expression for the temperature at which the rate is a maximum.

14.3 The viscosity of air is 1.72×10^{-5} kg/ms at 0^0C and 4.80×10^{-5} kg/ms at 1000^0C.

Treating air as a single-component gas of molecular weight, derive from the above formula an expression for "the" diameter of air molecules, and evaluate d for $T = 273^0$K and $T = 1273^0$K.

14.4 Derive equations (14.3-12) and (14.3-13), listing all assumptions.

The quantity

14.5 E,

14.6 S,

14.7 R_A,

14.8 $\exp(-E/\mathcal{R}T)$,

is called the:

A steric factor.

B Arrhenius term.

C volumetric reaction rate.

D rate of collision with sufficient energy to precipitate reaction.

E activation energy.

The mass rate per unit volume of collisions between A and B molecules in a mixture increases with:

14.9 the pressure of the mixture,

14.10 the absolute temperature of the mixture,

14.11 the mass fraction of A in the mixture,

14.12 the mass fraction of B in the mixture,

raised to the power:

A 1;

B 2;

C $\frac{1}{2}$;

D $-\frac{1}{2}$;

E $-3/2$.

The symbol used for:

14.13 Avogadro's number,

14.14 the number of molecules of A per unit volume,

14.15 Boltzmann's constant,

14.16 the universal gas constant,

14.17 the number of collisions between molecules A and B in unit time,

is:

A k;

B R;

C Z;

D n_A;

E N.

The value of:

14.18 Avogadro's number,

14.19 the mass of a hydrogen atom,

14.20 the diameter of a molecule,

is, in S.I. units, roughly:

A 6×10^{-26};

B $(1/6) \times 10^{-26}$;

C 6×10^{-10};

D 3×10^{-10};

E 6×10^{26};

14.21 All the following statements are true about bi-molecular collision-controlled reactions, except:

A any "activated-complexes" that are formed decompose in a time that is very short compared with that which elapses between collisions.

B therefore it is likely that a reaction which is collision-controlled at one pressure will become decomposition-controlled when the pressure increases greatly.

C if the concentrations of the reactants are doubled, so is the volumetric reaction rate.

D if the activation energy is zero, the volumetric reaction rate will decrease as the temperature increases.

E if the pressure is doubled, the volumetric reaction rate is quadrupled.

The molecular interaction represented by the equation:

14.22 $C_2H_6 + 3\frac{1}{2}O_2 \rightarrow 2CO_2 + 3H_2O$,

14.23 $O_2 + H \rightarrow OH + O$,

14.24 $H + H + M \rightarrow H_2 + M$,

14.25 $A + B \rightarrow (AB)^*$,

14.26 $A^* + M \rightarrow C + D + M$,

describes:

A one step in a chain reaction involving radicals.

B the formation of an activated complex.

C an impossibility.

D the three-body re-association of radicals.

E the decomposition of an activated molecule.

14.27 The "equivalent hard-sphere" diameter of a molecule,

14.28 The average energy of collisions,

14.29 The viscosity of a gas,

14.30 The number of molecules in a kg mole,

A increases approximately in proportion to the absolute temperature T.

B increases in proportion to a power of T slightly in excess of $\frac{1}{2}$.

C diminishes slightly as T increases.

D is independent of T.

E increases approximately in proportion to T^2.

14.31 Differentiation of the Arrhenius term allows all the following deductions, except:

A the fractional increase in the value of the term for unit increase in T is $E/(\mathcal{R}T^2)$.

B if E is 1.6×10^8 J/kg mole and T is 316^0K, a 1^0K rise in T increases the term by about 20%.

C with the same value of E, but with T equal to 3160^0K, a 1^0K rise in T enlarges $\exp(-E/\mathcal{R}T)$ by 0.2%.

D reactions having large activation energies are less sensitive to changes in temperature than those with small values.

E if the Arrhenius term were to be approximated by the expression αT^β, with α and β as constants, β should be given the value $E/\mathcal{R}T$.

14.32 It is necessary to determine activation energies experimentally

because

there is no way of predicting them from existing knowledge of the structure of molecules.

14.33 The steric factor can be calculated from knowledge of molecular structure

because

most molecules are hard spheres.

14.34 The value of the Arrhenius term is usually much greater than unity

because

E often lies in the range from 4×10^7 to 2×10^8 J/kg mole, and T is usually between 300 and $3{,}000^0$K for combustion reactions.

14.35 The volumetric rate of a bi-molecular, collision-controlled reaction increases in proportion to the square of the pressure

because

the velocities of the molecules are not all the same at any given temperature.

14.36 The curve of volumetric reaction rate versus temperature exhibits a maximum at the temperature of the fully-burned gas, T_b,

because

at least onc of the reactants is wholly consumed at the condition of maximum reaction rate.

14.37 The rate of a collision-controlled reaction is never zero, unless one of the reactants has been consumed

because

$\exp(-E/\mathcal{R}T)$ equals zero only at the absolute zero of temperature.

14.38 The rate of diminution of the mass fraction of a reactant is inversely proportional to the square of the pressure, for a collision-controlled reaction

because

the volumetric reaction rate is directly proportional to the square of the pressure for such reactions.

14.39 The rate of a chemical reaction increases rapidly with temperature, until stopped by disappearance of reactants,

because

the mass of species A colliding with species B per unit volume and time is proportional to $T^{-3/2}$.

14.40 The curve of R versus T is usually concave upward at low T, even when the reactants are consumed as the temperature rises

because

$E/\mathcal{R}T$ is usually much greater than unity.

14.41 A qualitative knowledge of the kinetic theory of gases allows one to expect all of the following except:

A the mean free path diminishes with increase in pressure.

B the molecular speed is independent of pressure.

C the frequency of collisions between molecules per unit volume is proportional to the square of the pressure.

D the rate of reaction per unit volume is proportional to the square of the pressure.

E the rate of reaction between a gas and the reactive walls of its container is proportional to the square of the pressure.

ANSWERS TO MULTIPLE-CHOICE QUESTIONS

Answer	Problem number (14's omitted)
A	6, 11, 12, 15, 23, 28, 32, 37, 40
B	8, 9, 16, 19, 25, 29, 35, 39
C	7, 17, 21, 22, 27
D	14, 20, 24, 30, 31, 34, 38
E	5, 10, 13, 18, 25, 33, 36, 41

CHAPTER 15

SPONTANEOUS IGNITION

15.1 ENGINEERING RELEVANCE OF THE PHENOMENON

A) WHAT IS MEANT BY "SPONTANEOUS IGNITION"

When a reactive mixture is formed, raised to a definite temperature and pressure, and then left alone, it may burst into flame after a certain time. "Burst into flame" means "rise rapidly in temperature", "emit visible radiation", "suffer rapid chemical transformation", etc. This phenomenon is often called "spontaneous ignition".

The word "spontaneous" distinguishes the ignition from that caused by external agencies such as sparks, heated walls, and pilot flames. Other words are: auto-ignition; self-ignition; they refer to the same thing.

B) RELEVANCE TO RECIPROCATING ENGINES

(i) Diesel engines

Liquid fuel is injected into air which has been raised in pressure and temperature by compression in the cylinder; part of it vaporises rapidly and mixes with the air. After a delay of a few milli-seconds, the mixture ignites; the ignition rapidly consumes the already vaporised and mixed fuel, and thereafter the rate of burning is controlled by the rate of injection and by the mixing characteristics of the two-phase, turbulent, fuel-air jet.

Remarks

(1) This process is not a very pure example of spontaneous ignition, because there is no sharp separation between the mixture-preparation and ignition phases; the mixture is not "left alone" during the

ignition-delay period. Nevertheless, the essential feature is present: the main temperature rise in the gas is accomplished by the reacting gas on its own; there is no input of high-temperature gas from an external source.

(2) When the ignition delay is long, the amount of fuel which is vaporised and mixed, ready to burn, is augmented by the continuing flow from the injector. Ignition, when it comes, makes a "big bang", which is the characteristic "knock" of Diesel engines. Therefore, in order to reduce the knock, the fuel should be highly reactive, so that the ignition delay will be short.

(3) Other means are sometimes adopted: the injection is effected in two stages, the first providing enough fuel to allow ignition, and the second providing the main fuel after ignition has been accomplished.

(4) High-speed automotive engines require more reactive fuels than slow-speed marine engines, because the thermodynamic penalty of a long ignition delay depends on the crank-angle change corresponding to this delay, rather than on the time.

(ii) Gasoline engines

The main flame is initiated by the spark, and spreads through the gaseous fuel-air mixture; it usually takes several milli-seconds to reach the most remote corners of the combustion chamber. The gas in these corners is hot, having been compressed both by the piston and by the expansion of the already-burned gas; it therefore tends to ignite spontaneously. If this spontaneous ignition has time to occur, before the main flame arrives, a very sharp pressure rise occurs; this is the origin of gasoline-engine knock, sometimes (confusingly) called "detonation".

Remarks

(1) Therefore, in order to reduce knock in a gasoline engine, one needs a fuel which is highly resistant to spontaneous ignition. The requirement is the opposite of that in the Diesel engine.

(2) Iso-octane has a long ignition delay, at any given pressure and temperature, and so is resistant to knock; n-Heptane ignites easily. The Octane-number scale is defined by reference to these two fuels.

(3) The occurrence of knock can also be reduced by making sure that the spark-initiated flame reaches all parts of the combustion chamber quickly. For example, it is better to put the spark near the middle than off to one side.

c) RELEVANCE TO RAM-JET ENGINES

In some designs of ram-jet for very high-speed flight, the inlet diffuser may provide sufficient compression and temperature rise for the fuel, injected into the air stream at the combustion chamber inlet, to vaporise and ignite before it has travelled more than a few feet downstream. If so, it will be possible to dispense with a bluff-body flame-holder, which is a cause of undesirable drag.

Remarks

(1) Although this is a steady-flow process rather than a transient one for an observer travelling with the ram-jet, it fits into the "spontaneous-ignition" definition from the point of view of a particle of fuel.

(2) It is true that, as in the case of the Diesel engine,

vaporisation and mixing are taking place throughout the ignition period; but, once again, these do not affect the main issue: the gas inflames as a result of its own efforts.

(3) In this case, obviously, fuels are required which have short ignition delays. The engine-designer can help by arranging that the pressure ratio of the diffuser is high; this causes high pressure and temperature, and low velocity.

15.2 FUNDAMENTAL CONSIDERATIONS

A) REACTION KINETICS

From Lecture 14, for a collision-controlled bi-molecular reaction, the rate obeys the equation:

$$R_A = K\, p^2\, m_A m_B\, T^{-3/2} \exp\{-E/(\mathcal{R}T)\} \qquad , (15.2\text{-}1)$$

where A and B are the two reactants.

Equations of this kind can be written for all the collision-controlled bi-molecular reaction steps in the whole combustion process; and, for those steps which are not collision-controlled or bi-molecular, other equations can be written.

In principle, the constants K and E, and any which appear in equations for the other types of reaction, are known or knowable constants. They ought to be tabulated like thermodynamic properties; but the research to allow this is still in its early stages.

B) THERMODYNAMICS (INCLUDING ATOMIC BALANCES)

As the temperature rises from the unburned-gas temperature T_o to the burned-gas temperature T_1, one can be sure that

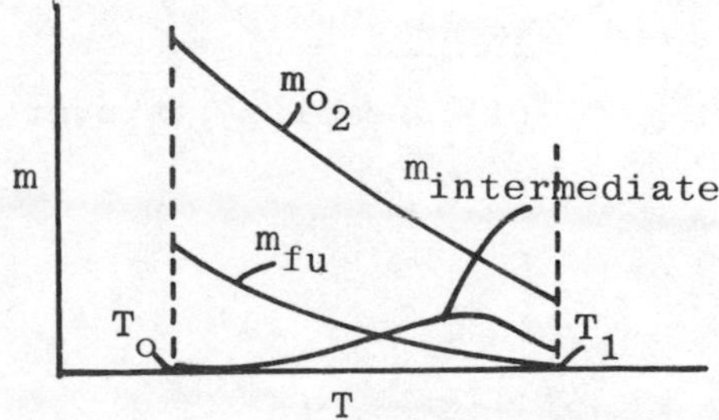

the fuel is used up, and the oxygen with it. Intermediates (radicals, decomposition products of the fuel) will increase in concentration in the lower-temperature range; but most of these will also have been consumed by the time T is reached.

Therefore the mass-fraction curves will be qualitatively as shown in the sketch.

Thermodynamics alone is unable, in general, to dictate the complete course of the curves; for these depend in part on the relative rates of competing reactions (i.e. reactions which consume the same reactants but with different products). However, one can be sure that the sketch is qualitatively correct.

c) HEAT TRANSFER, MASS TRANSFER, PHASE CHANGES

Although the processes in this sub-title occur in most practical spontaneous-ignition phenomena, they are not essential. There will therefore be no need to call on the relevant theories.

15.3 MATHEMATICAL MODEL

A) DESCRIPTION

(i) The reaction

To procure simplicity of analysis, while still retaining the main qualitative features of the situation, it will be supposed that the whole reaction can be accomplished in a single bi-molecular step, as described in Chapter 14, viz:

$$\text{fuel} \quad + \quad \text{oxygen} \quad \rightarrow \quad \text{product.}$$

Remarks

(1) Despite its over-simplicity, this supposition

allows many practical combustion phenomena between hydrocarbons and air to be understood, and even predicted.

(2) The equation for the rate of consumption of fuel is therefore (15.2-1), with $A \equiv fu$ and $B \equiv ox$. Of course, K and E are constants which must be determined experimentally, or estimated from fundamental considerations.

(ii) The reactants

Because there is now only one reaction path, thermodynamics is sufficient for the determination of the relation of m_{fu} and m_{ox} to T, provided that conditions are adiabatic and the pressure or volume is held constant.

It will be assumed that the molecular weights of fuel, air and products are equal, and that the specific heats are also equal and independent of temperature. The consequence is that the relations between m_{fu}, m_{ox} and T are linear as in Section 14.3 above.

Of course, more complex thermodynamic properties can be handled also, if need be; but there is no point in retaining troublesome but inessential details after such bold simplifications of the reaction kinetics have already been made.

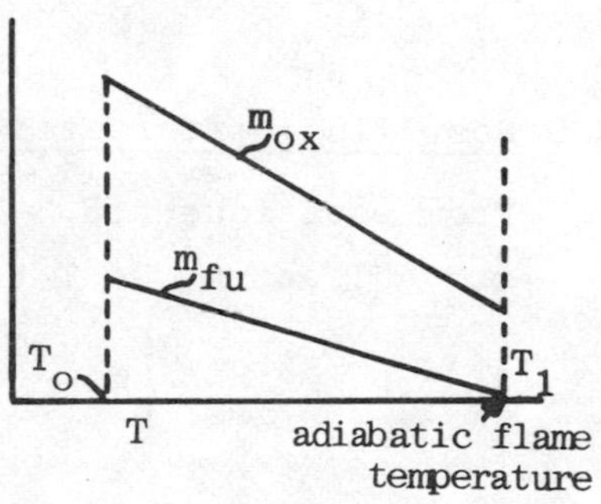

(iii) Other conditions

As implied above, it will be supposed that the gaseous system is adiabatic, and of constant pressure or volume.

At time t = 0, the temperature is at T_o, and the concentrations are $m_{fu,o}$ and $m_{ox,o}$. The task is to calculate how T, m_{fu} and m_{ox} vary thereafter with time.

B) EQUATIONS

(i) Differential equation for temperature

Because m_{fu}, m_{ox} and T are linked, it is necessary to write a differential equation for only one of them. For a constant-specific-heat mixture, chemical reaction has the same effect on temperature as an external heat addition; so the 1st Law of Thermodynamics can be written as:

$$c\rho \frac{dT}{dt} = R_{fu} H \qquad . \ (15.3\text{-}1)$$

Here c is either the constant-pressure or constant-volume specific heat, as appropriate.

(ii) Ideal-gas law

It is necessary to use :

$$\frac{p}{\rho} = \frac{\mathcal{R}}{M} T \qquad . \ (15.3\text{-}2)$$

(iii) Resulting differential equation for T

From equations (15.3-1), (15.2-1) and the linear $m_{fu} \sim T$ and $m_{ox} \sim T$ relations, one finds :

$$\frac{dT}{dt} = p\, T^{-\frac{1}{2}} \frac{HK\mathcal{R}}{cM} \left\{ m_{fu,o} - \frac{c(T-T_o)}{H} \right\} \left\{ m_{ox,o} - \frac{sc(T-T_o)}{H} \right\} \times$$

$$\times \exp\{-E/(\mathcal{R}T)\} \qquad , \ (15.3\text{-}3)$$

or, in terms of ρ:

$$\frac{dT}{dt} = \rho T^{\frac{1}{2}} \frac{HK\mathcal{R}^2}{CM^2}\left\{m_{fu,o} - \frac{c(T-T_o)}{H}\right\}\left\{m_{ox,o} - \frac{sc(T-T_o)}{H}\right\} \times$$

$$\times \exp\{-E/(\mathcal{R}T)\} \quad . \quad (15.3\text{-}4)$$

(iv) Important observation

Whichever equation is used, the right-hand side is a function of temperature alone, for a given problem; therefore the equation can certainly be solved by numerical means. It is of the form: $\frac{dy}{dx} = f(y)$, which can be expressed as: $x = \int_o^y \frac{1}{f(y)} dy + \text{const.}$

c) SOLUTION

(i) Qualitative features

If the RHS of (15.3-3) or (15.3-7) is symbolised by $r(T)$, since it is known already that this function has the shape shown in the sketch, the reciprocal, $1/r$, must also be as shown.

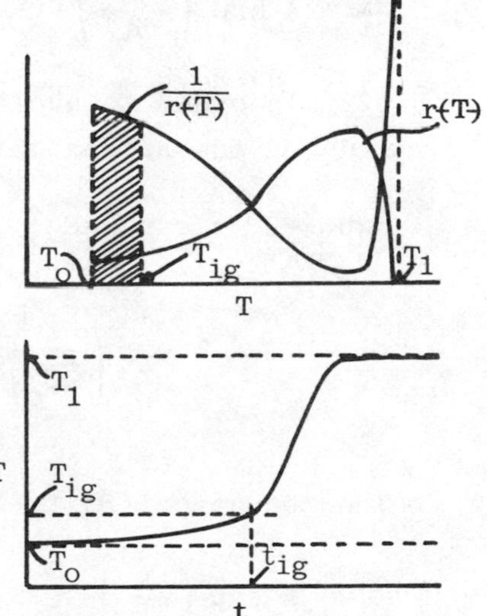

Hence, or otherwise, it may be deduced that the $T \sim t$ curve has the form of the sketch on the right.

Once the temperature has begun to rise significantly, this sketch implies, the further rise to near T_1 occurs very rapidly; therefore, in order to define the ignition delay precisely, it is convenient to take it as the time at which τ, defined as $(T-T_o)/(T_1-T_o)$ has the value of 0.1, say; what value happens to be chosen will not be very important. Thus, the ignition delay is given by the shaded area of the top sketch; T_{ig} is $T_o + 0.1\ (T_1 - T_o)$. The task is now to determine t_{ig}, and to see how it is related to p, K, etc.

(iii) Approximations

With the use of numerical methods and a digital computer, the T $\sim$ t relation can be easily determined, without further approximations. For present purposes, however, it is useful to proceed analytically; this necessitates some approximations, specifically:

(1) In the range $T_o \leqslant T \leqslant T_{ig}$, the terms $T^{\frac{1}{2}}$, $\{m_{fu} - c(T - T_u)/H\}$ and $\{m_{o_2} - s\ c\ (T - T_u)/H\}$ vary much less than the exponential term. They will therefore be regarded as constant. Hence :

$$\frac{dT}{dt} \approx A \exp\left(-\frac{E}{\mathcal{R}T}\right) \quad . \quad (15.3\text{-}5)$$

(2) In order to avoid numerical integration, use is made of the approximation:

$$\exp\left(\frac{-E}{\mathcal{R}T}\right) = \exp\left(\frac{-E}{\mathcal{R}T_o}\right) . \exp\left\{\frac{-E}{\mathcal{R}T_o} \ . \ \frac{(T_o - T)}{T}\right\}$$

$$\approx \exp\left(\frac{-E}{\mathcal{R}T_o}\right) . \exp\left\{\frac{E(T - T_o)}{\mathcal{R}T_o^2}\right\} \quad . \quad (15.3\text{-}6)$$

Then equation (8) takes the integrable form:

$$\frac{dT}{dt} \approx A \exp\left(\frac{-E}{\mathcal{R}T_o}\right) \cdot \exp\left\{\frac{E(T - T_o)}{\mathcal{R}T_o^2}\right\} \quad . \quad (15.3\text{-}7)$$

(iii) The solution

It is now a simple matter to solve (15.3-7) in the

required terms; the result is:

$$t_{ig} = \frac{1}{A} \exp\left(\frac{E}{\mathcal{R}T_o}\right) \cdot \frac{\mathcal{R}T_o^2}{E}\left[1 - \exp\left\{\frac{-E(T_{ig}-T_o)}{\mathcal{R}T_o^2}\right\}\right] . \quad (15.3\text{-}8)$$

(iv) Preliminary discussion

When $E(T_{ig}-T_o)/(\mathcal{R}T_o^2)$ exceeds unity, the second exponential term has little influence (this is a confirmation of a remark in Section 15.3 C (i)). A final approximation can therefore be made, giving :

$$\boxed{t_{ig} \approx \frac{\mathcal{R}T_o^2}{AE} \cdot \exp\left(\frac{E}{\mathcal{R}T_o}\right)} . \quad (15.3\text{-}9)$$

Here it should be recalled that A is defined by:

$$A \equiv H \frac{pT_o^{-\frac{1}{2}}}{c} \frac{K\mathcal{R}}{M} m_{fu,o}\, m_{ox,o} \text{ for constant-pressure ignition} , \quad (15.3\text{-}10)$$

and by: $$A \equiv H \frac{\rho T_o^{\frac{1}{2}}}{c} K \left(\frac{\mathcal{R}^2}{M}\right) m_{fu,o}\, m_{ox,o} , \quad (15.3\text{-}11)$$

for constant-volume ignition.

15.4 DISCUSSION

A) MAIN INFLUENCES ON THE IGNITION DELAY

(i) Pressure:

The equations show that $t_{ig} \propto (\text{initial pressure})^{-1}$. Evidently, therefore, the use of high compression ratios in engines makes spontaneous ignition easier. However, this is not the main effect of compression ratio, which is manifested through the accompanying temperature rise.

(ii) Temperature:

- From (15.3-9),

$$\frac{dt_{ig}}{t_{ig}} = \frac{2dT_o}{T_o} - \frac{dA}{A} - \frac{E}{\mathcal{R}T_o^2}\,dT_o \quad , \quad (15.4\text{-}1)$$

i.e. $$\frac{1}{t_{ig}}\frac{dt_{ig}}{dT_o} = \frac{1}{T_o}\left(2 - \frac{E}{\mathcal{R}T_o} \pm \frac{1}{2}\right)$$

(N.B. + for constant-pressure; - for constant volume).

- Typically, $E/\mathcal{R}$ is of the order of 20,000 ^{0}K, so $E/(\mathcal{R}T_o)$ lies between 100 (for T_o = 200) and 10 (for T_o = 2000^{0}C). So the $E/(\mathcal{R}T_o)$ term usually dominates the right-hand side.

- Example: T_o = 500 ^{0}K; $E/\mathcal{R}$ = 20,000 ^{0}K; constant-volume ignition.

  Then $$\frac{1}{t_{ig}}\frac{dt_{ig}}{dT_o} = \frac{1}{500}\left(2 - 40 - \frac{1}{2}\right)$$

  $$= -.077 \quad . \quad (15.4\text{-}2)$$

  So a 13 ^{0}C increase in temperature will reduce the ignition delay to 1/e of its initial value.

(iii) Concentration:

The greater are $m_{fu,o}$ and $m_{ox,o}$, the smaller is t_{ig}. Therefore, in order that the ignition delay should be small, inert mixture components (e.g. nitrogen) should be eliminated. The stoichiometric mixture ratio has no special significance or merit.

B) SOME EXPERIMENTAL DATA

(i) From Mullins (1953)

- The experiments were conducted by injecting finely atomised or gaseous fuels into a stream of heated air. The ignition delay was deduced from the distance downstream of the injection at which the flame was formed.

- For most fuels (hydrocarbons) it was found that

  $$t_{ig} \propto p^{-1} \quad .$$

  This confirms that the dominant reactions are bi-molecular and collision-controlled.

- The fuel concentration had little influence on the ignition delay within a fairly narrow (2:1) range of flows. Possibly the mixture was not, in any case, quite uniform in composition and temperature within the stream.

- Large differences were observed between various fuels, as shown below.

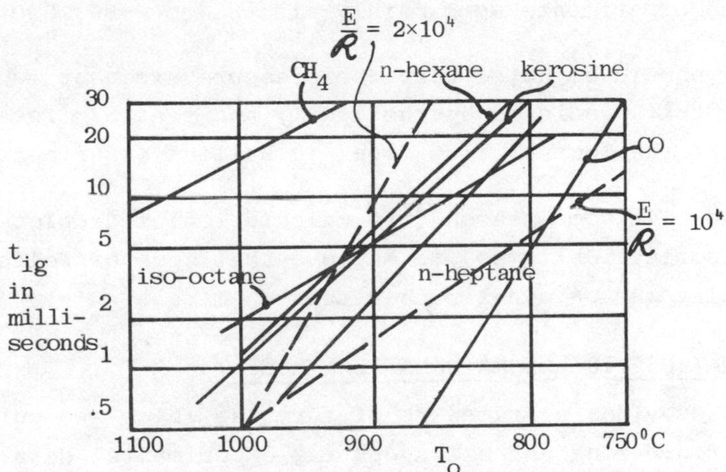

- Note the manner of representation. The vertical scale is logarithmic; the horizontal one is linear in $1/T_o$. So $t_{ig} \propto \exp(E/\mathcal{R}T_o)$ will give a straight line of slope proportional $E/\mathcal{R}$. Two broken lines serve as examples.

  It can be deduced that the experimental data can be described fairly well by eq. (15.3-9); $E/\mathcal{R}$ lies in the range: 10^4 to 2×10^4.

(ii) From Wolfer (1938)

- The experiments were transient. Compressed air was heated in a pressure vessel; then liquid fuel was injected, and the pressure-time variation was measured. The ignition delay was deduced by observing the instant at which the pressure began to rise steeply.

- Results. For a commercial Diesel fuel, Wolfer found that his measurements fitted the formula:

$$t_{ig} = \frac{.44}{p^{1.19}} \exp(4560/T) \text{ milliseconds} \qquad (15.4\text{-}3)$$

where p is in atmospheres, and T is in ^{0}K.

The experiments were mainly with $8 \leqslant p \leqslant 30$, $590 \leqslant T \leqslant 780$.

- It should be noted that the pressure exponent is near to that predicted by the model, and that $E/\mathcal{R}$ is about half the lowest value found in Mullins experiments.

There is some reason (via calculation of droplet-vaporisation times) to suppose that some of Wolfer's delay was physical rather than chemical.

c) WHAT USE IS THE MATHEMATICAL MODEL?

(i) It provides a framework of formulae which are suitable for ordering and extrapolating experimental data.

(ii) It provides understanding; ie it enables one correctly to predict the directions and orders of magnitude of the changes to ignition delay that follow changes in conditions.

(iii) It represents the first step on the road to a complete prediction procedure. This involves the exact (numerical) solution of many simultaneous differential equations of the same kind as (15.3-3 or 4). It is easy to make the computations with a digital computer; but certainty about the chemical-kinetic constants of the individual reaction steps is still lacking.

15.5 REFERENCES

1. GAYDON A G (1967)
"The use of shock tubes for studying fundamental combustion processes".
Eleventh Symposium on Combustion.
The Combustion Institute, Pittsburgh, pp 1-10.

2. MARTINENGO A, MELCZER J & SCHLIMME E (1965)
"Analytical investigations of stable products during reaction of adiabatically compressed hydrocarbon-air mixtures".
Tenth Symposium on Combustion.
The Combustion Institute, Pittsburgh, pp 323 - 330.

3. MULLINS B P (1953)
"Studies on the spontaneous ignition of fuels injected into a hot air stream".
Fuel, Vol 32, p 211.

4. WOLFER H H (1938)
"Der Zundverzug in Dieselmotor".
VDI Forschungsheft 392.

EXERCISES TO FACILITATE ABSORPTION OF MATERIAL IN CHAPTER 15

MULTIPLE-CHOICE PROBLEMS

15.1 All the following are spontaneous-ignition phenomena, except:

A auto-ignition.

B gasoline-engine "knock".

C ignition by mixing of unburned mixture with combustion products.

D the happenings in the "chemical shock tube", just behind the shock wave.

E the inflammation of a haystack as a consequence of exothermic chemical reactions in the interior.

15.2 All the following statements are true of ignition in Diesel engines, except:

A the process is a transient one.

B as a rule, all the fuel has been injected into the cylinder before ignition starts.

C although the fuel is injected in the liquid phase, the fuel-air-ratio pattern during injection is similar in qualitative respects to that in a turbulent single-phase jet.

D "Diesel knock" is caused by the sudden inflammation of fuel which was injected, vaporised and mixed with air earlier in the cycle.

E one way to reduce "Diesel knock" is to use a fuel which ignites easily.

15.3 All of the following measures are likely to reduce the tendency of a gasoline engine to knock, except:

A placing the spark plug as near as possible to the

centre of the combustion chamber.

B increasing the speed of flame propagation by increasing the turbulence level.

C increasing the speed of flame propagation by employing a mixture ratio close to stoichiometric.

D increasing the compression ratio of the engine.

E lowering the temperature of the air flowing into the engine.

Of the following set of completions (A, B, C, D, E), the right one for the sentence beginning:

15.4 in the ram-jet engine for high-speed flight,

15.5 in the gasoline engine,

15.6 in the Diesel engine,

is:

A the ignition delay should be long, so as to avoid excessive noise and vibration.

B the ignition delay should be long, so as to avoid development of excessive pressures.

C the ignition delay should be short, so that the flame does not become excessively long.

D it is desirable to have a low pressure in the combustion chamber, so that the pressure drop is also low.

E an easily igniting fuel makes the least noise.

15.7 All the following statements are true except:

A iso-octane would be a bad fuel for a Diesel engine.

B n-heptane would be a good fuel for a ram-jet.

C "detonation" is the name properly used for the high-velocity propagation of flame in a duct, involving a strong shock wave followed by a spontaneous-ignition zone.

D the process in gasoline engines which is sometimes called "detonation" involves the passage of a shock wave, followed by a zone of spontaneous ignition.

E gasoline-engine knock is more likely to occur at low engine speeds because then the spark-initiated flame travels slowly.

15.8 The equation which expresses the rate of consumption of reactant A per unit volume, when it engages in a collision-controlled reaction with reactant B, is:

A $R_{fu} = K\,p^2\,m_{fu}\,T^{-3/2}\exp(-E/\mathcal{R}T)$.

B $R_{fu} = K\,p^2\,m_{fu}\,m_{O_2}\,T^{3/2}\exp(E/\mathcal{R}T)$.

C $R_{fu} = K\,p\,m_{fu}\,m_{O_2}\,T^{-3/2}\exp(E/\mathcal{R}T)$.

D $R_{fu} = K\,p^2\,m_{O_2}\,T^{3/2}\exp(E/\mathcal{R}T)$.

E $R_{fu}\ K\,p^2\,m_{fu}\,m_{O_2}\,T^{-3/2}\exp(-E/\mathcal{R}T)$.

In the following sketch, the curve marked: 15.9, 15.10, 15.11, 15.12, could represent:

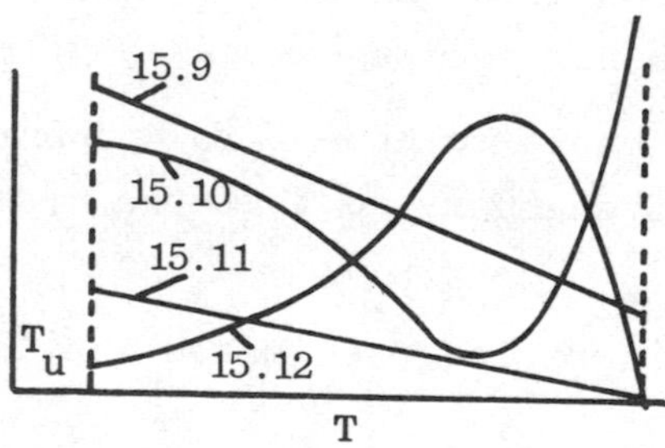

A the fuel mass fraction.

B the oxygen mass fraction.

C the mass fraction of combustion products.

D the volumetric rate of reaction.

E the reciprocal of the rate of reaction.

15.13 The differential equation for the rate of rise of temperature, in the single-step model of spontaneous ignition is:

A $c \frac{dT}{dt} = R_{fu}H$

B $c \frac{dT}{dt} \quad R_{fu}H\rho$

C $\rho \frac{dh}{dt} = R_{fu}H$

D $\frac{c}{\rho} \frac{dT}{dt} = R_{fu}H$

E $c \, \rho \, \frac{dT}{dt} = R_{fu}H$

15.14 All the following statements are true about the method of solving the differential equation of the single-step spontaneous-ignition model, except:

A the variations of m_{fu} and m_{ox} are implicitly neglected.

B it was therefore not essential for us to require that m_{fu} and m_{ox} vary linearly with temperature.

C the term $T^{\frac{1}{2}}$ is also treated as constant.

D if the activation energy were very small, the approximations would not be permissible.

E if the activation energy were very small, the ignition delay would probably be too long for practical interest in any case.

15.15 Gasoline-engine knock becomes more severe immediately the accelerator pedal is pressed down,

<u>because</u>

the engine speed then increases.

15.16 For multi-step ignition processes, thermodynamics is unable to predict the $m_{fu} \sim T$ relation,
<u>because</u>
such processes cannot be adiabatic.

15.17 The rate of a single-step collision-controlled chemical reaction is proportional to $\exp\{E/(\mathcal{R}T)\}$,
<u>because</u>
only collisions with sufficient kinetic energy of motion along the normal result in re-arrangements of atoms.

15.18 The equation, $R_{fu} = K\, p^2\, m_{fu}\, m_{ox}\, T^{-3/2} \exp\{-E/(\mathcal{R}T)\}$, is a true representation of the real fuel $\sim$ oxygen reactions,
<u>because</u>
such reactions usually proceed by way of a single step.

15.19 In the equation of the last problem, the quantity K which is appropriate to a real fuel $\sim$ oxygen reaction cannot be determined from collision-rate theory,
<u>because</u>
the equation is based on a severe over-simplification of the actual chemical-kinetic phenomena.

15.20 Provided K and E are taken from experimental data, the equation of the last-but-one problem is found to describe many real fuel-oxygen reaction phenomena adequately,
<u>because</u>
such phenomena are the results of a very large number of individual steps.

15.21 In the single-step model of fuel-oxygen ignition, the mass fractions of fuel and oxygen are linearly related,
<u>because</u>
fuel and oxygen are consumed by the chemical reaction in fixed mass proportions.

15.22 In the single-step model of fuel-oxygen ignition, the mass fraction of oxygen is linearly related to the temperature,
<u>because</u>
the chemical reaction is exothermic.

15.23 For a stoichiometric mixture, the volumetric reaction rate is proportional to $(T_1 - T)^2$ as T approaches T_1 according to the adiabatic single-step model,
<u>because</u>
the Arrhenius term ceases to be valid in this case.

15.24 A general analytical solution can be obtained for the time-temperature variation of the adiabatic, single-step, ignition reaction,
<u>because</u>
the integral to be evaluated is $\int \rho/(R_{fu}H)dT$.

15.25 The ignition delay is always defined as the time for T to attain T_{ig} where $(T_{ig} - T_o)/(T_1 - T_o)$ equals 0.1,
<u>because</u>
t_{ig} depends strongly on the value of $(T_{ig} - T_o)/(T_1 - T_o)$.

15.26 $\text{Exp}(-E/\mathcal{R}T)$ can be put approximately equal to $\exp(-E/\mathcal{R}T_o)$. $\exp\{E(T-T_o)/\mathcal{R}T_o{}^2\}$,
<u>because</u>
the two expressions have equal values when T is close to T_o, which is true of the critical period of ignition.

15.27 The ignition delay is proportional to the reciprocal of the pressure, for a collision-controlled bi-molecular ignition process,
<u>because</u>
t_{ig} is not very sensitive to the way in which T_{ig} is defined.

15.28 The ignition delay diminishes when the concentration of

inert diluent in the unburned mixture is diminished,
<u>because</u>
the adiabatic flame temperature is then increased.

15.29 The ignition delay is a minimum, for a fixed concentration of diluent, when the fuel/oxygen ratio is stoichiometric,
<u>because</u>
this mixture has the highest adiabatic flame temperature.

15.30 The ignition reaction of iso-octane in air can be presumed to have a lower activation energy than that of methane in air,
<u>because</u>
the ignition delay of iso-octane is lower than that of methane, at any given temperature.

15.31 t_{ig} is plotted on a logarithmic scale against T_o on a reciprocal one,
<u>because</u>
then the points can be expected to lie on a nearly-straight line.

15.32 If methane were used as a fuel for a spark-ignition engine, knock would probably be severe,
<u>because</u>
its ignition delay is several times as great as that of conventional fuels.

15.33 Carbon monoxide would be a good fuel for a compression-ignition engine, could the problem of supplying it to the engine be easily solved,
<u>because</u>
its activation energy is rather low.

15.34 Spontaneous ignition is employed as the method of starting and stabilising the flames in gas-turbine combustion chambers,
<u>because</u>
kerosine-air ignition delays at the pressure and temperature of the compressor outlet are below 1 milli-second.

15.35 Tetraethyl lead is added to fuels for spark-ignition engines,
<u>because</u>
it renders the fuel less resistant to spontaneous ignition.

15.36 If tetraethyl lead were added to the fuels of London taxis, their engines would probably become less noisy,
<u>because</u>
"Diesel knock" is the result of too long an ignition delay.

ANSWERS TO MULTIPLE-CHOICE QUESTIONS

Answer	Problem number (15's omitted)
A	5, 11, 19, 21, 26, 31
B	2, 9, 15, 20, 22, 27, 28, 30
C	1, 4, 16, 23, 33, 35
D	3, 7, 12, 17, 24, 29, 32, 36
E	6, 8, 10, 13, 14, 18, 25, 34

CHAPTER 16

THE STIRRED REACTOR

16.1 ENGINEERING RELEVANCE

A) REACTION-RATE LIMITATIONS OF STEADY-FLOW FLAMES

- Gas turbines contain combustion chambers, between compressor outlet and turbine inlet, in which fuel and air burn in a steady turbulent diffusion flame.

  Gas turbines for aircraft may have to operate, at high altitudes, with low pressures in the combustion chambers. It is found that the efficiency of combustion deteriorates under these conditions; if the pressure falls low enough, the flame may even be extinguished.

- Even at ground-level conditions, efficiency deterioration and extinction may result if, in order to reduce the size or weight of the engine, the combustion chamber is made too small.

- These effects, as will be seen, are consequences of the limited rates at which chemical reactions can proceed. Extinction comes about when "the chemical kinetics cannot keep pace with the supply of reactants".

B) RESEARCH ON THE RELEVANT PROPERTIES OF FUELS

- When the above phenomena had been recognised as chemical-kinetic in origin (this was in the early 1950's; until then, deficient atomisation had been often blamed), the question arose: what tests can be made of the available fuels, in order to distinguish the good from the bad?

  The gasoline engine had its Octane number, and the Diesel engine its Cetane number (a measure of ease of ignition). What "number" would be appropriate for the gas turbine?

Small quantities of additives (e.g. tetraethyl lead) were known to affect the performances of fuels for gasoline engines. Could the extinction of a gas-turbine flame be prevented by the addition of small quantities of some "dope" to the fuel?

The experiments of Mullins (see Chapter 15) were part of the attempt to answer these questions. As it turned out, they proved not to be very useful for that purpose, because spontaneous ignition is affected by reaction rates at the low-temperature end of the range, and extinction, as will be seen, by those near the high-temperature end. If the complete reaction-rate expression could be characterised by just two constants, e.g. K and E of Chapter 14, it would not matter in what temperature range they were measured; but, since the bi-molecular collision-controlled model fits reality only roughly, extrapolation over several hundreds of degrees K is not permissible.

New experiments, which did produce data which allowed "ranking" of fuels, and an assessment of the upper limit to combustion-chamber performance, were devised by Longwell, Frost and Weiss (1953). The apparatus was a refractory sphere, into the centre of which fuel and air were injected through many small holes. The exhaust products escaped through fewer, larger, holes in the shell. The gases, which were highly turbulent as a consequence of the high

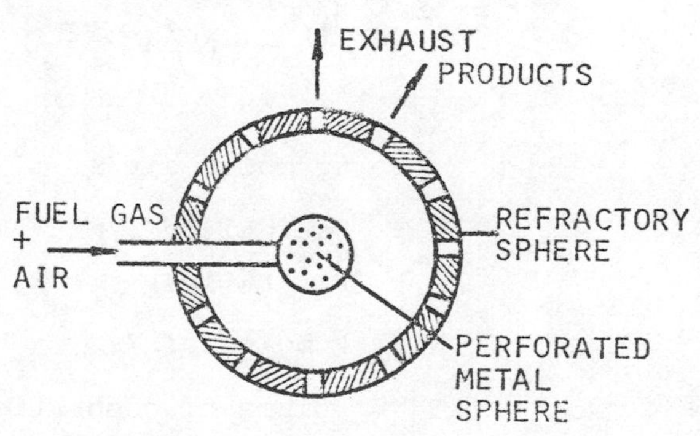

injection velocities, burned within the sphere. The flow pattern shown in the sketch, with its system of elongated recirculation eddies, caused the temperatures and concentrations within the spherical enclosure to be almost uniform. The apparatus thus approximated to the ideal of a "fully-stirred, adiabatic, steady-flow reactor", which is the mathematical model which will be discussed below.

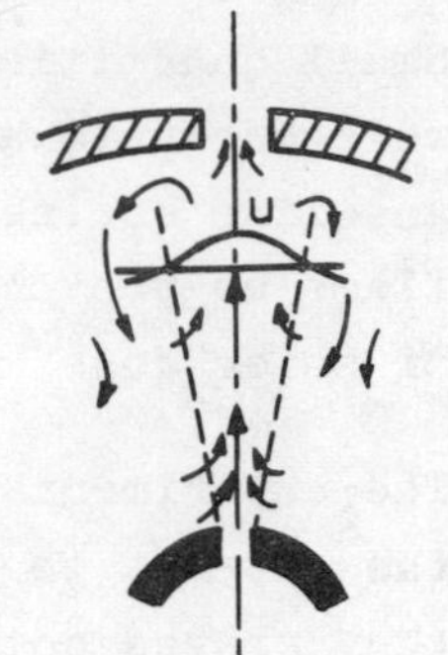

Some experimental results (Longwell and Weiss, 1955).

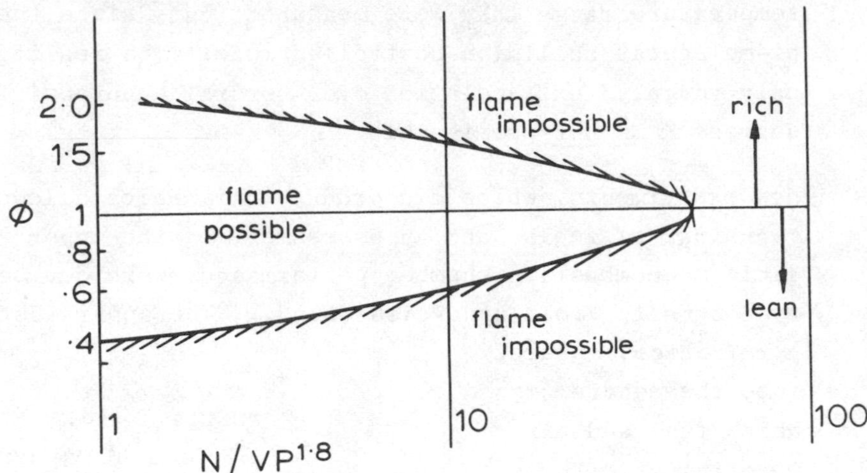

kg moles air/s. m^3 $(atm)^{1.8}$

$\phi \equiv$ (actual fuel/air ratio) ÷ (stoichiometric fuel/air ratio).

$N \equiv$ g moles air/s.

$V \equiv$ volume of combustion chamber in litres.

$P \equiv$ pressure in combustion chamber in atm.

ϕ is called the "equivalence ratio".

The experimental technique was to supply fuel and air to the combustion chamber, to ignite the flame, and then to increase either the fuel or the air flow rates until the flame was "blown out", i.e. extinguished. Each extinction point helped to distinguish the "flame-possible" from the "flame-impossible" region on the diagram. This diagram holds for various hydrocarbon fuels, at an inlet temperature of 400 ^{0}K.

One of the most important findings of the investigation was that only very small differences existed between the extinction behaviour of the practically interesting fuels; and additives such as tetraethyl lead were without significant effect.

Note that the blow-out rate is proportional to the chamber volume, and to the pressure to the power 1.8.

c) THE CONNEXION BETWEEN THE "LONGWELL BOMB" AND PRACTICAL COMBUSTION SYSTEMS

- The stirred reactor is used only in the laboratory; the pressure drop in the gas flowing through it is too high, and its shape too awkward, for use in gas turbines. In order for the data obtained with it to be applicable to practical combustion systems, quantitative theories must be devised which connect observations in the one to performance in the other.

- In this lecture, the theory for the stirred reactor will be provided.

- Because flow conditions in gas-turbine combustion chambers are three-dimensional, and must be described by partial differential equations in which appear still-unknown functions of the turbulence, no complete theory can be given in these lectures; but dimensional analysis can provide some useful results. The development of a complete theory is still the subject of research.

16.2 FUNDAMENTALS

A) REACTION KINETICS

There is nothing to add to what has already been said in Chapters 14 and 15. In the mathematical model, it will again be supposed that the reaction can be described as a collision-controlled bi-molecular one. Longwell, Frost and Weiss were among the first to make this assumption; and it is still among the most sophisticated chemical-kinetic notions in use among gas-turbine engineers.

B) THERMODYNAMICS

There is little to add here either. One is concerned with a steady-flow process in a reactor which can be taken as adiabatic; so the relations between temperature, fuel concentration and oxygen concentration are the same as for the constant-pressure reaction considered in lecture 14.

For the single-step reaction, with uniform and constant specific heats, the equations are :

$$m_{fu} = m_{fu,o} - c\,(T - T_o)/H \qquad , \quad (16.2\text{-}1)$$

$$m_{ox} = m_{ox,o} + s\,(m_{fu} - m_{fu,o}) \qquad . \quad (16.2\text{-}2)$$

Does the fact that mixing occurs between burned and unburned gases affect the validity of these equations? No, because, when mixing without reaction occurs between a gas with $m_{fu,I}$, T_I and another with $m_{fu,II} T_{II}$, the resultant properties are $m_{fu,III}$, T_{III}, given by:

$$m_{fu,III} = \varepsilon\, m_{fu,I} + (1 - \varepsilon)\, m_{fu,II} \qquad , (16.2\text{-}3)$$

$$T_{III} = \varepsilon\, T_I + (1 - \varepsilon)\, T_{II} \qquad ; (16.2\text{-}4)$$

so:

$$m_{fu,III} - c(T_{III} - T_o)/H = \varepsilon\, \{ m_{fu,I} - c(T_I - T_o)/H\} +$$

$$+ (1 - \varepsilon)\, \{ m_{fu,II} - c(T_{II} - T_o)/H\} \qquad . (16.2\text{-}5)$$

But both the curly brackets contain contents with value zero, from equation (16.2-1); so the left-hand side must also equal zero, no matter what is the value of the mixture ratio ε. Therefore $m_{fu,\varepsilon}$ and T_ε also obey equation (16.2-1).

c) HEAT TRANSFER, FLUID MECHANICS, PHASE CHANGE

Once again, though important in the general case, these processes are without influence in the ideal stirred reactor.

16.3 MATHEMATICAL MODEL

A) DESCRIPTION

(i) The reaction

Once again, attention is restricted to a model reaction in which fuel and oxygen react together in a single bi-molecular collision-controlled step, characterised by two empirically determined constants K and E, which appear in the reaction-rate equation:

$$R_{fu} = K\, p^2\, m_{fu} m_{ox}\, T^{-3/2} \exp(-E/\mathcal{R}T) \qquad . (16.3\text{-}1)$$

Introduction of equations (16.2-1) and (16.2-2) into this relation allows the reaction rate R_{fu} to be expressed as a function of temperature alone as before. As explained in earlier chapters, this function has the qualitative shape shown in the sketch. Note that the maximum lies nearer to T_1 than T_o.

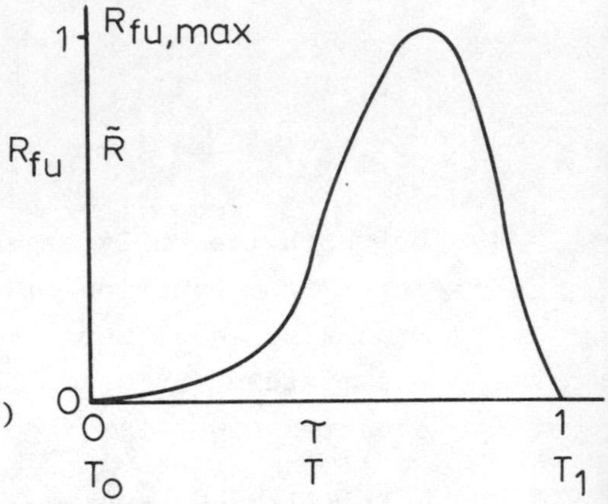

The reaction rate can be expressed conveniently in terms of the maximum value, $R_{fu,max}$, and the non-dimensional temperature rise, τ, defined by :

$$\tau \equiv (T-T_o)/(T_1-T_o) \quad . \quad (16.3\text{-}2)$$

If the non-dimensional reaction-rate $\tilde{R}$ is defined by:

$$\tilde{R} \equiv (R_{fu}/R_{fu,max}) \quad , \quad (16.3\text{-}3)$$

the $\tilde{R} \sim \tau$ relation is as shown in the sketch.

Note that, in defining $\tilde{R}$, R_{fu} is divided by the maximum value which that quantity can have for the given values of: pressure p, inlet temperature T_o, and inlet concentrations of fuel and oxygen, $m_{fu,o}$ and $m_{ox,o}$.

(ii) The flow conditions

The unburned gas flows steadily, at rate $\dot{m}$ (kg/s) into the reactor of volume $V(m^3)$. The turbulence level is so high that the same temperature and concentration values prevail throughout the reactor; and consequently they prevail in the outlet stream also.

The walls of the reactor are adiabatic, and non-catalytic; they therefore do no more than constrain the flow to stay in a prescribed volume.

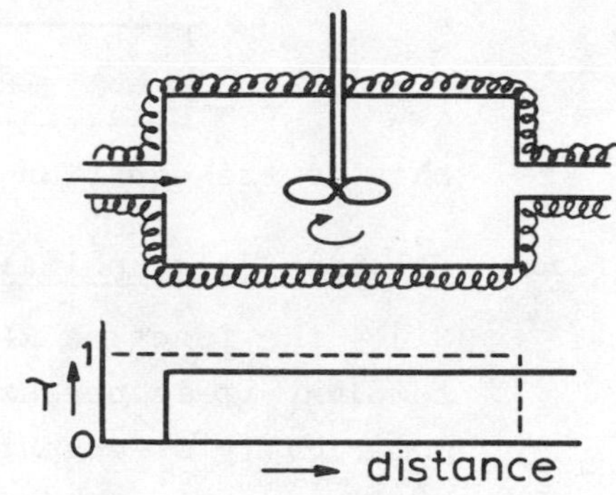

These conditions can be represented symbolically by the sketch.

(Remark: If it seems implausible that there should be a discontinuity of temperature in the gases at the inlet plane, but none at the outlet, one may imagine oneself sitting in a full bath with the hot tap running, and the water overflowing to waste. Stir vigorously. What is the temperature of the overflow? That in the bath? Or that of the hot tap ?)

B) ANALYSIS

(i) Equations

Reaction rate: $R_{fu} = R_{fu,max}\,\tilde{R}(\tau)$; (16.3-3)

Fuel balance: $\dot{m}\,(m_{fu,o} - m_{fu}) = R_{fu}\,V$; (16.3-4)

Steady-flow energy equation (or (16.2-1)):

$(m_{fu,o} - m_{fu}) = \tau\, c(T_1 - T_o)/H$. (16.3-5)

(ii) Expression in non-dimensional form

Define L by:

$$L \equiv \frac{\dot{m}\, c(T_1 - T_o)}{H\, R_{fu,max}\, V} \qquad . \; (16.3\text{-}6)$$

L is sometimes known as the "chemical loading" of the reactor. It is non-dimensional. Note that the numerator has the dimensions of heat per unit time; and so does the denominator.

Insertion in (16.3-3, 4 and 5), and elimination of all dimensional quantities, leads to:

$$\boxed{L\,\tau = \tilde{R}(\tau)} \quad ; \ (16.3\text{-}7)$$

this is the equation that has to be solved.

(iii) Solution by graphical means

Since the function $\tilde{R}(\tau)$ is algebraically rather complex, it is necessary to solve (16.3-7) either numerically or graphically. Here the latter means is used.

The sketch shows the right-hand side as a single curve, the left-hand side as a set of straight lines through the origin, with L as parameter. L is indeed proportional to the slope of the line. Solutions to (16.3-7) are represented on the sketch as the intersection of a straight line with the curve.

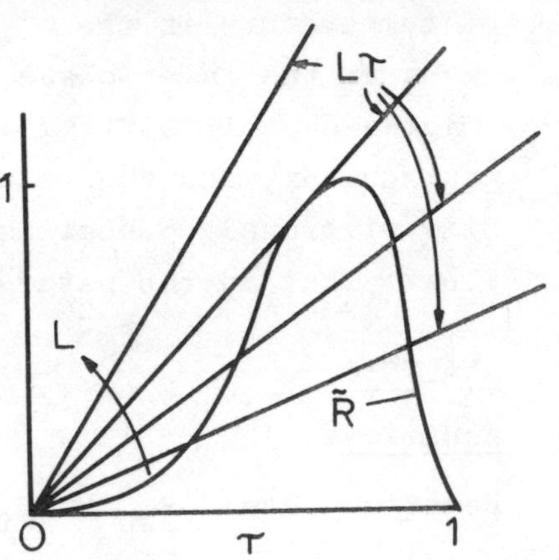

c) DISCUSSION

(i) The number of solutions

In general, there are three intersections of a straight line with the reaction-rate curve; there are therefore three solutions to (16.3-7), for a fixed value of the loading L.

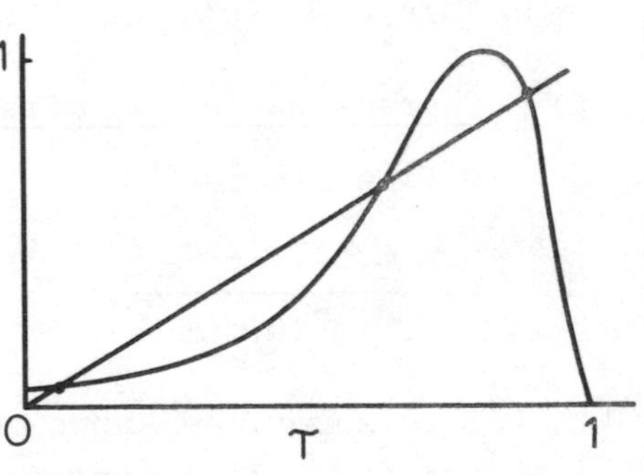

The left-hand solution lies very close to the origin, because $\tilde{R}$ is extremely small where $\tau = 0$. In this case, the gas leaving the reactor is in practically the same (unburned) state as that which enters it.

The right-hand solution lies between $\tau = 1$ (the thermodynamic equilibrium) and a value of τ which is very slightly less than that which corresponds to the maximum reaction rate ($\tilde{R} = 1$); the gases are hot.

However, if the slope of the straight line is too great (i.e. $L > L_{critical}$ in sketch), neither the right-hand nor the middle solution is real; so no hot-gas solution is possible.

It follows that the gases in the reactor can burn only if the chemical loading of the reactor is less than or equal to a critical value; if the reactor is running at a lower L than this, and the "hot" (i.e. right-hand) solution prevails, a sufficiently great increase in the mass flow rate, i.e. in L, eliminates the possibility of flame. This is the explanation of the blow-out (extinction) phenomenon which is observed experimentally.

(ii) Stability of the solutions

For $L < L_{crit.}$, three solutions are possible, according to a steady-state analysis; but are they all stable to small disturbances from the condition of balance? The answer is that the two outer ones ("cold" and "hot") are stable; the middle one is not. The latter point is proved as follows; and the former can be proved in the same way.

Suppose the reactor is operating in the middle-solution condition; then suppose that a momentary interruption in the fuel supply causes the reactor

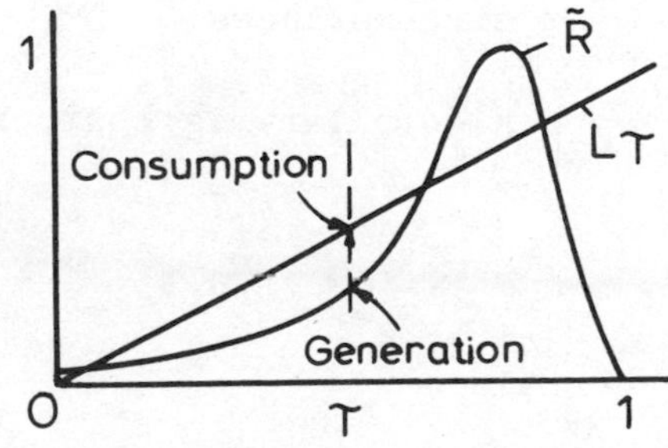

temperature to fall somewhat. The result is, as the sketch shows, that the rate of reaction ("heat generation") falls off more rapidly than the left-hand side of equation (16.3-7) ("heat consumption"). Therefore the reactor temperature falls still more; the decline is not stopped until the left-hand ("cold") solution is attained.

Similarly, if a momentary increase in the fuel supply raises the reactor just above the middle-solution one, the "heat-generation" term outweighs the "heat-consumption" one. Then the temperature is bound to rise further until the right-hand ("hot") solution is attained.

Even though two solutions are stable, they cannot both exist simultaneously. Which is adopted by the reactor depends on the starting conditions. If cold reactants are supplied to a cold reactor, no flame will form until an ignition source is provided. These unsteady-state phenomena cannot of course be handled by the steady-state equations.

(iii) The loading at extinction

The critical value of the chemical loading L, $L_{crit.}$, intersects the $R(\tau)$ curve just to the left of the maximum. Since τ there is rather less than unity, and the maximum R equals unity, L_{crit} is somewhat above 1.0. One may take :

$$L_{crit.} \approx 1.3 \qquad , \text{(16.3-8}$$

as fairly close to the truth for many practical circumstances.

A corollary is that, at the point of extinction, the

value of τ, which can also be regarded as the combustion efficiency of the reactor, is a little below 0.8. The variation of combustion efficiency with loading can indeed be deduced from the foregoing analysis to obey the relation shown in the sketch.

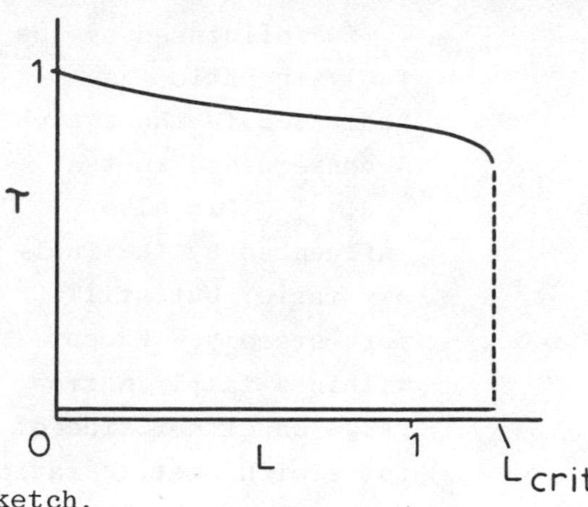

16.4 PRACTICAL IMPLICATIONS

A) WHAT INFLUENCES EXTINCTION

(i) Equation (16.3-8) in physical terms

Reference to the definitions implies:

$$\text{at extinction: } \dot{m} \approx 1.3 \frac{H \; R_{fu,max} \; V}{c(T_1 - T_o)} \quad . \quad (16.4\text{-}1)$$

The main implications are that flames may be blown out by:- increasing the mass flow rate; diminishing the size of the reactor; or diminishing the value of $R_{fu,max}$.

(ii) What influences $R_{fu,max}$

Because the maximum value of R occurs at around $\tau = 0.8$, as stated above, $R_{fu,max}$ can be deduced from :

$$R_{fu,max} \approx Kp^2 \left\{ m_{fu,o} - \frac{0.8c(T_1-T_o)}{H} \right\} \left\{ m_{ax,o} - \frac{0.8c(T_1-T_o)}{H/c} \right\}$$

$$\exp\left\{ \frac{-E/\mathcal{R}}{T_o+0.8(T_1-T_o)} \right\} \times \left\{ T_o + 0.8(T_1-T_o) \right\}^{-3/2} \quad . \quad (16.4\text{-}2)$$

This implies that the main factors influencing ($R_{fu,max}$) are :

- pressure p ($R_{fu,max} \propto p^2$),
- adiabatic flame temperature T_1 (very steep because of the Arrhenius term).

T_1 is influenced by the fuel/air ratio, as indicated in the sketch. A consequence is that $R_{fu,max}$ is also influenced by the fuel-air ratio, but still more steeply. Except within a fairly narrow range on either side of the stoichiometric ratio, the maximum reaction rate is very low indeed.

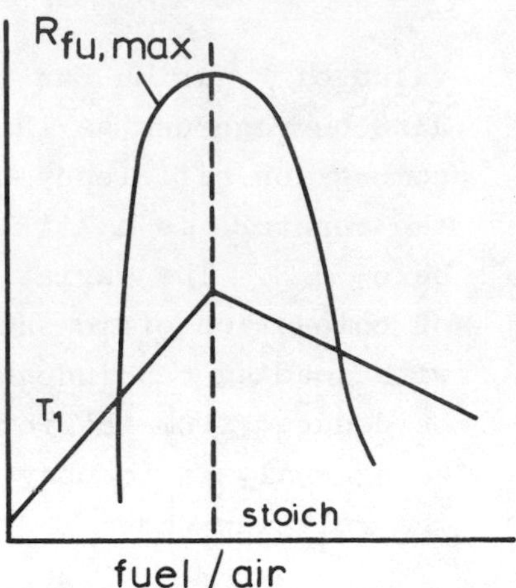

(iii) The influence of pressure

The value for fixed m_{fu}, m_{ox}, T at inlet is :

$$\frac{\dot{m}_{crit.}}{V} \propto p^2 \qquad . (16.4\text{-}3)$$

Therefore, since $\dot{m}$ is proportional to ρ V A, where $\rho \equiv$ density, V $\equiv$ velocity in inlet duct, A $\equiv$ cross-sectional area of inlet duct, and since $\rho \propto p$:

$$V_{crit.} \frac{A}{V} \propto p \qquad . (16.4\text{-}4)$$

So the blow-out velocity is proportional to the first power of the pressure.

(iv) The influence of the surface-to-volume ratio

Equation (16.4-4) shows that, if the linear size of the reactor increases, at fixed pressure and inlet velocity, the blow-out velocity rises linearly in proportion to V/A, i.e. to the linear dimension. Big flames are harder to blow out than small ones.

B) COMPARISON WITH EXPERIMENT

(i) The Longwell-Weiss data

The experimental results reported earlier become fully understandable in the light of the mathematical model. A deviation lies in the pressure exponent: 1.8 in the experiments, 2 in the model. The explanation may lie in experimental inaccuracy; but it is actually probable that the reaction is not wholly collision-controlled.

Of course, the data go further than the analysis. They allow values to be ascribed to K and E. Longwell and Weiss recommended 1.75×10^8 J/kg mole for the latter.

It is of great practical interest that the available fuels differ little from each other in maximum reaction rate.

(ii) Gas-turbine combustion chambers

Only qualitative agreement can be noted at present, because of the absence of a theory for practical reactors. Practical experience at high altitudes, and with small combustion chambers, is now easy to understand; and changing fuels is found indeed to have little influence on the efficiency or stability of flames in practice as well as in the "stirred" reactor.

Even though there is no complete theory for the practical combustion chamber, it should be expected that the chemical loading L, because it is a dimensionless group, would come into such a theory. Perhaps a real reactor differs from a fully-stirred one only in exhibiting extinction at a value of L which depends on the geometry of the chamber. Experience confirms this reasonable expectation.

(iii) Other experimental data

The stirred reactor has been employed, since its invention

by Longwell et al, by many investigators of combustion chemistry. Its advantage, as compared with other reactor systems (e.g. the Mullins spontaneous-ignition apparatus, or the laminar propagating flame), is that only <u>algebraic</u> equations are involved in its analysis, even when multiple chemical reactions are employed; all other reacting systems involve differential equations. Of course, practical realisations of stirred reactors exhibit non-uniformities of composition even there; and these reintroduce the need for differential equations. However such effects are still secondary.

Recent publications on stirred reactors include those of:- Miles (1964); Jain and Spalding (1966); Jenkins, Yumlu and Spalding (1967); Jones and Prothero (1968); Evangelista, Shinnar and Katz (1969); Bowman, Pratt and Crowe (1973); Pratt and Bowman (1973).

16.5 REFERENCES

1. BOWMAN B R , PRATT D T & CROWE C T (1973)
"Effects of turbulent mixing and chemical kinetics on nitric oxide production in a jet-stirred reactor".
Fourteenth Symposium (International) on Combustion, The Combustion Institute, pp 819 - 830.

2. EVANGELISTA, J J, SHINNAR R & KATZ S K (1969)
"The effect of imperfect mixing on stirred combustion reactors".
Twelfth Symposium (International) on Combustion, The Combustion Institute, pp 901 - 912.

3. LONGWELL J P, FROST E E & WEISS M A (1953)
"Flame stability of bluff-body recirculation zones".
Industrial and Engineering Chemistry, Vol. 45, p 1629.

4. LONGWELL J P & WEISS M A (1955)
"Heat release rates in hydrocarbon combustion".
I Mech E/ASME Joint Conference on Combustion, pp 334-340.

5. JAIN V K & SPALDING D B (1966)
"The effects of finite recirculation in a stirred reactor".
Combustion and Flame, Vol. 10, pp 37 - 43.

6. JENKINS D R, YUMLU V S & SPALDING D B (1967)
"Combustion of hydrogen and oxygen in a steady-flow adiabatic stirred reactor".
Eleventh Symposium (International) on Combustion, 779, The Combustion Institute.

7. JONES A & PROTHERO A (1968)
"The solution of the steady-state equations for an adiabatic stirred reactor".
Combustion and Flame II, No. 3.

8. MILES G A (1964)
"Performance analysis of the exothermic, adiabatic well-stirred reactor".
Ph.D. Thesis, Massachusetts Institute of Technology.

9. PRATT D T & BOWMAN B R (1973)
"PSR - A computer program for calculation of combustion reaction kinetics in a micromixed perfectly-stirred reactor".
Circular 43, Engineering Extension Service, Washington State University.

EXERCISES TO FACILITATE ABSORPTION OF MATERIAL IN CHAPTER 16

MULTIPLE-CHOICE PROBLEMS

16.1 The efficiency of a gas-turbine combustion chamber tends to diminish at high altitude, and the flame may even be extinguished,
mainly because
the low rate of flow of fuel through the atomiser causes the droplets to become rather large.

16.2 It was once believed that spontaneous-ignition experiments would permit the quantitative characterisation of fuels for gas turbines,
partly because,
additives such as tetraethyl lead were found to have a significant effect on the efficiency and stability limits of gas-turbine combustion chambers.

16.3 The flame-stabilisation process in a gas turbine differs qualitatively from that in a steady-flow spontaneous-ignition rig,
because
in the former process, the incoming fresh gases undergo their initial temperature rise as a result of mixture with recirculating burned gas, and are not left to heat up through their own reaction.

16.4 Longwell, Frost and Weiss employed a water-cooled steel sphere as their reaction vessel,
because
a refractory-lined sphere would have been too quickly damaged by the hot gases.

16.5 If the rate of fuel supply to a flame-filled "Longwell bomb" is steadily increased, without change of air-supply rate, extinction is bound to occur,
because
below an equivalence ratio of 0.4 no flame is possible.

16.6 If a stoichiometric propane-air mixture were supplied to a "stirred reactor" ("Longwell bomb") at the rate of 10^3 kg/s per m^3 of reactor volume, and the inlet temperature and the pressure were 400 ^{0}K and 1 atm respectively, the flame would be extinguished,

because

the maximum value of $N/Vp^{1.8}$ is about 50 kg mole air per m^3 per second.

16.7 If a mixture of kerosine vapour and air, in the ratio 1 to 60 by mass, were supplied to a stirred reactor at 400 ^{0}K and 2 atm pressure, it would be impossible to sustain a flame at any rate of flow,

because

the stoichiometric ratio is 1 to 15, and no appreciable reaction rate can be sustained at an equivalence ratio of 0.25.

16.8 A high-octane gasoline is likely to exhibit a comparatively low value of $N/Vp^{1.8}$ for extinction in a stirred reactor,

because

experience with spark-ignition engines shows such fuels to be relatively unreactive.

16.9 Spherical combustion chambers are not used in gas turbines,

because

they have a large volume/surface ratio.

16.10 The oxygen concentration is connected with the gas temperature by: $m_{ox} = m_{ox,o} + c(T-T_o)s/H$, in the model employed in the text,

because

the reaction proceeds in a single step, and the specific heats of all the mixture components are equal and independent of temperature.

16.11 The plot of R_{fu} versus T exhibits a maximum as T

approaches T_1,

<u>because</u>

the $T^{-3/2}$ term eventually diminishes the reaction rate faster than exp{-E/(T)} can increase it.

16.12 The gases inside a Longwell-Weiss stirred reactor are very nearly uniform in temperature and concentration,

<u>mainly because</u>

the incoming fuel and air are injected through a large number of holes.

16.13 If the stirred reactor did not have adiabatic walls, it would still be possible for m_{fu} and T to be connected quantitatively by thermodynamics alone provided that the magnitude of the heat loss were known,

<u>because</u>

the enthalpy of the gases in the reactor would differ by a known amount from that in the supply pipe.

16.14 The fuel concentration in the gases flowing out of the reactor is always zero, for equivalence ratios below unity,

<u>because</u>

it is impossible for fuel to pass through a flame, where oxygen is present in excess, without being completely consumed.

16.15 The value of the chemical loading at extinction is lowest for reactions with low activation energy,

<u>because</u>

these have $\tilde{R} \sim \tau$ curves with maxima lying nearest to $\tau = 0$.

16.16 The high-temperature intersection of the straight line with the $\tilde{R} \sim \tau$ curve represents a stable equilibrium,

<u>because</u>

if the temperature were to rise momentarily, the increase

in reaction rate would cause a still greater rise.

16.18 Stoichiometric mixtures have the greatest values of $R_{fu,max}$,
<u>chiefly because</u>
they have the highest values of T_1.

16.19 Longwell's finding that the mass flow rate at extinction is proportional to $(\text{pressure})^{1.8}$ cannot be regarded as valid,
<u>because</u>
theory proves that the exponent must be 2 for a collision-controlled reaction.

16.20 If a gas-turbine combustion chamber is replaced by a set of smaller, geometrically-similar combustion chambers having the same frontal area, performance at high altitude is likely to deteriorate,
<u>because</u>
the chemical loading will have been increased.

16.21 The maximum "volumetric heat release rate" of a hydrocarbon-air mixture is about 10^6 k J/m^3s at 1 atm and 400 ^0K inlet temperature,
<u>because</u>
for a typical hydrocarbon 1 kg of fuel combines with about ½ moles of air, the maximum value of $N/Vp^{1.8}$ is about 20 kg mole air/s m^3 $(atm)^{1.8}$, and the calorific value of the fuel is around 10^4 kJ/kg.

16.22 The combustion chamber of a gas turbine usually consists of an inner "flame tube" into which fuel and air are mixed roughly in the ratio 1:15, and an outer annulus carrying the remainder of the air (about 45 times the amount of fuel),
<u>because</u>
fuel and air will burn at an appreciable rate only when the mixture ratio is fairly close to stoichiometric.

ANSWERS TO MULTIPLE-CHOICE PROBLEMS

Answer	Problem number (16's omitted)
A	3, 7, 13, 18, 20, 22
B	1, 5, 9, 12
C	2, 11, 16, 21
D	6, 8, 10, 15, 19
E	4, 14, 17

CHAPTER 17

FLAME STABILISATION BY BLUFF BODIES

17.1 ENGINEERING RELEVANCE OF THE PHENOMENON

A) WHAT IS MEANT BY THE TITLE

(i) A "bluff body" is a solid object of such a shape that, when it is suspended in a fluid stream, a re-circulatory flow is formed in its immediate wake. Examples are:- a sphere; a circular-sectioned cylinder with its axis at right angles to the direction of flow; a cone with its axis parallel to the direction of flow; a V-sectioned "gutter" with its edges at right angles to the flow direction; etc.

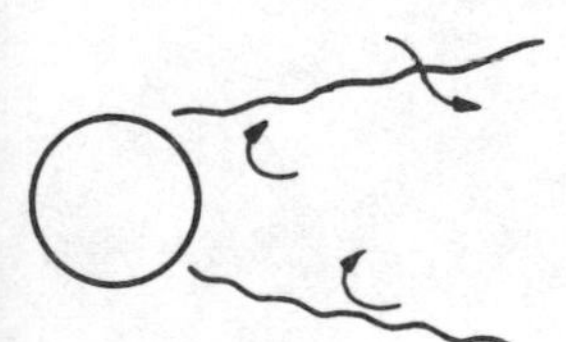
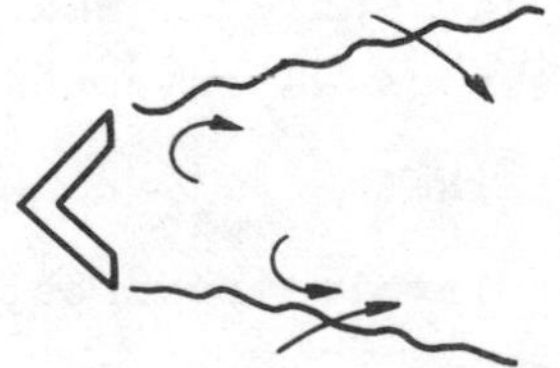
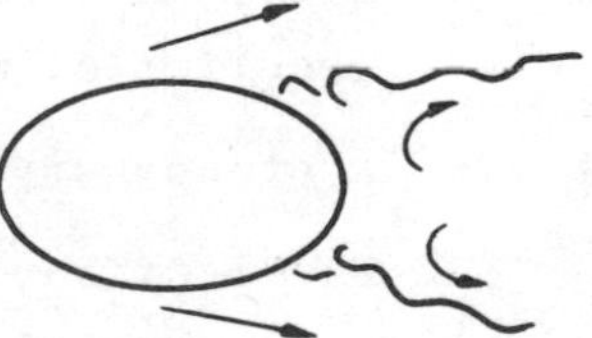

FIG. 17.1 RECIRCULATING FLOW BEHIND SPHERE, GUTTER AND STRUT.

(ii) If the fluid stream is a gas capable of sustaining an exothermic reaction, moving at an average velocity greatly in excess of the speed of laminar flame propagation of the mixture, a steady flame can exist only if a flame-stabilising device is provided in the stream. The most common of these is the "bluff body".

The essential feature is the recirculation found in the immediate wake of the body; this carries upstream, and mixes with fresh, unburned gas, the products of combustion of fuel and oxidant which were injected earlier.

Another name for "bluff-body flame-stabiliser" is "baffle". "Flame-holder" is also common.

(iii) Other flame-stabilising devices may be used. These include:-

- a jet of hot gas injected into the stream in the direction of flow;
- a jet of gas at the same temperature as the on-coming stream, at such an angle to it that re-circulation is provoked;
- a stream-lined surface, heated by external means to a high temperature;
- a stream-lined catalytic surface, heated by the reaction of main-stream gases in contact with it;
- a steady stream of sparks.

The last three do not involve recirculation.

Some of these devices are sketched below.

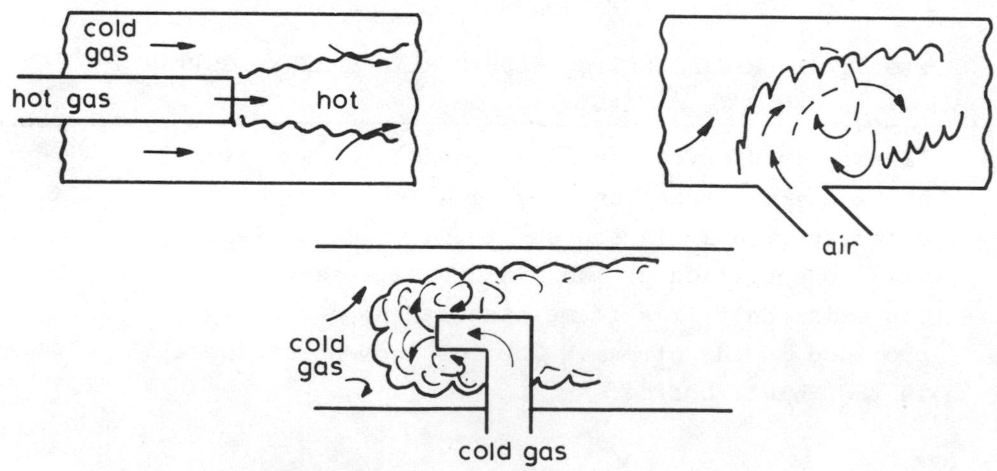

FIG. 17.2 FLAME STABILISATION BY JET INJECTION

(iv) If the whole fluid stream is combustible, the stabilisation region is normally followed by a propagation region, in which the flame spreads from the bluff-body stabiliser to the wall. If the fuel-air ratio of the stream is non-uniform, this propagation may be of only limited extent. Fig. 17.3 illustrates this.

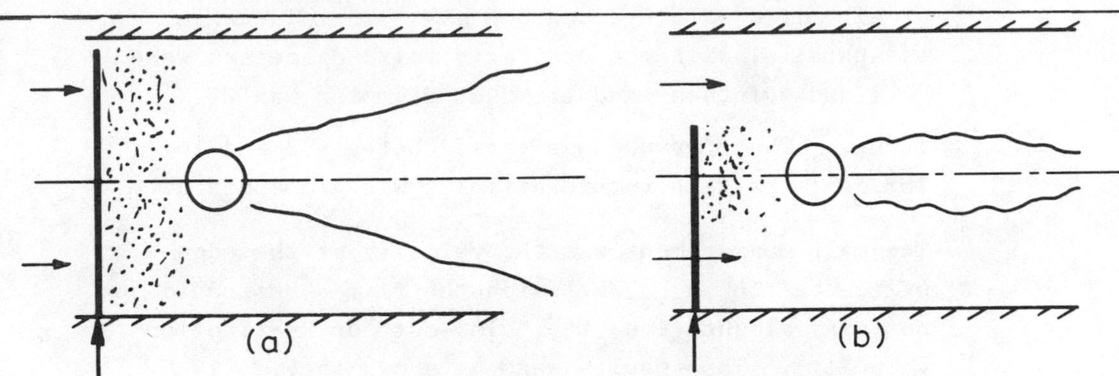

FIG. 17.3 FLAME PROPAGATION DOWNSTREAM OF A BAFFLE, (a) FOR A UNIFORM FUEL-AIR MIXTURE, (b) WHEN THE FUEL IS CONFINED TO A CENTRAL REGION.

B) WHERE BLUFF-BODY FLAME STABILISATION IS USED

(i) The "after-burner" (also called "reheat") systems of aircraft gas turbines employ flame stabilisation by baffles.

Similar devices are to be found in ram-jet engines.

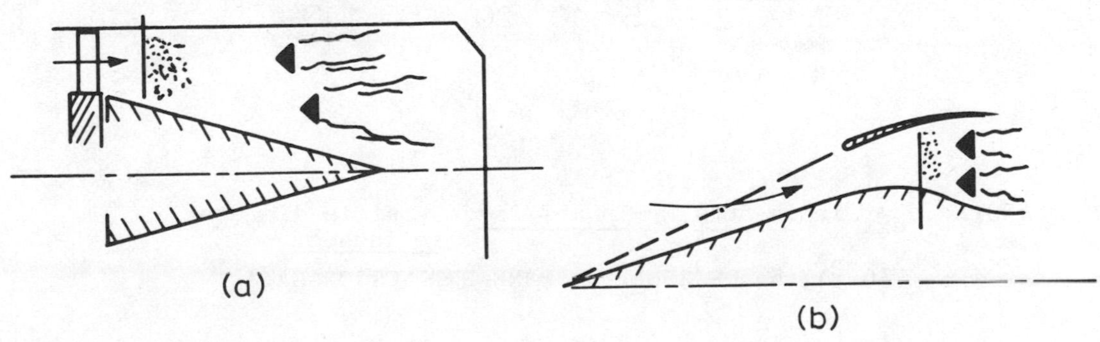

FIG. 17.4 (a) "AFTER-BURNER"; (b) "RAM-JET" SYSTEMS

(ii) In one sense, the main combustion chamber of a gas turbine employs flame stabilisation by a bluff body; but in this case the baffle area is much greater than the free area of the duct.

(c) SOME EXPERIMENTAL DATA

(i) De Zubay (1950) suspended discs in steady streams of air mixed with hydrocarbon gas. The disc axis was parallel with the duct axis; disc diameters were $\frac{1}{4}$, $\frac{1}{2}$ and 1 inches, and the duct diameter was $2\frac{3}{4}$ inches. The pressure was varied between 3 and 15 lbf/in^2; the temperature at inlet was atmospheric.

The main measurement was the velocity at the edge of the baffle, u_{ext}, at which the flame just could not stay alight (i.e. the "blow-out" or "extinction" velocity). The results were as shown in Fig. 17.5.

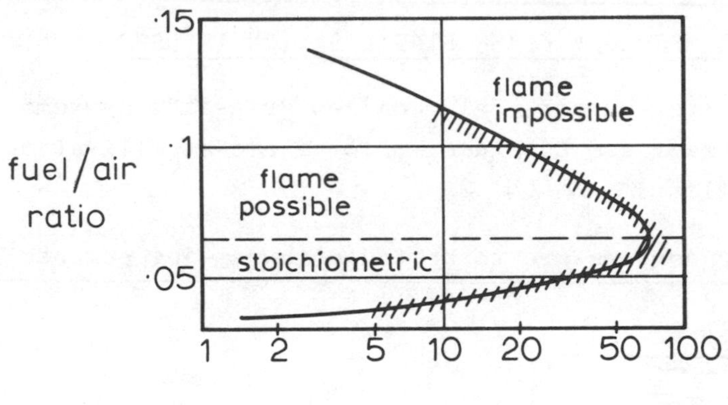

Shading indicates region occupied by experimental points

NOTE: u_{ext} is in ft/s, p in lbf/in^2, D(baffle diameter) in inches.

FIG. 17.5 DE ZUBAY's FLAME-EXTINCTION DATA

Thus, large diameter, high pressure, and a mixture ratio near stoichiometric gave the highest extinction velocity. At 1 atm pressure, with the 1/4 inch diameter baffle, the extinction velocity was 300 ft/s; at the same pressure and D = 1 inch, u_{ext} was 970 ft/s.

(ii) Barrère and Mestre (1954) performed similar blow-out tests on cylindrical flame-holders of various cross-sectional shape and size, suspended in a duct of rectangular cross-section. The pressure was atmospheric; the fuel was propane, mixed with air; and the upstream mixture temperature was 290^0K.

In the experiments which gave the results shown in the sketch, the baffle width was 5mm and the width of the duct was 30mm.

Baffle 1 was a flat strip with its broad side normal to the stream. Baffle 2 was a 90^0 angle, with its apex upstream. Baffle 3 was a circular cylinder.

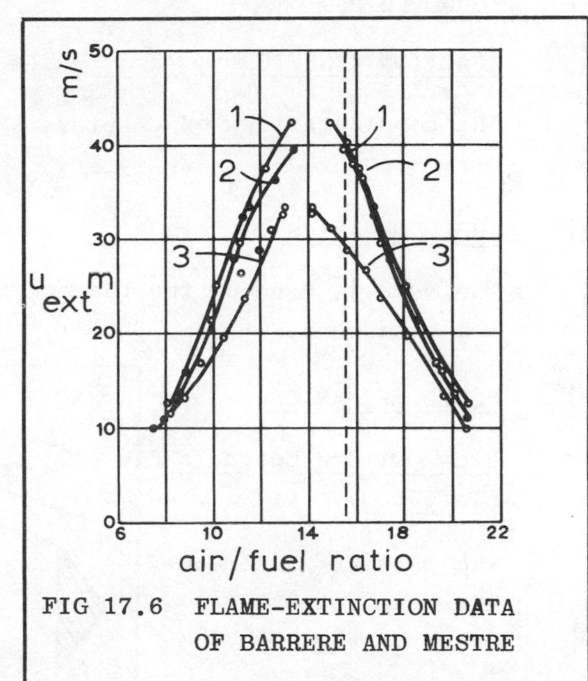

FIG 17.6 FLAME-EXTINCTION DATA OF BARRERE AND MESTRE

Fig. 17.6 shows that Baffle 1 gives higher extinction velocities than Baffle 3. The differences between Baffles 1 and 2 are probably not significant in view of the scatter in the experimental data.

The diagram also shows that, in this case, the maximum values of u_{ext} lie on the rich side of the stoichiometric mixture ratio. This behaviour is often exhibited by high-molecular-weight fuels and small-width baffles; the explanation is given in Section 17.4(b)(ii) below.

(D) THE PRESENT TASK

The purposes of the following discussion are:- to understand the above findings in the light of chemical kinetics; to predict what value u_{ext} will attain in other circumstances; to consider whether there are means whereby u_{ext} can be significantly increased, without serious penalty.

17.2 FUNDAMENTALS

(A) REACTION KINETICS

The considerations of Chapters 14 and 15 apply here also.

(B) THERMODYNAMICS

The same is true of the thermodynamic aspects of the phenomena.

(C) FLUID MECHANICS

(i) Flow pattern behind a bluff body

The pattern of stream-lines is roughly as shown in Fig. 17.7. The reverse-flow region is usually between two and five body widths long. The fluid within the closed stream lines is highly turbulent at the velocities which are encountered in flame-stabilisation practice.

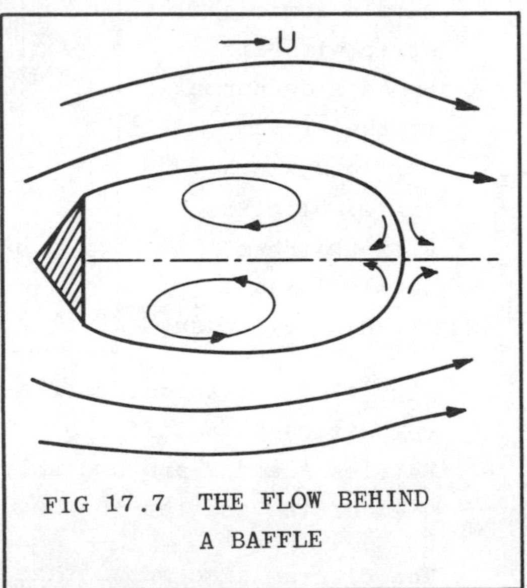

FIG 17.7 THE FLOW BEHIND A BAFFLE

(ii) The drag coefficient of a bluff body

This quantity, C_D, is defined by:

$$C_D \equiv F/(\tfrac{1}{2}\rho u^2 A) \quad ,(17.2\text{-}1)$$

where $F \equiv$ force on body, $A \equiv$ cross-sectional area of body, $\rho \equiv$ fluid density upstream, $u \equiv$ fluid velocity upstream; it usually has a value of about unity, at Reynolds numbers greater than 10^4. The exact value depends on the shape of the body.

Reynolds number must here be defined as $\rho u D/\mu$, where D is the baffle width, and μ is the viscosity of the approach-stream gases.

(iii) Mean shear stress in the mixing layers enclosing the recirculating flow

A momentum balance on the control volume of Fig. 17.8 gives, for a supposed uniform pressure:

$$C_D \cdot \frac{1}{2}\rho u^2 D \approx S \ell j \quad , (17.2\text{-}2)$$

where $j = 2$ for a plane flow, 4 for an axi-symmetrical flow, and $S \equiv$ shear stress.

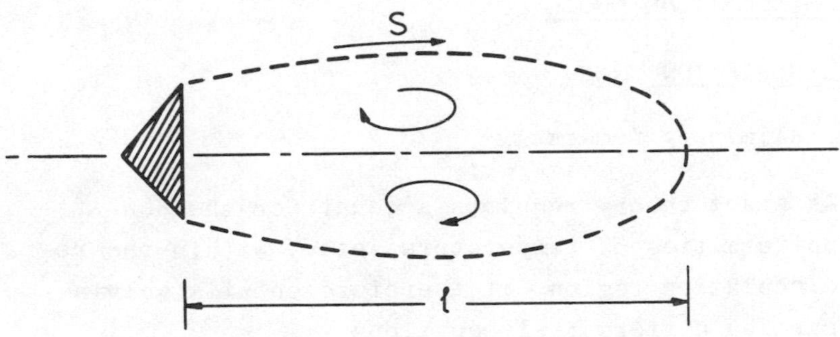

FIG. 17.8 CONTROL VOLUME FOR FORCE BALANCE

Hence,

$$\frac{S}{u} = \frac{C_D}{2j} \cdot \frac{D}{\ell}\rho u \quad .(17.2\text{-}3)$$

(iv) Reynolds Analogy

It is reasonable to suppose that the Reynolds Analogy between momentum and heat transfer is approximately valid for a turbulent mixing layer. So the heat-transfer process across the mixing layer, which may be characterised by a heat-transfer coefficient α, will approximately obey the relation:

$$\frac{\alpha}{c} = \frac{S}{u} = \frac{C_D}{2j} \cdot \frac{D}{\ell} \rho u \qquad ,(17.2\text{-}4)$$

where c stands for the specific heat of the gas.

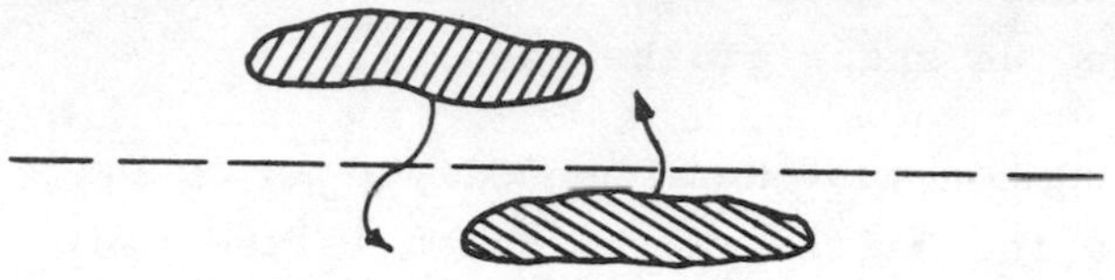

exchange rate g kg/m^2S

FIG. 17.9 TURBULENT EXCHANGE ACROSS CONTROL-VOLUME BOUNDARY

Mass transfer will be similarly related; the whole turbulent-mixing-layer process can thus be treated as a mass exchange across the closed stream-line at a rate per unit area of g, where:

$$g = \frac{\alpha}{c} = \frac{S}{u} = \frac{C_D}{2j} \frac{D}{\ell} \rho u \qquad .(17.2\text{-}5)$$

17.3 MATHEMATICAL MODEL

(A) DESCRIPTION

(i) Preliminary remark

An exact theory requires account for the non-uniformities of temperature, etc., within the re-circulation region; it therefore entails solving partial differential equations.

Instead, for the sake of ease, only order-of-magnitude predictions will be made here.

(ii) Flow pattern

The region enclosed by the closed stream line which springs from the point of separation on the bluff body will be treated as a region in which turbulence is so intense that the temperature and concentration are uniform. Its volume is approximately $D\ell$ per unit length, for a plane flow, and $\pi D^2 \ell/4$ for an axi-symmetrical flow.

The surface area of the region is 2ℓ per unit length in the first case, and $\pi D\ell$ in the second.

(ii) Exchange between the recirculation region and the main stream

The heat-transfer rate per unit area will be: $\alpha(T_r - T_u)$, where T_r is the temperature of the recirculation region, and T_u is the temperature of the surrounding, unburned, stream.

Similarly, the exchange rate of fuel, from the stream to the recirculation region will be, per unit area: $g(m_{fu,u} - m_{fu,r})$.

Heat transfer to the baffle, and by radiation to the surroundings will be supposed absent.

(iii) The reaction

It will be presumed that the reaction is the same single-step collision-controlled model reaction as has been used above already; the rate can thus be represented by a function $\tilde{R}(\tau)$ as before.

(B) ANALYSIS

(i) Equations

The fuel balance equation is:

$$(m_{fu,u} - m_{fu,r})\, g\, j\, \ell = R_{fu,r}\, D\ell \qquad .(17.3\text{-}1)$$

This equates the supply of fresh fuel by turbulent exchange to the rate of consumption of fuel by chemical reaction.

Further, equation (17.2-5) connects the supply rate, with C_D, ρ, u, etc; and, from Chapter 16, equation

(16.3-3:

$$R_{fu}/R_{fu,max} = \tilde{R}(\tau) \qquad ,(17.3\text{-}2)$$

where τ is the reactedness. The latter quantity is related to m_{fu} and T by the relation expressing uniformity of enthalpy:

$$\begin{aligned} H(m_{fu,u} - m_{fu,r}) &= c(T_r - T_u) \quad , \\ &= c(T_b - T_u)\,\tau \end{aligned} \qquad .(17.3\text{-}3)$$

(ii) Reduction to a single equation

Straightforward manipulation now leads to:

$$\frac{C_D}{2} \cdot \frac{D}{\ell} \cdot \frac{\rho u\ c\ (T_b - T_u)}{H\ R_{fu,max}\ D} \cdot \tau = \tilde{R}(\tau) \qquad .(17.3\text{-}4)$$

It is now useful to define a "baffle loading", L_b, by:

$$L_b \equiv \frac{\rho u\ c\ (T_b - T_u)}{H\ R_{fu,max}\ D} \qquad .(17.3\text{-}5)$$

This has the significance of:

$$\frac{\text{fuel supply rate per unit stream area}}{\text{maximum volumetric consumption rate} \times \text{width}}.$$

Then equation (17.3-4) becomes:

$$\boxed{\left\{\frac{C_D}{2}\,\frac{D}{\ell}\,L_b\right\}\tau = \tilde{R}(\tau)} \qquad .(17.3\text{-}6)$$

(iii) Solution

Equation (17.3-6) is obviously similar to equation (16.3-7); the place of L in that equation is here taken by $\{(C_D/2)(D/\ell)\,L_b\}$. Therefore, everything that in Chapter 16 was said about L now applies to the quantity in curly brackets. In particular, it can be deduced that extinction occurs when L_b takes the value $L_{b,ext}$, given by:

$$\boxed{L_{b,ext} \approx 1.3\,\frac{2}{C_D}\,\frac{\ell}{D}} \qquad .(17.3\text{-}7)$$

(c) DISCUSSION

(i) The magnitude of $L_{b,ext}$

Since $C_D \approx 1$, and $\ell/D \approx 3$, it can be deduced that:

$$L_{b,ext} \approx 8 \qquad .(17.3\text{-}8)$$

Actually, larger values are to be expected because it is found that C_D is smaller, and ℓ/D larger, for a baffle with a flame behind it than for a uniform-density flow. One reason is that the shear stresses and mass-exchange rates are proportional to the gas density; and this is greatly diminished in the flame region. However, the analysis has contained several approximations; it is reasonable to expect only that $L_{b,ext}$ lies in the range from, say, 4 to 40. Experiment must provide more exact information.

(ii) The main implications of the result

The definition of L_b (9), together with the finding that the quantity has a definite value (which is not yet known precisely) at extinction, implies that the extinction velocity, u_{ext}:

is proportional to the baffle width D,

and to the peak reaction rate of the mixture.

The last implication suggests that u_{ext} will be greatest for the stoichiometric fuel-air ratio; further, since $\dot{m}''_{fu,max} \propto p^2$ and $\rho \propto p$, the extinction velocity will be proportional directly to the pressure.

Equation (11) implies that $L_{b,ext}$ (and so u_{ext}) will be greatest when ℓ/D is large, and C_D small. The first of these is likely to be caused by a baffle shape that causes the fluid to be flung away from the symmetry axis, as indicated in the sketch.

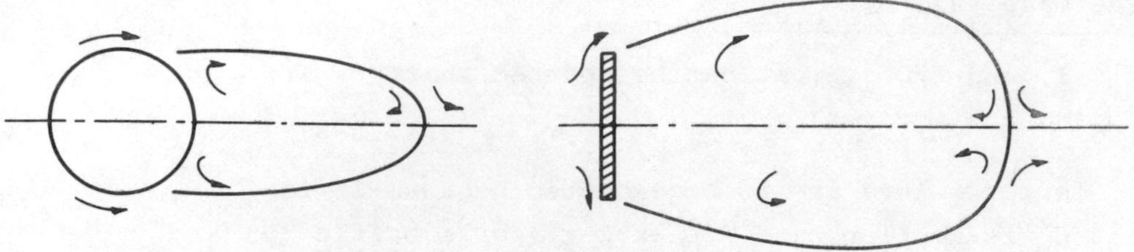

FIG. 17.10 INFLUENCE OF BAFFLE SHAPE ON LENGTH OF RECIRCULATION REGION.

17.4 PRACTICAL IMPLICATIONS

(A) COMPARISON WITH THE EXPERIMENTAL DATA OF DE ZUBAY

(i) De Zubay found that, at fixed fuel-air ratio, u_{ext} was proportional to $p^{0.95}$ and $D^{0.85}$.

The first result is in close agreement with the prediction for a collision-controlled reaction; the second is not far from the prediction and may be explained by the fact that the enlargement of D, in De Zubay's experiment, caused the presence of the duct wall, neglected in the present analysis, to have an appreciable effect.

(ii) A quantitative comparison can be made by the use of the Longwell-Weiss data for the quantity $R_{fu,max}$.

From Section 16.1, a stoichiometric mixture can burn air at the rate 58 moles/l s at 1 atm pressure. Therefore, $R_{fu,max} \approx \frac{58\times29}{15} = 1.12 \times 10^2 \text{kg/m}^3\text{s}$.

With $L_{b,ext}$ defined by (9), and given the value 10; and with $H = 10^4$ kcal/kg, $T_b - T_u = 2000$°K, $c = 0.25$ kcal/kg°K, $\rho = 1$ kg/m³, $D = \frac{1}{4}'' = 6.35 \times 10^{-3}$ m, there results:

$$u_{ext} = 143 \text{ m/s} = 470 \text{ ft/s}.$$

This is of the same order as, but greater than, the extinction velocity deduced directly from De Zubay's correlation in Section 17.1. The agreement is as good as can be expected in view of the uncertainty about C_D, ℓ/D, and the completeness of mixing in the wake region.

(iii) The effect of fuel-air ratio

The Longwell-Weiss data imply that the reaction rates, and so the extinction velocities, are very small when the equivalence ratio lies outside the range: $.4<\phi<2.0$. Since the stoichiometric fuel/air ratio is above .067, this means that u_{ext} will be very small for fuel/air ratios above .133 and below .0267. The data of De Zubay confirm this prediction fairly well.

(B) COMPARISON WITH THE DATA OF BARRÈRE AND MESTRE

(i) Order of magnitude of the blow-out velocity

Fig. 17.6 shows a peak velocity of about 45 m/s for a baffle width not much below that of De Zubay's smallest. So the change of geometry appears to have halved the value of u_{ext}; however, De Zubay reports the velocity in the annular gap around the baffle, not that upstream.

Because there is no precise information about how L/D and C_D change with the geometry, and because no reasonable estimate can be made of the effect of departures from uniformity of temperature and concentration in the recirculation region, order-of-magnitude agreement between predictions and experiment must suffice.

(ii) Influence of fuel-air ratio

Although the u_{ext} curves of Barrère and Mestre have a form similar to that of the Longwell-Weiss and De Zubay data, their peaks lie on the "rich" side of stoichiometric. This shift is usually found to disappear at high Reynolds number; it is a result of the low laminar diffusion coefficient of the large (slowly-moving) fuel molecules, which causes the mixture ratio in the wake region to be lower than that in the main stream. At high Reynolds numbers, the laminar transport effects are completely swamped by turbulent ones; but this had probably not yet occurred in the Barrère-Mestre experiments.

(iii) Influence of baffle shape

The largest value of u_{ext} is given by the strip, the lowest by the circular-section cylinder. This tendency is consistent with the theory, in so far as it can be expected that the p/D ratio of the recirculation region is greater for the former baffle than for the latter.

(C) PRACTICAL USE OF THE THEORY

(i) Negative aspects

It is of very great use that the flame-stabilisation process is seen as nothing but the interaction of high-temperature reaction kinetics with the aerodynamics of the recirculation. The following can therefore be excluded from consideration, as stabilisation-promoting devices:-

- the addition of small quantities of "dope" to the fuel (because these are found not to influence stirred-reactor behaviour);
- The supply of heat by sparks to the recirculation region at a rate which is small compared with the "thermal turnover" of that region (i.e. $H\ R_{fu,max} \times$ volume of region);
- the supply of additional fuel to the recirculation region when the mixture ratio there is already stoichiometric;
- other suggestions by well-intentioned but unenlightened inventors.

(ii) Positive uses; to improve design

Flame stability can be improved by limiting the rate of exchange of fluid between the recirculation region and the main stream; for example the baffle might be flanked by side-pieces for this purpose.

In one sense, the flame tube of the main combustion chamber of a gas turbine is just such a recirculation region with limited access of fresh gas; of course, the mixture ratio is there not uniform, so the need for "protection" of the flame is especially great.

For the flame stabiliser of an after-burner system, this "protection" is never practised; for, though it would increase stability, it would reduce the supply of hot gas from the recirculation region to cause propagation of flame through the remainder of the mixture.

(iii) Positive uses; to aid development (modelling)

High-altitude test facilities are expensive to construct and run. The recognition that extinction occurs at a definite value of a non-dimensional criterion (L_b) allows predictions of full-scale performance to be based on small-scale tests. Thus, since $R_{fu,max}\, D/\rho$ is proportional to pD for a bi-molecular reaction, it is common to adopt the "pD scaling rule"; this implies that, to simulate behaviour at 1/2 atm pressure, a 1/2 size model should be tested at 1 atm pressure.

This practice allows development to be carried on more swiftly and cheaply. By good fortune, it ensures Reynolds-number similarity also.

Another technique, which avoids the building of a scale model, is to modify $R_{fu,max}$ by diluting the mixture, for example with steam. It is necessary to determine the "rate-of-exchange" between pressure and steam, either by tests directly on a baffle stabiliser or by measurements with a "Longwell bomb".

17.5 REFERENCES

1. BARRÈRE M & MESTRE A (1954)

"Stabilisation des flammes par des obstacles".

In "Selected combustion problems", Butterworth's, London.

2. BRAGG S L & HOLLIDAY J B (1956)

"The influence of altitude operating conditions on combustion-chamber design".

Selected Combustion Problems, Vol. 2, Butterworth's, London, pp 270-295.

3. LEFEBVRE A H & HALLS G A (1959)

"Simulation of low combustion pressures by water injection".

Seventh Symposium (International) on Combustion, Butterworths, London, p 654.

4. De ZUBAY E A (1950)

"Characteristics of disk-controlled flame".

Aero. Digest, July.

EXERCISES TO FACILITATE ABSORPTION OF MATERIAL IN CHAPTER 17

MULTIPLE-CHOICE PROBLEMS

17.1 A bluff body makes a convenient flame stabiliser,

because

it creates recirculatory flow in its immediate wake which continuously mixes hot burned gas with fresh combustible mixture.

17.2 There is no need to use a flame stabiliser in the "reheat duct" of an aircraft jet engine,

because

the gases into which the fuel is injected, coming directly from the turbine, are hot enough for spontaneous ignition to be sufficiently rapid.

17.3 The experiments of de Zubay imply that the blow-out velocity increases in nearly direct proportion to the diameter of the baffle,

because

the fuel-air ratio for extinction was found to depend on the quantity $(u\ p^{-.95}\ D^{-.85})$, regardless of the fuel type or inlet temperature.

17.4 It is probable that, if Barrère and Mestre had operated their apparatus at pressures in excess of atmospheric, the range would have become narrower.

because

it is unlikely that the different shapes of their baffles would radically alter the behaviour of the flames.

17.5 The peak values of u_{ext} in the Barrère-Mestre experiments occurred with slightly rich mixture ratios, probably

because

the laminar diffusion coefficient of the fuel was appreciably higher than that of the other gases in the mixture.

17.6 The drag coefficient of a bluff body is usually fairly independent of Reynolds number, when this is high,

because

the point of breakaway of the boundary layer from the surface does not then vary with Reynolds number.

17.7 According to the Reynolds Analogy, the quantities g, α/c_p and S/u are numerically equal for a turbulent mixing process,

because

mass, heat and momentum are all exchanged by means of the transfer and inter-mingling of "parcels" of fluid.

17.8. The neglect of normal-pressure forces in the momentum balance leads to no error,

because

the pressure is uniform over the whole control surface.

17.9 The enthalpy in the immediate wake of a flame stabiliser exceeds that in the main stream,

because

the temperature is higher in the wake than in the stream.

17.10 If the Reynolds Analogy were not obeyed, and g were less than α/c_p for the boundary of the recirculation zone, the enthalpy there would be lower than that in the main stream,

because

the supply of fresh fuel would not be in balance with the outward heat transfer.

17.11 The "baffle loading" L_b contains ℓ rather than D in its definition,

because

the length of the recirculation zone is not usually a known quantity.

17.12 The "shear-stress-coefficient" for the boundary of the recirculation zone, $S/\rho u^2$, is of the order of 0.1,

because

laminar viscosity has little effect on the processes in the wake at high Reynolds numbers.

17.13 The value of $L_{b,ext}$ for a given baffle will be independent of the activation energy of the reaction in question,

because

L_b is a dimensionless quantity.

17.14 L_b must be expected to depend upon the Reynolds number, in general,

because

the collision rate between molecules and the viscosity of the gas are connected by a relation derivable from the kinetic theory of gases.

17.15 The blow-out velocity for a baffle is directly proportional to the pressure, according to the mathematical model,

because

the volumetric rate of a collision-controlled reaction is directly proportional to pressure.

17.16 Although the local fuel-air ratio in gas-turbine combustion chamber is not uniform, it is likely that the "pD scaling rule" will still apply provided that the overall fuel-air ratio is kept constant,

because

at high Reynolds numbers, with fine atomisation and negligible heat losses to the walls, there is no reason why the dimensionless loading should depend upon anything but the overall fuel-air ratio.

17.17 The model and full-scale phenomena are not exactly similar when pD scaling is observed,

because

the Reynolds numbers will differ.

17.18 It is likely, in the absence of other information that a combustible mixture which has a low ignition delay will have a high blow-out velocity on a baffle,

because

spontaneous ignition is dependent on the values of R_{fu} at low τ and baffle stabilisation on the values of R_{fu} at high τ.

17.19 An increase in the upstream temperature T_u is likely to increase the velocity of the stream at which the flame is extinguished, mainly

because
the definition of the loading has u in the numerator, and the density ρ is inversely proportional to the absolute temperature.

17.20 Even when a theory shows that there are very few ways open for the improvements of the relevant engineering equipment, its possession is still valuable,

because
it discourages waste of time and money in hopeful but misguided endeavour.

ANSWERS TO MULTIPLE-CHOICE PROBLEMS

Answer	Problem number (17's omitted)
A	1, 7, 10, 16, 20
B	6, 12, 14, 18, 19
C	3, 5, 15, 17,
D	4, 9, 11, 13
E	2, 8

CHAPTER 18

PROPAGATION OF A LAMINAR FLAME THROUGH A PRE-MIXED COMBUSTIBLE GAS

18.1 ENGINEERING RELEVANCE OF LAMINAR FLAME PROPAGATION

(A) WHERE PROPAGATION IS DESIRED

Many heating appliances burn gaseous fuel, pre-mixed with air, in laminar jets emerging from small-diameter orifices. These jets are simultaneously diffusion flames and pre-mixed flames, as shown in Fig. 18.1. Concern is here with the inner, pre-mixed flame; this is conical in shape, the sine of the angle being equal to the ratio of the velocity of the unburned gases to the speed of propagation.

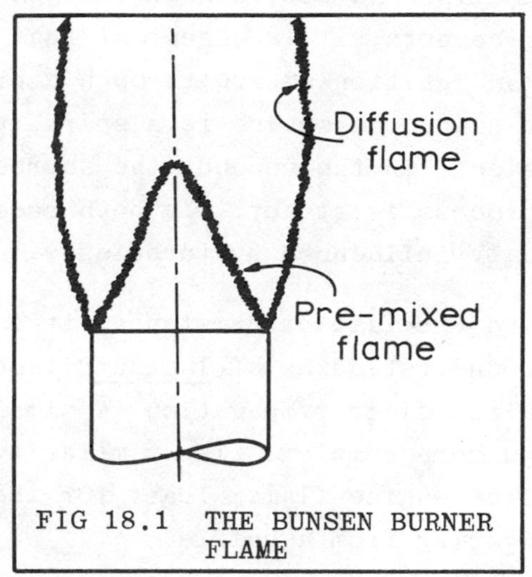

FIG 18.1 THE BUNSEN BURNER FLAME

When designing the burner, the engineer needs to be able to predict the shape of the flame; and for this he needs to know the flame speed.

(B) WHERE PROPAGATION MUST BE AVOIDED

Sometimes combustible mixtures must be conveyed in ducts without burning there. If there is a source of ignition downstream, the velocity in the vicinity of this source should be well above the flame-propagation speed; otherwise, the flame will "flash back", often with disastrous results. Flash-back is easy to produce with a Bunsen burner by operating it with a wide-open air hole and low gas flow; the first allows the mixture ratio to be near stoichiometric; the second allows the velocity in the burner tube to fall below the propagation speed.

(Strictly speaking, the flash-back phenomenon is rather more complex than the above discussion implies; for, near the walls of the tube, the flow velocity is zero; so there is always <u>some</u> part of the duct in which the flame-propagation speed exceeds the flow velocity. However, to take account of non-uniformities in velocity is beyond the present scope; the simple discussion at least explains the main facts adequately.)

(C) PROPAGATION PHENOMENA THAT ARE NOT LAMINAR

In gasoline engines, and in ram-jets and gas-turbine after-burners, it is essential that flame should propagate from an ignition source through a pre-mixed gas. In the first case, the source is a spark, and the process is transient; in the second, the source is a baffle, and the process is steady. In both cases the flow is usually strongly influenced by turbulence.

Despite the last circumstance, it is useful to have a clear understanding of <u>laminar</u> flame propagation, because turbulent flame propagation is similar in some respects, though more complex. The similarity is greatest for the gasoline-engine flame, least for the steady flame propagating from a baffle.

(D) LAMINAR FLAME SPEED AS AN INDICATION OF THE REACTION-KINETIC PROPERTIES OF MIXTURES

Laminar flame speeds are easy to measure (e.g. by observing the cone angle of a Bunsen burner); if they can be linked with such quantities as $R_{fu,max}$, the Bunsen burner may be a useful means of distinguishing between more and less reactive fuels. It is certainly cheaper to build than the "Longwell bomb", and it requires much smaller quantities of the fuel which is being tested. The latter is an important point when only a small quantity of the newly-produced fuel is available.

It will be shown that there is a connexion between S and $R_{fu,max}$. This is the main purpose of the lecture.

18.2 FUNDAMENTALS

(A) REACTION KINETICS) The information needed under these headings is the same as for Chapters 16 and 17.

(B) THERMODYNAMICS)

(C) TRANSPORT PROCESSES

Whereas, in the stirred reactor and in the baffle-stabilised flame, the fresh gas was caused to react by being mixed with combustion products, in the laminar flame the same effect is produced by the conduction of heat and diffusion of matter from one side of the flame to the other. The difference is one of scale rather than kind; for, in a gas, heat conduction and diffusion are brought about by molecular inter-mingling.

Use will therefore have to be made of:

(i) Fourier's heat-conduction law:

$$\dot{q}'' = -\lambda \frac{dT}{dx} \quad ; \quad (18.2\text{-}1)$$

(ii) Fick's diffusion law:

$$G_{diff,fu} = -\mathcal{D}\rho \frac{dm_{fu}}{dx} \quad . \quad (18.2\text{-}2)$$

Another fact to be used is that, for gases of fairly uniform molecular weight, the thermal conductivity and the diffusion coefficient are related by:

$$\frac{\lambda}{c} \approx \mathcal{D}\rho \quad . \quad (18.2\text{-}3)$$

(D) MATHEMATICS

The distributions of temperature and concentration in a one-dimensional flow are in question; therefore the differential equations to be solved are ordinary ones.

Because of the strong non-linearity of the reaction-rate function, the equations are non-linear, and therefore impossible to solve by analytical means.

An approximate solution technique will be used; this involves: (i) guessing plausibly the shapes of the profiles of the dependent variables; (ii) determining the values of the unknown parameters in these profiles by reference to integral forms of the basic differential equations. This technique is worth remembering, because it often allows fairly accurate solutions to be obtained without recourse to a computer.

18.3 MATHEMATICAL MODEL

(A) DESCRIPTION

(i) Frame of reference

The distributions of T and m_{fu} along a direction normal to the flame are considered; the distance x in this direction is measured from a point which is fixed relative to the flame.

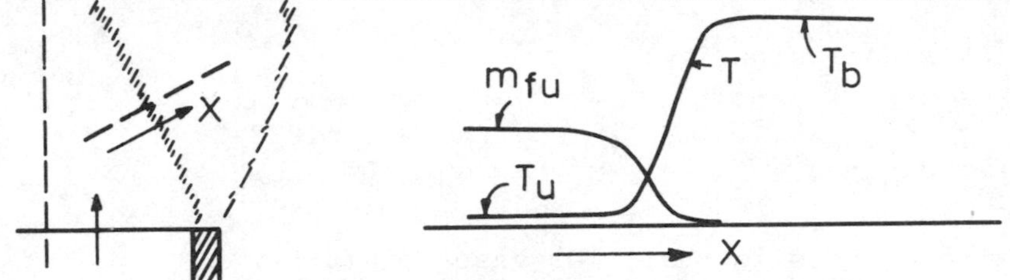

FIG. 18.2 REPRESENTATION OF THE COORDINATE SYSTEM AND OF THE PROFILES THROUGH THE FLAME

At $x = -\infty$, the gas will be taken to be in its unburned state; at $x = +\infty$, it will be in its fully-burned (equilibrium) state, with m_{fu} equal to zero (for weak mixtures).

The last-mentioned restriction is made simply for definiteness; it is not essential.

The situation has thus been "idealised": in reality, for example, the point where x equals $+\infty$ lies in the atmosphere, where the temperature is low. No appreciable error results from neglect of this point, if the flame is thin (which it is).

(ii) Properties of the mixture

The fuel and oxidant will be supposed to take part in a single-step exothermic reaction.

The specific heats and molecular weights will be taken as independent of temperature, and the same for all mixture components. This permits definition of enthalpy as:

$$h \equiv cT + H\, m_{fu} \qquad , (18.3\text{-}1)$$

as in earlier chapters.

λ/c will be taken as exactly equal to $\mathcal{D}\rho$ at all concentrations and temperatures; the symbol Γ will be used (units kg/ms). Thus:

$$\frac{\lambda}{c} = \mathcal{D}\rho = \Gamma \qquad . (18.3\text{-}2)$$

There is no need for Γ to be independent of temperature or concentration.

(iii) Other features of the model

Heat transfer by radiation is taken as negligible. Conditions are steady.

(B) EQUATIONS AND BOUNDARY CONDITIONS

(i) Conservation equations

Mass velocity: G = const. . (18.3-3)

Fuel: $$G\frac{dm_{fu}}{dx} - \frac{d}{dx}\left(\mathcal{D}_{fu}\,\rho\,\frac{dm_{fu}}{dx}\right) = -R_{fu} \qquad . (18.3\text{-}4)$$

Oxygen: $$G\frac{dm_{ox}}{dx} - \frac{d}{dx}\left(\mathcal{D}_{ox}\,\rho\,\frac{dm_{ox}}{dx}\right) = -R_{fu}s \qquad . (18.3\text{-}5)$$

Steady-flow energy equation:

$$G \frac{dh}{dx} - \frac{d}{dx} \left(\lambda \frac{dT}{dx} + \mathcal{D}_{fu} \rho \, H \frac{dm_{fu}}{dx} \right) = 0 \qquad . \ (18.3\text{-}6)$$

(ii) Consequences of the transport-property relation relation (18.3-2)

From (18.3-4) and (18.3-5):

$$G \frac{d}{dx} \left(m_{fu} - \frac{m_{ox}}{s} \right) - \frac{d}{dx} \left\{ \Gamma \frac{d}{dx} \left(m_{fu} - \frac{m_{ox}}{s} \right) \right\} = 0 \quad . \ (18.3\text{-}7)$$

From (18.3-6):

$$G \frac{dh}{dx} - \frac{d}{dx} \left(\Gamma \frac{dh}{dx} \right) = 0 \qquad . \ (18.3\text{-}8)$$

These two equations are source-free.

(iii) Boundary conditions

Far upstream ($x = -\infty$), the quantities m_{fu}, m_{ox}, T and h have known values, indicated by subscript u.

Far downstream ($x = +\infty$), the absence of interactions with the surroundings ensures that h equals h_u and $(m_{fu} - m_{ox}/s)$ equals $(m_{fu} - m_{ox}/s)_u$. Moreover, m_{fu} will be zero for a weak mixture, and m_{ox} will be zero for a rich one.

(C) SOLUTION OF THE MATHEMATICAL PROBLEM

(i) The uniformity of h and $m_{fu} - m_{ox}/s$.

Integration of equations (18.3-7) and (18.3-8), with insertion of the boundary conditions, yields:

$$h = h_u \qquad , \ (18.3\text{-}9)$$

and : $$m_{fu} - m_{ox}/s = (m_{fu} - m_{ox}/s)_u \qquad . \ (18.3\text{-}10)$$

The details of the derivation are as follows. Let ϕ stand for either $m_{fu} - m_{O_2}/s$ or h. Then integration of either (18.3-7) or (18.3-8) leads to:

$$G\phi - \Gamma \frac{d\phi}{dx} = \text{const} = G\,\phi_u \qquad , \; (18.3\text{-}11)$$

wherein the far-upstream boundary condition has been invoked.

Further integration yields:

$$\phi - \phi_u = \text{const.}\ \exp\,(G \int \Gamma^{-1}\,dx) \qquad ; \; (18.3\text{-}12)$$

and, since ϕ is finite at large x, it follows that the constant must equal zero. Hence:

$$\phi - \phi_u = 0 \qquad . \; (18.3\text{-}13)$$

This is what had to be proved.

As a consequence, variables m_{fu}, m_{ox} and T are connected together in precisely the same way as for the stirred reactor; and R_{fu} can therefore be treated as a function of just one of: m_{fu}, m_{ox}, T; and there is need to solve a differential equation for only one of these variables.

(ii) Choice of equation

Because the temperature variation is the most obvious physical characteristic of the flame, and because its use does not require distinction between rich and weak mixtures, T will be used as the variable. Then from (18.3-4) and (18.3-6):

$$\boxed{G\,\frac{dT}{dx} - \frac{d}{dx}\left(\Gamma\,\frac{dT}{dx}\right) = \frac{H}{c}\,R_{fu}} \qquad . \; (18.3\text{-}14)$$

(iii) Solution by "profile method"

Let the temperature profile be the linear one shown in Fig. 18.3; and let the thickness of the flame region be δ.

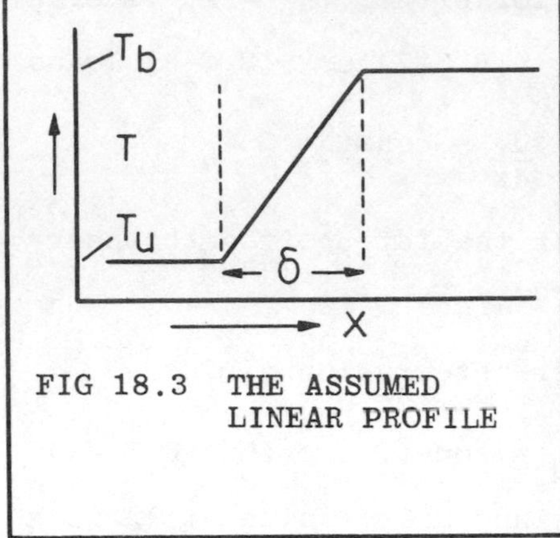

FIG 18.3 THE ASSUMED LINEAR PROFILE

Integration of (18.3-14), from $-\infty$ to $+\infty$, leads to:

$$G(T_b - T_u) = \frac{H}{c}\int_{-\infty}^{+\infty} R_{fu}\, dx$$

$$= \frac{H}{c}\,\frac{\delta}{(T_b-T_u)}\int_{T_u}^{T_b} R_{fu}\, dT$$

i.e. $G(T_b - T_u) = \frac{H}{c}\,\frac{\delta\,\overline{R_{fu}}}{(T_b-T_u)}$, say. (18.3-15)

The conduction term vanishes, because $\frac{dT}{dx}$ equals 0 at both boundaries.

Integration from $x = -\infty$ to the mid-point of the temperature profile, on the other hand, gives:

$$\frac{G(T_b - T_u)}{2} - \Gamma_{\frac{1}{2}}\,\frac{(T_b - T_u)}{\delta} = \frac{H}{c}\int_{-\infty}^{x_{\frac{1}{2}}} R_{fu}\, dx \approx 0\ . \quad (18.3\text{-}16)$$

The setting of the reaction-rate integral to zero is justified because very little chemical reaction takes place at the low temperatures which prevail to the left of the half-way point.

Combination of (8.3-15) and (8.3-16) now leads to:

$$G = \left[2\, H\, \Gamma_{\frac{1}{2}}\, \overline{R_{fu}}/\{c(T_b - T_u)\}\right]^{\frac{1}{2}} \quad , (18.3\text{-}17)$$

and

$$\delta = \left[2\, \Gamma_{\frac{1}{2}}\, c(T_b - T_u)/(H\, \overline{R_{fu}})\right]^{\frac{1}{2}} \quad . (18.3\text{-}18)$$

These equations provided information about the desired quantities: the flame speed ($S \equiv G/\rho$), and the flame thickness, δ.

(D) DISCUSSION

(i) It is seen from (8.3-17) that the flame speed increases in proportion to the square root of the temperature-average reaction rate. It also increases with the thermal conductivity (= Γc), again to the power 1/2.

(ii) Since S is the same as G/ρ and ρ is proportional to p, if R_{fu} is proportional to p^n, the flame speed S is proportional to $p^{\frac{n}{2}-1}$.

If n equals 2, as is true for a bi-molecular collision-rate-controlled reaction, S is independent of pressure. If n equals 1.8, as suggested by the Longwell-Weiss experiments, S is proportional to $p^{-.1}$; i.e. the flame speed decreases as the pressure rises.

(iii) Equation (8.3-18) shows that the flame thickness δ is proportional to $p^{-n/2}$; so low-pressure flames are thick.

18.4 RELATION TO EXPERIMENT

(A) EFFECTS OF VARIOUS EXPERIMENTAL CONDITIONS ON FLAME SPEED

(i) The adiabatic flame temperature

The average reaction rate, like the maximum value for a given mixture, is strongly influenced by the highest temperature in the flame, T_b. The two main influences on this are the mixture ratio, and the initial temperature, T_u. Fig. 18.4 illustrates these influences.

It is evident that the greatest value of the flame speed occurs for the stoichiometric mixture, and that the flame speed has appreciable velocities only for a narrow range of mixture ratios, similar to that for stability of a stirred-reactor or bluff-body flame.

The data accord qualitatively with the predictions of the mathematical model.

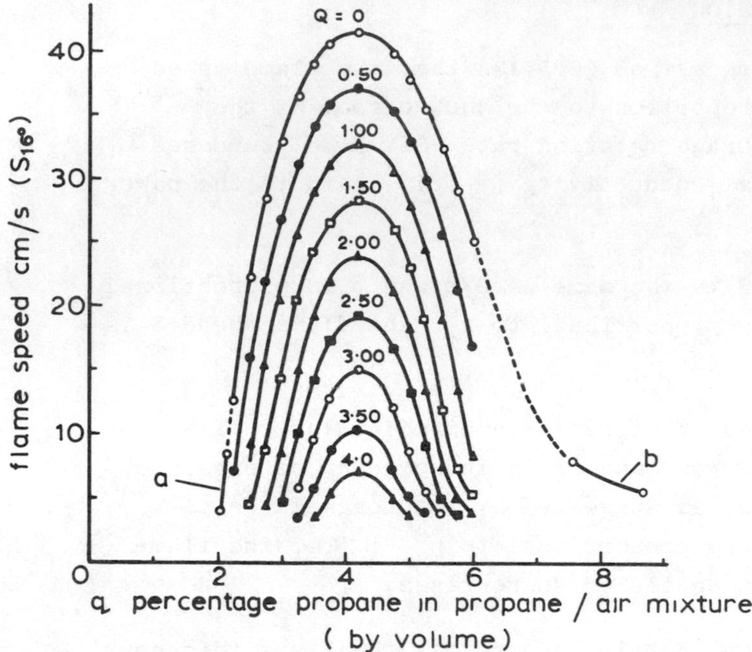

Flame speed as a function of mixture ratio for different amounts of heat removed from the flame, Q cal/ml. propane. Curves a, b, Egerton & Thabet (1952).

Curves from Botha and Spalding (Proc. Roy. Soc. A, Vol. 225, pp 71-96).
$p = 1$ atm.
$Q \equiv$ heat extracted from flame to lower initial temperature from 16°C.
$S_{16} \equiv S$ relative to gas at 16°C.

FIG. 18.4 MEASUREMENTS MADE ON A COOLED-POROUS-PLUG BURNER

(ii) The pressure

Experimental data show that flame speed tends to diminish as pressure rises, for most mixtures, but the effect is slight. This is in accordance with the expectations set out in Section 8.3(d).

(iii) Differences between fuels

Most hydrocarbons, when mixed with air, have flame speeds of the same order of magnitude, viz. 0.3 m/s for stoichiometric mixtures at atmospheric temperature and pressure.

This behaviour is like that in a stirred reactor, and unlike that in a spontaneous-ignition apparatus.

(B) FURTHER SIGNIFICANCE OF THE FINDINGS

(i) It may be expected that turbulence will increase flame speeds greatly over the laminar values, because the effective value of Γ for a turbulent mixture is 100 or 1000 times that for a laminar one. Qualitatively, this expectation is borne out in practice; but the quantitative connexions are not strong, chiefly because it is rarely possible to create a one-dimensional turbulent flame.

(ii) If the shape of the $R_{fu} \sim T$ curve is fixed, the temperature-average value of R_{fu} will be a fixed fraction of its maximum value; and in any case the fraction is not likely to vary much. It follows that flame-speed data can be used to provide predictions of $R_{fu,max}$ Two factors have prevented this practice from being often used:

(1) S is influenced by transport properties (Γ) as well as reaction-kinetic ones; so flame speed does not separate the two sorts of properties so well as the stirred reactor.

(2) A slightly greater acquaintance with, and confidence in, combustion theory is needed for its use than most combustion engineers possess.

EXERCISES TO FACILITATE ABSORPTION OF MATERIAL IN CHAPTER 18

MULTIPLE-CHOICE PROBLEMS

18.1 The pre-mixed flame of a Bunsen burner is roughly conical in shape,
because
the velocity of the gases emerging from the pipe is fairly uniform, and the propagation speed is a constant for a given mixture.

18.2 The quantities λ/c and $\mathcal{D}\rho$ are usually almost equal for a gaseous mixture,
because
both have the dimensions mass/(length × time).

18.3 In the immediate upstream part of a laminar pre-mixed flame, the mass flow rate of fuel exceeds $G\ m_{fu,u}$,
because
the diffusional flux is there finite, and acts in the same direction as the convective flux.

18.4 It is essential to define enthalpy as $cT + H\ m_{fu}$ rather than $cT + (H/s)\ m_{ox}$,
because
$cT + H\ m_{fu}$ is uniform through the flame when $\lambda/c = \mathcal{D}\rho$.

18.5 The reaction-rate properties of a hydrocarbon-air mixture can be measured more easily in a Bunsen burner than in a Longwell-Weiss stirred reactor,
because
laminar-flame propagation is a steady-flow process.

18.6 The "profile" method of solving equation (18.3-14) is advantageous,
because
it allows a numerical integration with two-point boundary conditions to be replaced by a quadrature.

18.7 The equation

$$G \frac{d}{dx}\left(m_{fu} - \frac{m_{ox}}{s}\right) - \frac{d}{dx}\left\{\Gamma \frac{d}{dx}\left(m_{fu} - \frac{m_{ox}}{s}\right)\right\} = 0$$

would cease to be valid if heat were lost by radiation from the gas,

<u>because</u>

the validity of the equation depends on the uniformity of enthalpy through the flame.

18.8 If $\mathcal{D}_{fu}\rho$ were very much less than λ/c, there would be a region in the flame in which the enthalpy exceeded h_u and h_b,

<u>because</u>

heat would be transferred by conduction to a region in which chemical reaction had not yet begun to diminish the concentration of fuel.

18.9 For a bi-molecular collision-controlled chemical reaction, the laminar flame speed increases linearly with pressure,

<u>because</u>

G is proportional to $(\overline{R_{fu}})^{\frac{1}{2}}$ and $\overline{R_{fu}}$ is proportional to p^2.

18.10 The <u>exact</u> solution of the differential equation is unlikely to yield the relation

$G = \text{const.}\ [H\ \Gamma_{\frac{1}{2}}\ \overline{R_{fu}}/\{c_p\ (T_b - T_u)\}]^{\frac{1}{2}}$,

<u>because</u>

this relation is not dimensionally homogeneous.

18.11 Flame speeds are greatest, as a rule, for stoichiometric mixtures,

<u>because</u>

these mixtures have the shortest spontaneous-ignition times.

18.12 If the upstream temperature of a combustible mixture is increased, the flame temperature will also be increased,

<u>because</u>

the spontaneous-ignition times fall with increase in temperature.

18.13 If fuel A has a laminar flame speed 1.1 times that of fuel B, at equal pressure, initial temperature and equivalence ratio, it is likely that fuel A will give blow-out velocities in bluff-body-stabilisation experiments that are 20% in excess of those for fuel B,

because

the flame speed is proportional to the square root of the volumetric reaction rate, while the blow-out velocity is proportional to this rate directly.

18.14 The flame-speed relation can be expressed as:

$G \approx [2\, \Gamma_{\frac{1}{2}}\, \overline{R_{fu}}\,(\delta\, m_{fu})]^{\frac{1}{2}}$,

where $\delta\, m_{fu}$ is the change in fuel fraction that is brought about by combustion,

because

in a single-step reaction the quantity of oxygen which disappears is proportional to the quantity of fuel consumed.

18.15 For a stoichiometric hydrocarbon-air mixture at atmospheric initial conditions, $\overline{R_{fu}}$ is of the order of 100 kg/m^3s,

because

ρ is about 1 kg/m^3, S is about .3m/s, $\delta\, m_{fu}$ is about .065, and $\Gamma_{\frac{1}{2}}$ is about 3×10^{-5} kg/m s.

18.16 For a stoichiometric hydrocarbon-air mixture at initially atmospheric conditions, the maximum volumetric-heat-release rate is about 1.6×10^6 J/m^3s,

because

the maximum is about four times the mean and the calorific value is about 4×10^7 J/kg.

18.17 Although the flames which propagate in gasoline-engine combustion chambers are usually turbulent, even laminar flames would travel fast enough to burn up all the fuel in the available time,

because

such flames would traverse chambers of the usual size in less than .1 s, and there is much more time than this available for flame travel.

18.18 If a flame is ignited at the centre of a spherical enclosure containing a combustible gaseous mixture, the flame will move at a velocity smaller than the laminar flame speed,

because

the pressure in the enclosure will increase and the densities of the burned and unburned gases are different, and change with time.

18.19 If a way could be found of doubling the thermal conductivity and diffusion coefficient of a combustible gas, its laminar flame speed would be increased by a factor of $\sqrt{2}$,

because

S is proportional to $\Gamma^{-\frac{1}{2}}$.

ANSWERS TO MULTIPLE-CHOICE PROBLEMS

Answer	Problem number (18's omitted)
A	1, 6, 8, 13, 15
B	2, 5, 12, 14
C	11, 19
D	3, 4, 9, 16, 18
E	7, 10, 17

CHAPTER 19

IGNITION BY PILOT FLAMES AND SPARKS

19.1 ENGINEERING RELEVANCE OF THE PHENOMENA

(A) SPARK IGNITION

(i) Desired ignition

Sparks, produced by electrical discharges, are used to ignite the fuel and air in:

gasoline engines (every cycle);
domestic boilers (every cycle);
gas turbines (at the start of operation only).

Usually it is easy to provide an energy of spark which greatly exceeds the minimum to ignite the gases; but it is still necessary to understand what are the characteristics of spark and mixture which most favour ignition.

An interesting question is: When the carburettor is adjusted so as to provide very weak mixtures, the sparks supplied by the electrical equipment of a gasoline engine, though satisfactory at normal mixture strengths, may fail to ignite the flame. Can this failure be remedied simply by supplying more energetic sparks?

The answer is: Only to a limited extent. Outside the "inflammability limits", the laminar flame speed may be taken as zero; so the required energy becomes infinite.

(ii) Undesired ignition

In coal mines, oil refineries, paint shops, and many kinds of process plant, combustible fuel-air mixtures may be present in the atmosphere. Sparks may be generated by:- the contacts of electrical equipment, iron-shod shoes striking on stone, the doffing of a nylon shirt, the stroking of a cat, etc. Since explosions are undesired in these situations, it is important to know under what circumstances the combination of a spark and a combustible mixture will lead to ignition.

(B) PILOT-FLAME IGNITION

(i) In a ram-jet

Ignition by a pilot flame has been described in Lecture 17. A jet of hot combustion products is injected into a stream of unburned fuel and air flowing in the same direction. Flame spreads from the former to the latter, as a consequence of heat conduction, diffusion, mixing, and chemical reaction.

The process is a steady-flow one.

(ii) In domestic gas-burning equipment

A small flame burns continuously, its fuel consumption being small enough to constitute no serious expense. When the equipment is to operate, the main supplies of fuel and air flow past the pilot flame, and are ignited by it.

This process is an unsteady one; and removal of the pilot flame after the main flame was ignited would have no effect on the stability of the latter.

(C) SIMILARITIES BETWEEN PILOT-FLAME AND SPARK IGNITION

(i) Successive temperature profiles after the passage of a spark through a combustible mixture are represented in Fig. 19.1.

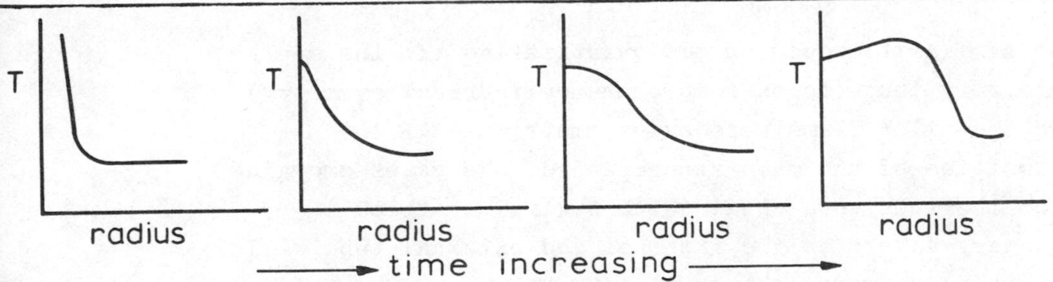

FIG. 19.1 TEMPERATURE-VERSUS-RADIUS PROFILES AT SUCCESSIVE TIMES AFTER THE PASSAGE OF A SPARK THROUGH A GASEOUS COMBUSTIBLE MIXTURE.

The sketches show how the spark creates a small region of high-temperature gas; conduction causes the peak to fall and its base to become smaller; but the fall is eventually checked by the exothermic effect of the chemical reaction; and, in the end, a steady flame propagates away from the ignition source.

(ii) Temperature profiles at successive stations downstream through the mixing region of a pilot flame are represented by the isotherms and temperature profiles of Fig. 19.2.

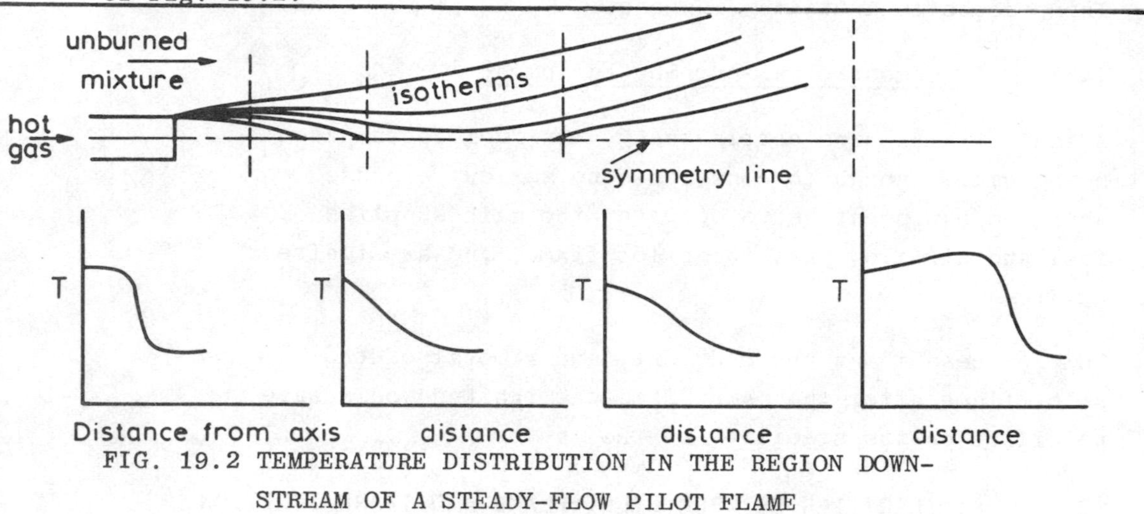

FIG. 19.2 TEMPERATURE DISTRIBUTION IN THE REGION DOWNSTREAM OF A STEADY-FLOW PILOT FLAME

(iii) There is an obvious qualitative similarity between the temperature profiles for the two situations, one unsteady in a single space dimension, the other steady in two space dimensions.

The similarity would become quantitative if: the spark were very long (so as to procure cylindrical symmetry) and the pilot flame were axi-symmetrical; if the velocities of the main stream and of the gases emerging from the tube were equal (with negligibly-thick boundary layers on the internal and external tube walls); if the gases were laminar in both cases; and if the spark had the appropriate energy distribution to provide, after its passage, a set of temperature and concentration profiles identical with those at the tube outlet. Of course, it would be very difficult to satisfy these conditions. The point is mainly important in suggesting that the two processes are mathematically similar, and can probably be handled by the same procedures of solution.

Indeed, the differential equations prove to be "parabolic", in both cases; and they may be solved by exact or approximate means which have the same form whether the second independent variable is time or longitudinal distance.

(D) THE PRESENT TASK

It is desired to understand how the variables which can be controlled by a designer influence the success or failure of ignition. If possible, quantitative relations are to be obtained, permitting predictions of such quantities as minimum spark energy, and minimum pilot-flame diameter. However, in the present work, only those mathematical techniques will be used which do not require extensive computations.

19.2 FUNDAMENTAL CONSIDERATIONS

The processes in question result from the interactions of chemical reaction, heat conduction, diffusion of matter, and the flow of fluids.

The single-step model of the chemical reaction will be used, and also the Fourier and Fick laws for the relevant transport processes. There is no need to introduce any new material under this heading. Indeed, it will have been observed that the same fundamental material is repeatedly being drawn upon for the mathematical models of the various phenomena. Although of course much detail can be added, the main territories of the scientific base of combustion have already been traversed.

19.3 MATHEMATICAL MODEL

(A) DESCRIPTION

(i) Transient or steady?

Analysis will be made of the transient temperature distribution in a one-dimensional space; so the two independent variables will be distance x, and time t.

However, it will be possible to interpret the results as pertaining to the particular two-dimensional steady-state situation in which the gas flows in a direction y, normal to x, at uniform velocity; it is necessary only to remark that: (1) for an observer travelling with the gas, there is a linear relation between the distance y and the time t; (2) if the flow velocity (strictly $c\rho uy/\lambda$ or $c\rho y^2/(\lambda t)$) is great enough, heat conduction in the y direction will be negligible compared with that in the x direction. The mathematical problems will therefore be formally identical in the two cases.

(ii) Type of symmetry

x will be the distance from a plane, for example a plane of symmetry; it is not the distance from a point (as in spherical symmetry) or a line (as in axial symmetry).

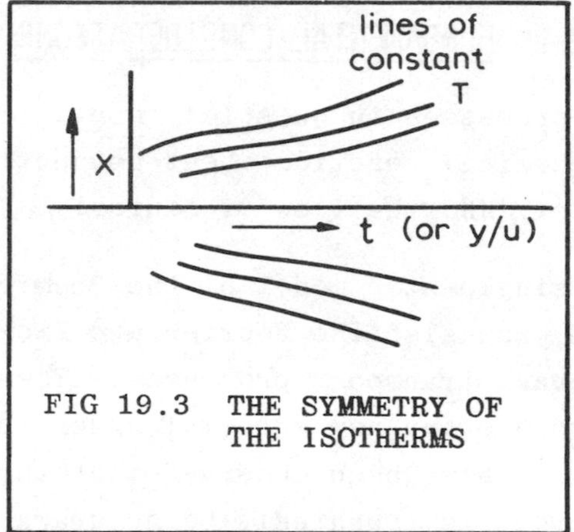

FIG 19.3 THE SYMMETRY OF THE ISOTHERMS

Radii of curvature can therefore be left out of the account; and igniting sparks must be imagined to spread out over plane areas which are wide compared with the thickness of the flames to which they give rise. This is done in the interests of simplicity of mathematical analysis, and in the expectation that the main features of the results will not be greatly dependent on the type of symmetry.

(iii) Transport processes

The fluid will be supposed laminar; so heat conduction obeys Fourier's law and diffusion obeys Fick's law, with the thermal conductivity and diffusion coefficient possessing their "molecular" values. There will be no need to require these quantities be independent of temperature; but λ/c will be put equal to $\mathcal{D}\rho$ for all components, in order for it to be possible to link the concentrations linearly with the temperature. It will also be convenient, though not essential, to suppose ρ = const., which is often close to the truth.

(iv) Reaction-kinetics

The single-step collision-controlled model of a reaction between fuel and air will be presumed. Together with the constancy of enthalpy, this supposition allows the reaction rate to be a function of temperature alone, as in Chapter 14.

(v) Initial and boundary conditions

The temperature and concentration profiles at the start will be chosen to ensure uniformity of enthalpy; then, as will be shown, the assumption $\lambda/c = \mathcal{D}\rho$ will entail that the enthalpy remains constant and uniform everywhere.

This starting-profile assumption marks a departure from the spark-ignition condition; for a spark definitely increases the enthalpy of the gas through which it passes. The assumption is needed if there is to be only one differential equation; analysis of the more realistic non-uniform-enthalpy start shows that the simple assumption introduces no seriously unrealistic features into the situation.

The particular initial profiles that are postulated at the start are shown in Fig. 19.4. They represent a slab of hot burned gas, suddenly inserted into a reservoir of unburned mixture.

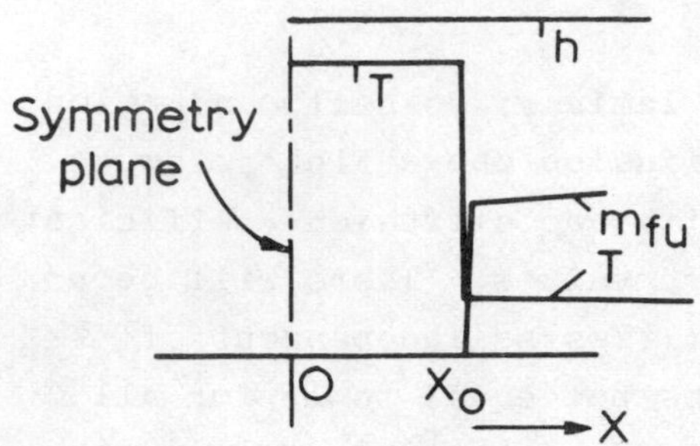

FIG. 19.4 INITIAL PROFILES OF T, h AND m_{fu}.

(B) EQUATIONS

(i) The differential equations are:-

Fuel conservation:

$$\rho\frac{\partial m_{fu}}{\partial t} = \frac{\partial}{\partial x}\left(\mathcal{D}_{fu}\rho\,\frac{\partial m_{fu}}{\partial x}\right) - R_{fu} \qquad (19.3\text{-}1)$$

Steady-flow energy equation:

$$\rho\frac{\partial h}{\partial t} = \frac{\partial}{\partial x}\left(\lambda\,\frac{\partial T}{\partial x}\right) + H\,\frac{\partial}{\partial x}\left(\mathcal{D}_{fu}\rho\,\frac{\partial m_{fu}}{\partial x}\right) \qquad (19.3\text{-}2)$$

Since $\lambda/c = \mathcal{D}_{fu}\rho = \Gamma$, say, and
$h \equiv cT + H\,m_{fu}$,

equation (19.3-2) can be written as:

$$\rho\frac{\partial h}{\partial t} = \frac{\partial}{\partial x}\left(\Gamma\,\frac{\partial h}{\partial x}\right) \qquad (19.3\text{-}3)$$

Further,since h is uniform at the start, so that $\partial h/\partial x$ equals zero then, $\partial h/\partial t$ must equal zero. Therefore, h never changes; and it follows that:

$$h = h_o \text{ for all } x \text{ and } t\,, \qquad (19.3\text{-}4)$$

where h_o is the value of h at the start, and is equal to $cT_u + H\,m_{fu,u}$.

Now that the validity has been proved, for this case, of the familiar linear relation between T and m_{fu}, there remains only one differential equation, of which the dependent variable can be either T or m_{fu}. The former is chosen here; the result is:

$$\boxed{\rho\frac{\partial T}{\partial t} = \frac{\partial}{\partial x}\left(\Gamma\,\frac{\partial T}{\partial x}\right) + \frac{H}{c}\,R_{fu}} \qquad . \; (19.3\text{-}5)$$

Of course, R_{fu} is a function of T alone.

(ii) Dimensionless form

This parabolic differential equation will not here be solved exactly; for this would necessitate the use of numerical methods, and would require a computer. Instead, an approximate method of solution will be used; this incorporates an assumption about the temperature-profile shape, and uses integral forms of the differential equation. First however the latter will be tidied up, as follows:

Let $$\tau \equiv (T - T_u)/(T_b - T_u) \qquad , \; (19.3\text{-}6)$$

to give: $$\rho\frac{\partial \tau}{\partial t} = \frac{\partial}{\partial x}\left(\Gamma\,\frac{\partial \tau}{\partial x}\right) + \frac{H}{c(T_b-T_u)}\,R_{fu} \qquad . \; (19.3\text{-}7)$$

Multiplication by Γ, and the observation that $\Gamma\rho$ is constant, lead to:

$$\Gamma_u\rho_u\,\frac{\partial \tau}{\partial t} = \Gamma\,\frac{\partial}{\partial x}\left(\Gamma\,\frac{\partial \tau}{\partial x}\right) + \left\{\frac{H}{c(T_b-T_u)}\right\}\Gamma R_{fu} \qquad . \; (19.3\text{-}8)$$

The aim of this simplification is to reduce the number of symbols in the equation (for ease of manipulation), and to ensure that those which remain are dimensionless (for ease of understanding). Flame-propagation theory, which shows the temperature average of the reaction rate to be important (Chapter 18), suggests definition of a dimensionless reaction rate ϕ:

$$\phi \equiv \frac{\Gamma R_{fu}}{\int_o^1 \Gamma R_{fu}\,d\tau} \qquad . \; (19.3\text{-}9)$$

Then, in preparation for the final stage, it is useful to write (19.3-8) as:

$$\frac{\Gamma_u \rho_u \ c(T_b - T_u)}{H \int_o^1 \Gamma R_{fu} \ d\tau} \cdot \frac{\partial\tau}{\partial t} = \Gamma \left\{ \frac{c(T_b - T_u)}{H \int_o^1 \Gamma R_{fu} d\tau} \right\}^{\frac{1}{2}} \frac{\partial}{\partial x} \left[\Gamma \left\{ \frac{c(T_b - T_u)}{H \int_o^1 \Gamma R_{fu} d\tau} \right\}^{\frac{1}{2}} \times \frac{\partial\tau}{\partial x} + \phi \right] \quad . \ (19.3\text{-}10)$$

Hence, with $\theta \equiv \dfrac{H \int_o^1 \Gamma R_{fu} d\tau}{\Gamma_u \rho_u \ c(T_b - T_u)} . t$, (19.3-11)

and $d\xi \equiv \dfrac{1}{\Gamma} \left(\dfrac{H \int_o^1 \Gamma R_{fu} d\tau}{c(T_b - T_u)} \right)^{\frac{1}{2}} dx$, (19.3-12)

there results:

$$\boxed{\frac{\partial\tau}{\partial\theta} = \frac{\partial\tau}{\partial\xi} + \phi(\tau)} \qquad (19.3\text{-}13)$$

(iii) <u>Integral forms of equation</u>

Integration of (19.3-13) from $\xi = 0$ to $\xi = \infty$ gives:

$$\frac{d}{d\theta} \int_o^\infty \tau \ d \ \xi = \int_o^\infty \phi \ d \ \xi \qquad , \ (19.3\text{-}14)$$

into which has been inserted the boundary-condition information that $d\tau/d\xi = 0$ at both bounds.

Integration from $\xi = \xi_m$ to $\xi = \infty$, on the other hand, gives:

$$\frac{d}{d\theta} \int_{\xi_m}^\infty \ d \ \xi + \tau_m \frac{d\xi_m}{d\theta} = - \left(\frac{\partial\tau}{\partial\xi}\right)_{\xi=\xi_m} + \int_{\xi_m}^\infty \phi \ d \ \xi \ ; (19.3\text{-}15)$$

here ξ_m is the ξ value of the point on the $\tau \sim \xi$ profile where $\tau = \tau_m$. The latter is an arbitrarily-chosen value, say 1/2.

These two equations can provide information about two properties of the temperature profile; so, if the <u>shape</u> of the profile is presumed to be known, its width and position can be deduced from these equations.

(Note: If a more flexible assumption is made, for example that the profile is a polynomial with several undetermined constants, it is necessary to employ more integral equations than two. However, it is possible to generate as many as are desired, simply by choosing different values for τ_m).

(iv) The linear-profile assumption

It is now supposed that τ is linear in ξ over a range of width (in terms of ξ) δ. The sketch (Fig. 19.5) explains.

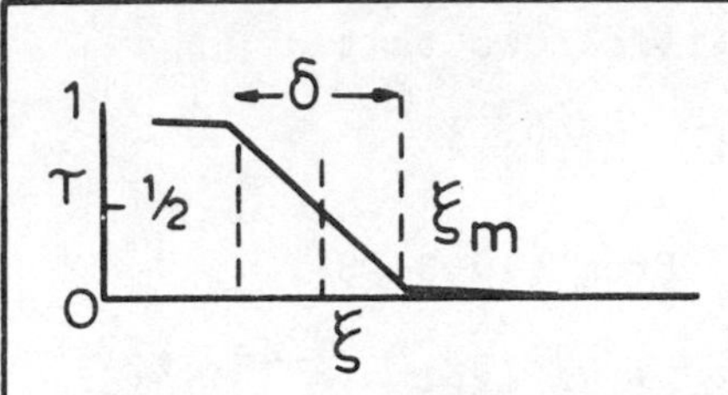

FIG 19.5 PRESUMED LINEAR PROFILE, FOR USE IN EVALUATION OF THE INTEGRAL AND THE SLOPE.

Further, τ_m is to be taken as 1/2; and ϕ is regarded as negligible for $\tau < \tau_m$.

Consequences of the assumptions are:

$$\int_0^\infty \phi \, d\,\xi = \int_0^\infty \phi/(d\tau/d\xi)\, d\xi = \delta \qquad ; \quad (19.3\text{-}16)$$

$$\int_{\xi_m}^\infty \phi \, d\,\xi = 0 \qquad . \quad (19.3\text{-}17)$$

Equations (19.3-14) and (19.3-15) now become:

$$\frac{d\xi_m}{d\theta} = \delta \qquad , \quad (19.3\text{-}18)$$

$$\frac{1}{4}\frac{d\delta}{d\theta} + \frac{1}{2}\frac{d\xi_m}{d\theta} = \frac{1}{\delta} \qquad . \quad (19.3\text{-}19)$$

(c) SOLUTION

(i) Substitution of (19.3-18) into (19.3-19) yields:

$$\frac{1}{4}\frac{d\delta}{d\theta} + \frac{1}{2}\,\delta = -\,\frac{1}{\delta}$$

i.e. $\frac{1}{8}\frac{d\delta^2}{d\theta} + \frac{1}{2}\delta^2 = 1$. (19.3-20)

(ii) The solution of this equation is:

$$\delta^2 = 2 + A\,e^{-4\theta} \quad , (19.3\text{-}21)$$

where A is an arbitrary constant. Since δ equals 0 at $\theta = 0$, it follows that:

$$\delta = \{2(1 - e^{-4\theta})\}^{\frac{1}{2}} \quad . (19.3\text{-}22)$$

(iii) From (19.3-18),

$$\xi_m = \xi_{m,o} + \int_o^\theta \{2(1 - e^{-4\theta})\}^{\frac{1}{2}}\, d\theta \quad . (19.3\text{-}23)$$

This cannot be expressed in closed form; but it can be evaluated numerically; and the form of $\xi_m \sim \theta$ variation which it implies must be as shown in Fig. 19.7.

(D) DISCUSSION

(i) Equation (19.3-22) shows that δ grows as shown in the sketch. In the steady state (large θ), its value is $2^{\frac{1}{2}}$.

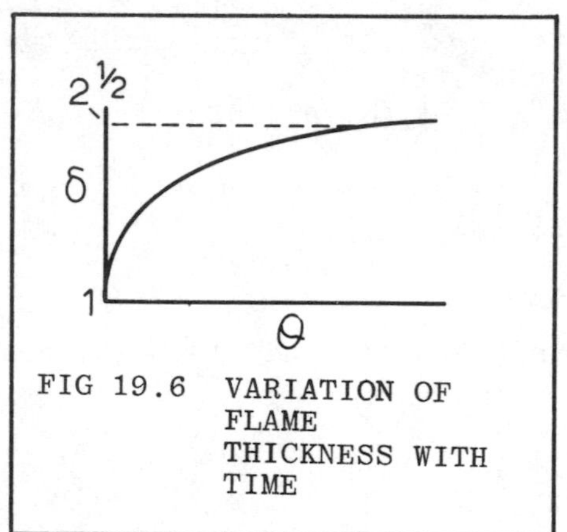

FIG 19.6 VARIATION OF FLAME THICKNESS WITH TIME

When this is interpreted in physical terms, it means:

profile thickness =

$$\overline{\Gamma}\left\{\frac{2c(T_b - T_u)}{H\int_o^1 \Gamma R_{fu} d\tau}\right\}^{\frac{1}{2}}$$

where $\overline{\Gamma}$ is an average value of Γ.

(ii) The movement of the flame

Fig. 19.7 shows how the head and foot of the flame, and its mid-point, vary with time. The curves are derived from equations (19.3-22) and (19.3-23).

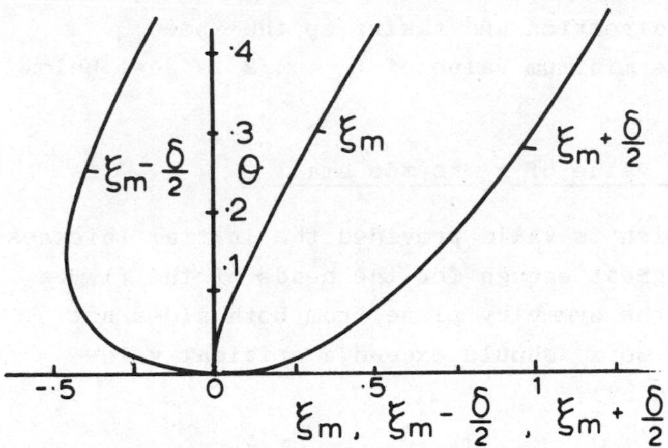

FIG. 19.7 RESULTS OF NUMERICAL INTEGRATIONS SHOWING HOW THE FLAME EDGES ($\xi_m - \delta/2$, $\xi_m + \delta/2$) AND CENTRE (ξ_m) VARY WITH TIME θ. (FROM: D B SPALDING, PROC. ROY. SOC. A, VOL. 245, PP 352-372, 1958).

Clearly, the mid-point moves to the right continuously, after making a slow start; the final slope of this curve is $2^{\frac{1}{2}}$, as implied by equation (19.3-18).

The foot of the flame also moves continuously to the right; it makes a rapid start, but finally takes up the same speed of advance as the mid-point. In physical terms, as the definitions imply:

$$S = \left(\frac{dx}{dt}\right)_{fast} = 2^{\frac{1}{2}}\, \Gamma_u \left\{\frac{c(T_b - T_u)}{H\int_o^1 \Gamma R_{fu} d\tau}\right\}^{\frac{1}{2}} \left\{\frac{H\int_o^1 \Gamma R_{fu} d\tau}{\Gamma_u \rho_u c(T_b - T_u)}\right\}$$

$$= \frac{1}{\rho_u}\left\{2\,\frac{H\int_o^1 \Gamma R_{fu} d\tau}{c(T_b - T_u)}\right\}^{\frac{1}{2}} \qquad . \ (19.3\text{-}23)$$

This is the same result as was obtained in Lecture 18 (eq. 18.3-17), except that the Γ appears in a slightly different place. (On this occasion, the non-uniformity of properties has been more scrupulously handled; and the profile has been taken as linear in ξ rather than in x.)

The head of the flame moves to the left at first, only later changing direction and taking up the speed: $d\xi/d\theta = 2^{\frac{1}{2}}$. The minimum value of $\xi_m - \delta/2$ is just below -0.5.

(iii) If the value of x_o is too small

The above solution is valid provided the initial thickness of the slab is great enough for the heads of the flames which approach the symmetry plane from both sides not to meet there. So x_o should exceed a critical value, $x_{o,crit}$, defined by:

$$x_{o,crit} \equiv \frac{1}{2}\,\Gamma_b\,\{c(T_b - T_u)/\left[H\int_o^1 \Gamma R_{fu}\,d\tau\right]\}^{\frac{1}{2}} \quad . \ (19.3\text{-}24)$$

It can be shown that $x_{o,crit}$ is $2^{-\frac{1}{2}}$ times the steady-state flame thickness, i.e. $2^{-\frac{1}{2}}\,\Gamma_b/(\rho u S)$.

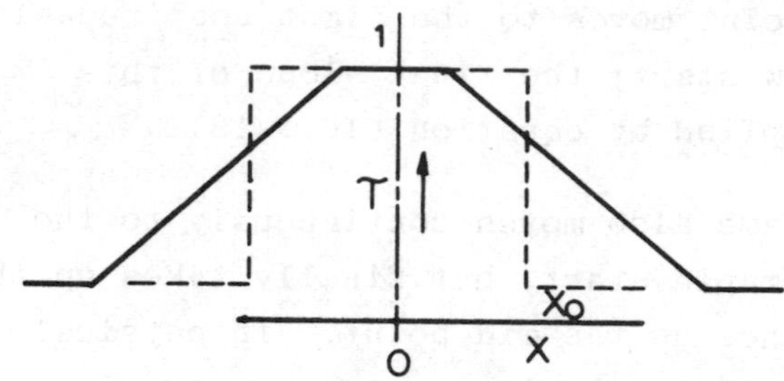

FIG. 19.8 ILLUSTRATION OF THE DOUBLE-SIDED HOT-GAS SLAB. IF x_o IS SMALL ENOUGH, THE TWO LINEAR PROFILES MAY MEET; AND THE FLAME CAN BE EXTINGUISHED WITHOUT PROPAGATING.

If x_o does not exceed $x_{o,crit}$, the two flames merge. The mathematical method will still be valid, but the equations will differ, to account for the fact that the highest value of τ in the flame may become less than unity; and it will also be necessary to prescribe the $R_{fu} \sim \tau$ relation more precisely than has been done so far.

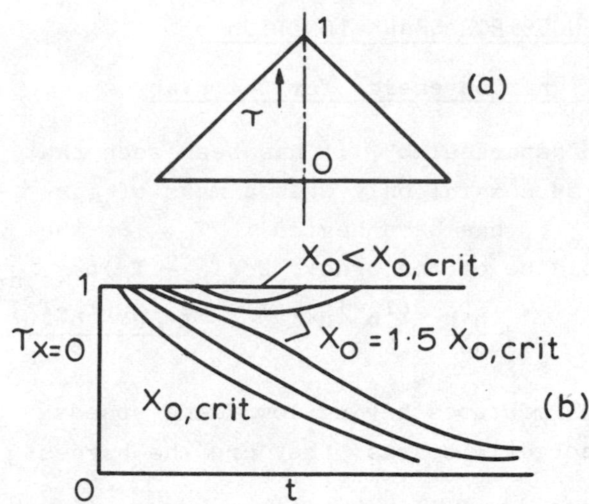

FIG. 19.9(a) ILLUSTRATION OF THE PRESUMED TEMPERATURE PROFILE, WHEN THE FLAMES STARTING ON THE TWO SIDES OF THE SLAB HAVE COALESCED.

(b) VARIATION OF THE CENTRE-PLANE TEMPERATURE $\tau_x = 0$ WITH TIME, FOR VARIOUS VALUES OF THE RATIO $x_o/x_{o,crit}$.

If x_o falls below $x_{o,crit}$ by only a small amount, the symmetry-plane temperature falls momentarily, but then recovers; the steady flames finally move off, leaving the fully-burned gas at the central plane.

However, if the deficit is too great ($x_o/x_{o,crit} < 0.7$, say), the central temperature falls steadily; i.e. the flame is extinguished; and ignition does not occur. There is therefore a minimum size of slab for ignition, approximately equal to $\Gamma_b/(\rho u S)$.

19.4 PRACTICAL CONSEQUENCES

(A) CONSEQUENCES FOR SPARK IGNITION

(i) Minimum ignition energy for the slab

For the idealised "spark slab", it has been seen that ignition will be successful only when a mass of gas equal to $2\ \rho_b\ x_{o,crit}$ has been heated to T_b. So the spark energy should be of the order: $2c(T_b - T_u)\rho\ x_{o,crit}$ per unit area, i.e. $2\ \dfrac{\Gamma_b\rho_b}{\rho uS}\ c(T_b - T_u)$, i.e. $2\ \dfrac{\Gamma_u c(T_b - T_u)}{S}$

Therefore, gaseous mixtures having low flame speeds require highly-energetic sparks: they are the hardest to ignite.

(ii) Minimum ignition energy for the spherically-symmetrical system

Although the case of flame propagation from a sphere of hot gas into cold surroundings has not been analysed here, the foregoing study suggests (and detailed analysis confirms) that in this case also the need is to create by means of the spark a hot-gas pocket of diameter greater than $2\ \Gamma_b/(\rho uS)$. (Probably the different geometry will require a modified numerical coefficient.) Then, since the volume of a sphere is $\pi d^3/6$, it can be concluded that the minimum ignition energy for a spherical flame is of order:

$$\frac{\pi}{6}\ c(T_b - T_u)\ \rho_b \left(\frac{\Gamma_b}{\rho uS}\right)^3$$

It follows that low-flame-speed gases will be very hard to ignite (energy $\alpha\ S^{-3}$); also, ignition will require more energy at low pressure (energy $\alpha\ \rho^{-2}$).

These results are confirmed experimentally.

(B) CONSEQUENCES FOR PILOT-FLAME IGNITION

(i) Minimum jet width for successful ignition

For the plane pilot jet, in uniform laminar flow, which is in effect the case which has been considered, it can be seen that the jet width must exceed a minimum size if flame propagation is to occur. Fig. 19.10 illustrates three possible situations.

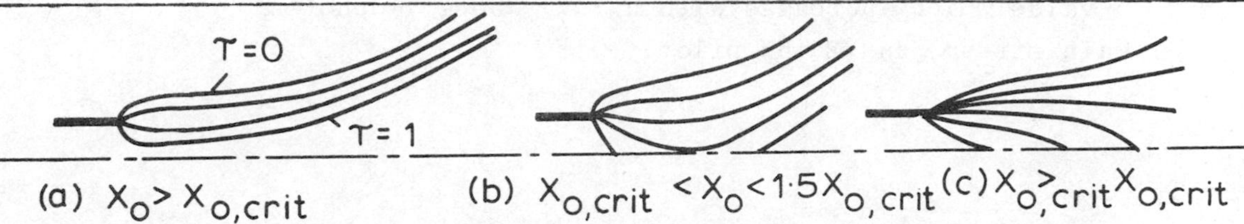

(a) $x_o > x_{o,crit}$ (b) $x_{o,crit} < x_o < 1.5x_{o,crit}$ (c) $x_o < x_{o,crit}$

FIG. 19.10(a) EASY IGNITION; THE HEAD OF THE FLAME DOES NOT PENETRATE TO THE SYMMETRY PLANE.

(b) MORE DÍFFICULT IGNITION; THE HEAD DOES PENETRATE TO THE SYMMETRY PLANE, BUT PROPAGATION DOES FINALLY OCCUR.

(c) EXTINCTION; THE FLAME DECAYS INTO A JET OF NON-REACTING GAS.

From equation (19.3-24), it can be seen that the width should exceed $\Gamma_b \left\{\dfrac{c(T_b - T_u)}{H \int_0^1 \Gamma R_{fu} \delta\tau}\right\}^{\frac{1}{2}}$. This is proportional to

$$\left\{\frac{\bar{\Gamma}\ c(T_b - T_u)}{H\ R_{fu,max}}\right\}^{\frac{1}{2}}.$$

It can be expected that large pilot-jet widths are required at low pressures; for $R_{fu,max}$ is approximately proportional to pressure squared.

(ii) Relevance to turbulent flow

In turbulent flow, the effective value of Γ is proportional ρu D. So $D \propto \left\{\dfrac{\rho u\ Dc(T_b - T_u)}{H\ R_{fu,max}}\right\}^{\frac{1}{2}}$.

Thus, extinction occurs when $\frac{\rho u\, c(T_b - T_u)}{H\, D\, R_{fu,max}}$ attains a definite value. This is precisely the quantity which proved to be significant for the extinction behaviour of a baffle-stabilised flame in Chapter 17.

When the pilot-flame and main-stream velocities differ, it may be expected that, at extinction, $\frac{\rho u\, c(T_b - T_u)}{H\, D\, R_{fu}}$ equals a value which increases with u_2/u_1, where 1 denotes main stream, and 2 the pilot.

EXERCISES TO FACILITATE ABSORPTION OF MATERIAL IN CHAPTER 19

MULTIPLE-CHOICE PROBLEMS

Spark ignition

19.1 Demonstrate that the differential equations governing the spark-ignition process can be expressed as:

$$\rho \frac{\partial m_{fu}}{\partial t} = \frac{1}{r^2} \frac{\partial}{\partial r} \left(\Gamma r^2 \frac{\partial m_{fu}}{\partial r} \right) - R_{fu} (m_{fu}, h),$$

$$\rho \frac{\partial h}{\partial t} = \frac{1}{r^2} \frac{\partial}{\partial r} \left(\Gamma r^2 \frac{\partial h}{\partial r} \right)$$

List all the simplifying assumptions which are implicit in these equations.

19.2 Demonstrate that, if the $h \sim r$ distribution has a fixed shape, viz.: $h - h_\infty = f(r/\delta)$, integration of the second equation yields:

$$\frac{d}{dt} \int_0^{\frac{1}{2}\delta} r^2 (h - h_\infty)\, dr - \left(\frac{\delta}{2}\right)^2 \frac{d\delta}{dt} \frac{1}{2} \rho_{\frac{1}{2}} (h_{\frac{1}{2}} - h_\infty) = \left(\Gamma r^2 \frac{\partial h}{\partial r}\right)_{\frac{1}{2}}$$

where the subscript $\frac{1}{2}$ denotes evaluation where $r = \delta/2$.

19.3 Derive a relation connecting the spark energy Q with $\int_0^\infty r^2\, h\, dr$.

The following problems consist of: statement, because, statement. Select A, B, C, D, or E, according to the following table:

	First statement	Second statement	Argument
A	true	true	true
B	true	true	false
C	true	false	-
D	false	true	-
E	false	false	-

19.4 Spark ignition can be considered as a special case of spontaneous ignition,

because

the gases ignite as a result of the self-heating produced by the chemical reaction in an isolated volume.

19.5 The gases in the location where a spark has just passed usually have a lower enthalpy than the surrounding unburned gases,

because

although their temperature is higher, some of the fuel will have been consumed.

19.6 Just after the passage of a spark, the direction of diffusion of fuel is radially inward,

because

the value of m_{fu} is lower near the centre than at larger radius.

19.7 The axial temperature profile in a pilot flame which is successful in causing flame propagation usually exhibits a minimum,

because

it is difficult in practice to ensure that the injection velocity is precisely the same as that of the surrounding stream of unburned gas.

19.8 Pilot flames are not often used as flame-stabilising devices in jet-engine after-burner systems,

because

they cause more drag than baffle stabilisers.

19.9 The equations governing spark ignition and propagation from pilots belong to the class known as parabolic,

because

they possess second derivatives with respect to one independent variable but only first derivatives with respect to the other.

19.10 The assumption $\lambda/c = \mathscr{D}\rho$ is fairly close to the truth for gases in which all molecules have approximately the same weight,

<u>because</u>

density is proportional to the reciprocal of the absolute temperature, λ is proportional to T raised to the power 3/4 approximately, and $\mathcal{D}$ is proportional to T raised to the power 7/4 approximately.

19.11 It requires more energy to ignite a stoichiometric mixture than one which is weak in fuel,

<u>because</u>

$(T_b - T_u)$ is a maximum for such mixtures, and the required spark energy is proportional to

$$\{c(T_b - T_u)/(H \int_o^1 \Gamma R_{fu} \, d\tau)\}^{3/2}.$$

19.12 Less energy is required to ignite a flame at high pressure than at low,

<u>because</u>

the energy is proportional to S^{-3} and S usually diminishes somewhat as the pressure rises.

19.13 The minimum spark energy for ignition diminishes as the initial temperature of the mixture rises,

<u>because</u>

the most temperature-sensitive term in the expression for the minimum energy is ρ_u S, and this diminishes as T_u rises.

19.14 A turbulent hot-gas jet injected into stagnant cold combustible gases can be expected to ignite them only if the injection velocity is below a value which is proportional to the jet diameter,

<u>because</u>

the entrainment rate of cold gas per unit axial length is proportional to the velocity times the diameter, while the rate of conversion of material per unit length by chemical reaction is proportional to diameter squared.

ANSWERS TO MULTIPLE-CHOICE QUESTIONS

Answer	Problem number (19's omitted)		
A	6,	9,	14
B	7,	10,	12
C	8,	13	
D	5,	11	
E	4		

CHAPTER 20

COAL-PARTICLE COMBUSTION

INTRODUCTORY NOTE

The combustion of liquid-fuel particles (i.e. droplets) was discussed in early chapters of this book; and the process was found to be physically controlled.

The combustion of solid-fuel (e.g. coal) particles is sometimes a physically controlled process; but chemical-kinetic factors often play an additional part. This is why discussion of the subject has been deferred to this stage of the book.

20.1 ENGINEERING-LEVEL CONSIDERATIONS

A) PRACTICAL OCCURRENCE

(i) Coal particles are burned, as "pulverised fuel" (PF), in power-station boilers. This manner of burning coal is preferred to combustion on grates despite the fact that the grinding process is expensive, because of the resulting flexibility of furnace design. The fuel is carried into the furnace by a stream of air.

(ii) Recently, attention has been given to a special kind of PF combustion, called "fluidized-bed" combustion. In this, the air velocity is just large enough to keep the coal particles "floating", but not high enough to sweep them from the combustion space.

Whereas, in a conventional PF furnace, the heat is transferred to the water-cooled furnace walls, in the fluidised-bed furnace, heat-transfer surfaces may be suspended as coils or tube banks within the combustion space.

(iii) In cement manufacture, where ash can mix harmlessly with the cement, pulverised coal is burned in a long rotating kiln.

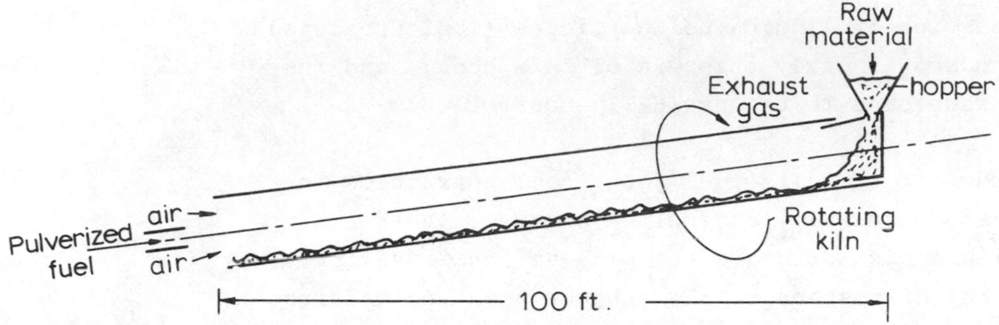

FIG. 20.1 DIAGRAMMATIC REPRESENTATION OF A ROTATING KILN FOR THE MANUFACTURE OF CEMENT.

(B) POWER-STATION USE; SOME FACTS FOR A TYPICAL LARGE INSTALLATION

(i) Boiler arrangement

Fig. 20.2 illustrates a typical installation.

The furnace space is a tall square-sectioned water-cooled space, into which pulverised coal and pre-heated air are blown.

The hot gases rise to the top, and pass out through banks of tubes which first superheat the steam and then pre-heat the water entering the boiler. They pass through the air pre-heater, before being finally exhausted to the chimney.

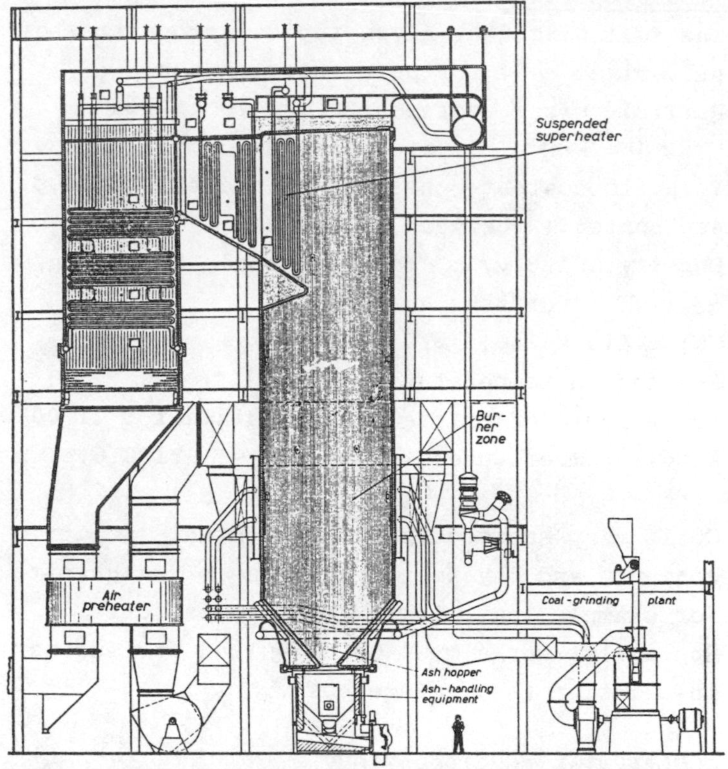

FIG. 20.2 ARRANGEMENT OF A PF-FIRED STEAM BOILER FOR POWER GENERATION. THE FURNACE IS OF THE "DRY-BOTTOM" TYPE, WHICH MEANS THAT ASH PARTICLES ARE EXTRACTED FROM THE BOTTOM AS A POWDER.

The task of the designer is to ensure that combustion is complete within the furnace, that the proper amount of heat is transferred, and that the ash is carried away without clogging the heat-transfer surfaces.

(ii) Fuel Properties

The following list gives typical properties of pulverized coal for power-station use:-

Particle size: between 30 and 70 micron (1 micron $= 10^{-3}$ mm).

Volatile content: between 30 and 40% by mass.

Ash content: between 16 and 24% by mass

Density: 1.3 g/cm^3 for pure coal, 2.6$\sim$5.2 g/cm^3 for ash.

Calorific value: 2.5×10^7 J/kg.

Ash fusion temperature: 1200°C.

Temperature of exit gases if adiabatic: 1700°C.

Actual temperature of exit gases: 1150°C.

Coals vary enormously in ash content, volatile content, and physical properties. Anthracite, for example, has little volatile matter. Specialist works must be consulted for further information (see references).

20.2 FUNDAMENTAL CONSIDERATIONS

A) THE THERMODYNAMICS OF COAL

Although coal has a volatile component, consisting of hydrocarbon gases which are released when the temperature of the coal is raised, this volatilisation is very different from that which characterises the vaporisation of a liquid fuel.

Hydrocarbon gases are released from coal as a consequence of a chemical reaction, not a simple phase change; and, when the temperature is lowered again, the gases do not condense so as to reconstitute the same coal substance.

The release of the gases may indeed be exothermic (i.e. heat-releasing) rather than endothermic (i.e. heat-absorbing), as is vaporisation. Moreover, there is no distinct temperature (like the boiling-point temperature) at which vapour release occurs; but the rate of release increases with rise of temperature, until no further volatiles are given off.

The substance which remains after the volatilisation process still contains much carbon. It is indeed what is known as "coke"; and this is a valuable fuel in its own right. It consists of pure carbon, mixed with incombustible ash.

B) THE PROBLEM TO BE CONSIDERED

There are still no satisfactory mathematical models of the volatilisation of coal and the consequent production of coke particles; and, if there were, they would almost certainly be too complex for presentation in an elementary text-book.

However, it is easy to make a quantitive study of the combustion of the coke particle itself; and, since this is often the step which takes longest, to do so is of practical interest to designers. Of course, for an anthracite which has a negligible content of volatiles in any case, the carbon-combustion phase is the only one of any importance.

In the remainder of this chapter, therefore, the problem considered is: what governs the burning rate of a carbon particle, and so determines the time between its injection into the furnace and the complete combustion of all the carbon in it?

The problem has much in common with that of the burning rate and time of a liquid-fuel droplet, dealt with in Chapter 6. However, whereas chemical kinetics exerted no influence there, the combustion of solid carbon is often kinetically influenced. Further, heat losses and gains by radiation can play a significant part, determining indeed whether chemical reaction can take place at all.

c) RELEVANT PHYSICAL AND CHEMICAL LAWS

(i) The simple carbon-oxygen reaction

Consistently with the practice of the remainder of the book, the oxidation of carbon will be regarded as obeying a chemical equation which causes C and O_2 to combine in a fixed proportion to form a unique oxide. But which oxide? Two equations are possible, as indicated already in section 6.2:

$$\begin{array}{llll} C + & O_2 \longrightarrow & CO_2 & ; \quad (20.2\text{-}1) \\ 12\text{kg} & 32\text{kg} & 44\text{kg} & \end{array}$$

and:

$$\begin{array}{llll} C + & \tfrac{1}{2}O_2 \longrightarrow & CO & \\ 12\text{kg} & 16\text{kg} & 28\text{kg} & . \quad (20.2\text{-}2) \end{array}$$

Which oxide is in practice dominant depends upon the temperature: CO_2 is formed at the lower temperature, and is indeed the desired product; CO is formed at higher temperatures, and is less desired because its formation is accompanied by a smaller heat release.

In the following analysis, therefore, CO_2, i.e. carbon dioxide, will be taken as the only oxidation product; and CO, i.e. carbon monoxide, will be regarded as absent. Therefore, the stoichiometric ratio s will take the value 32/12, i.e. 2.667.

Other features of the simple chemically reacting system will not be needed; Part 1 (see section 6.2) suffices.

(ii) Physical process

Because carbon is non-volatile, it is necessary to consider only the diffusion of one of the reactants, namely oxygen. It will appear that the rate at which oxygen can diffuse to the carbon surface will often set a limit to the maximum rate of combustion.

Fick's Law (Chapter 2) is of course the relevant formulation by which the diffusion rate can be computed.

(iii) Chemical-kinetic processes

When oxygen reaches the carbon-particle surface, it may not react with it chemically. Certainly it will not do so if the surface temperature is too low; and the rate of reaction is found to depend both on the temperature and the local oxygen concentration in accordance with a law of Arrhenius type, which can be written approximately as:

$$G_C = Kpm_{ox,o} \exp(-E/(RT)) \qquad (20.2\text{-}3)$$

Here G_C (kg/m^{2}s) stands for the rate of oxidation of carbon per unit area; K is a constant; p stands for the local gas pressure; $m_{ox,o}$ is the mass fraction of O_2 in the gases immediately adjacent to the carbon surface; E is the activation energy; R is the Universal Gas Constant; and T_o is the absolute temperature of the carbon surface.

A formula of this type can be derived on the presumption that the reaction rate is governed by the product of the rate at which molecules strike the surface with the fraction of the collisions which are sufficiently energetic. The argument, and the accompanying simplifications, qualifications and elaborations, are similar to those introduced in Chapter 14, in connexion with the bi-molecular gas-phase reaction; but no extensive discussion will be provided in the present case.

20.3 MATHEMATICAL MODEL OF CARBON-PARTICLE COMBUSTION

A) MOTIVATION

The purpose of the analysis is to provide formulae, similar to those for the burning droplet of Chapter 7, for the burning rate and time of a particle of solid carbon, suspended in an atmosphere containing oxygen.

Such formulae can aid the designer of furnaces who, even if he does not know precisely how long each injected particle remains in the furnace (because of the complexity of the flow patterns) can at least set a realistic lower limit to this time.

B) DESCRIPTION OF THE MATHEMATICAL MODEL

(i) Origin

The first published study of the present problem was made by Nusselt (1924), in a paper which anticipated much later work. He derived a formula similar to equation (20.3-6) below.

The influence of chemical-kinetic limitations was explored later by Tu, Davis and Hottel (1934).

(ii) Nature

The particle is supposed to be a perfect solid sphere; and the field of concentration around it is also taken as point-symmetrical. The quasi-steady-state assumption is made, as in Chapter 3.

Oxygen diffuses to the surface; carbon dioxide diffuses away; and the heat of combustion is partly conducted and partly radiated to the surroundings.

The conditions near the particle must therefore be qualitatively as indicated in the sketch of Fig. 20-3.

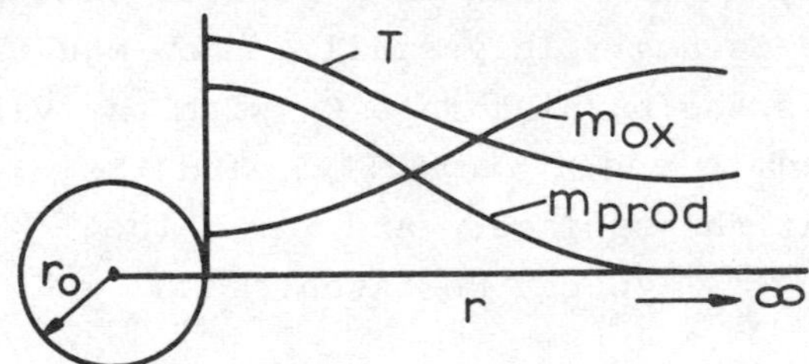

Fig. 20-3. Variations of temperature T and of mass fractions of oxygen and product, m_{ox} and m_{prod}, in the vicinity of a burning carbon particle.

The transport properties of the gas, namely Γ_{ox}, Γ_{prod}, and Γ_h, will be taken as uniform for convenience, so as to permit the deduction of closed-form solutions of the equation.

(iii) Discussion

Spherical symmetry will prevail in practice whenever the particle is very small, and when any relative velocity between it and the supporting gas has had time to die away under the influence of friction. When, however the relative velocity remains large (as in a fluidised bed), the departures from point symmetry will be such as to increase the burning rate.

This aspect of the model is so similar to that of the burning-droplet model as to require no further discussion. Indeed, the solid-carbon particle can be regarded in some respects as a special case of a burning droplet, as will now be demonstrated.

c) THE BURNING-RATE EQUATION

The special feature of carbon as a fuel is that its vapour pressure is negligibly small. Now equation (7.2-5) connects the burning rate G_o with the values of ϕ at the surface and at infinity, and the gradient of ϕ at the surface; and for ϕ the expression $m_{fu} - m_{ox}/S$ can be taken.

But m_{fu}, the mass fraction of carbon vapour, is negligibly small, as already stated. Therefore ϕ can be taken as $-m_{ox}/S$; and equation (7.2-5) becomes:

$$\frac{G_o r_o}{\Gamma_{ox}} = \ln \left[1 + \frac{(m_{ox,o} - m_{ox,\infty})}{\left\{-\Gamma_{ox}\,(dm_{ox}/dr)_o/G_o\right\}} \right] \quad . \ (20.3\text{-}1)$$

It is necessary to give a significance to the quantity in the $\{\ \}$ bracket in this equation. Obviously $\Gamma_{ox}\,(dm_{ox}/dr)_o$ is the diffusion rate of oxygen to the surface; and this is related to the total rate of transfer of oxygen there by:

$$G_{tot,ox} = G_o m_{ox,o} - \Gamma_{ox}\left(\frac{dm_{ox}}{dr}\right)_o \quad , \quad (20.3\text{-}2)$$

where the first term on the right-hand side is the convective contribution.

Now $G_{tot,ox}$ must equal $-s$ times G_C, the reaction rate of carbon; and G_C is also the same thing as G_o, because only carbon is being transferred. Hence:

$$\left\{-\Gamma_{ox}(dm_{ox}/dr)_o/G_o\right\} = -(s + m_{ox,o}) \quad ; \quad (20.3\text{-}3)$$

and so equation (20.3-1) reduces to:

$$\boxed{\frac{G_o r_o}{\Gamma_{ox}} = \ln\left[1 + \frac{m_{ox,\infty} - m_{ox,o}}{s + m_{ox,o}}\right]} \quad . \quad (20.3\text{-}4)$$

This equation permits the burning rate G_o to be computed whenever the oxygen concentration at the surface is known. If, because the carbon is highly reactive chemically, this should fall to zero, the rate becomes:

$$m_{ox,o} = 0 : \frac{G_o r_o}{\Gamma_{ox}} = \ell n\left[1 + \frac{m_{ox,\infty}}{s}\right] \quad . \quad (20.3\text{-}5)$$

Further, since $m_{ox,\infty}$ is usually equal to 0.232 or less, and s equals 2.67, the logarithmic expression on the right-hand side can be expanded to yield the approximate relation:

$$\frac{G_o r_o}{\Gamma_{ox}} = \frac{m_{ox,\infty}}{s} \quad . \quad (20.3\text{-}6)$$

This is the relation derived by Nusselt (1924).

D) THE INFLUENCE OF CHEMICAL KINETICS

A question now to be considered is: What governs the value of $m_{ox,o}$, the oxygen concentration at the surface? The answer is that it is settled by the simultaneous action of equation (20.2-3), which represents the chemical-kinetic influences, and equation (20.3-4) which represents the physical ones.

An algebraic solution of the two equations, if equation (20.3-4) is approximated by an expansion of the logarithm, gives:

$$\frac{G_o r_o}{\Gamma_{ox}} \approx \frac{m_{ox,\infty} - m_{ox,o}}{s + m_{ox,o}} \quad . \qquad (20.3\text{-}7)$$

Now G_C in equation (20.2-3) and G_o in (20.3-7) are identical in meaning. Hence $m_{ox,o}$ can be obtained from the combination of these equations which yields:

$$\frac{Kpm_{ox,o} \exp\left\{-E/(RT)_o\right\} r_o}{\Gamma_{ox}} = \frac{m_{ox,\infty} - m_{ox,o}}{s + m_{ox,o}} \quad . \qquad (20.3\text{-}8)$$

This is a quadratic in $m_{ox,o}$, which may be solved in the usual way. However, the relative importances of physics and chemistry are best seen by re-writing the equation as:

$$\boxed{\frac{m_{ox,o}}{m_{ox,\infty}} = \left[1 + (s + m_{ox,o})\frac{Kpr_o}{\Gamma_{ox}} \exp\left(\frac{-E}{RT}\right)\right]^{-1}} \quad . \qquad (20.3\text{-}9)$$

Inspection of this equation permits the following conclusions to be drawn:

- Because $m_{ox,o}$ is much smaller than s, its appearance on the right-hand side has no influence on the qualitative behaviour of that term.
- When the particle radius r_o is very small, the second term in the square bracket is also small. Then $m_{ox,o}/m_{ox,\infty}$ must be close to unity: the surface and infinite-distance concentrations of

oxygen nearly coincide.

- In the latter case, equation (20.3-4) is not useful for computing the burning rate; but equation (20.2-3) <u>is</u> useful: the process is <u>kinetically controlled</u>.

- If r_o is not too small, and the surface temperature is large, the second term in the bracket of equations (20.3-9) becomes large. The consequence is that $m_{ox,o}/m_{ox,\infty}$ tends to zero. This can be expressed in words as: the oxygen diffusing to the surface is consumed by carbon as soon as it arrives.

- In this case, it is equation (20.2-3) which is useless as a determinant of the combustion rate; but G_o can be easily evaluated from equation (20.3-4). The combustion process is <u>physically controlled</u>.

- At intermediate values of the quantity

$$(s + m_{ox,o}) \frac{Kpr_o}{\Gamma_{ox}} \exp\left(-\frac{E}{RT_o}\right) ,$$

both physical and chemical factors influence the burning rate.

- A generally convenient expression for the burning rate may be obtained by combining (20.3-9) and (20.2-3) to give:

$$\boxed{G_o = m_{ox,\infty} \quad \left\{Kp \exp\left(\frac{-E}{RT_o}\right)\right\}^{-1} + \frac{(s+m_{ox,o})}{\Gamma_{ox}}}$$

(20.3-10)

In this equation, the terms

$\{Kp \exp(-E/RT)\}^{-1}$ and $r_o \dfrac{(s+m_{ox,o})}{\Gamma_{ox}}$

can be regarded as being respectively chemical and physical "resistances" to combustion. The larger resistance always dominates the process.

E) THE PARTICLE BURNING TIME

The variation of radius with time is given by:

$$\frac{dr_o}{dt} = \frac{-G_o}{\rho_C} \quad , \qquad (20.3\text{-}11)$$

wherein ρ_C stands for the density of carbon.

A soluble differential equation for r_o can be obtained by combining this with (20.2-11); and this can be solved analytically if the particle temperature can be taken as uniform and the variation of $m_{ox,o}$ on the right-hand side of (20.3-10) can be neglected. The result is a quadratic relation between t and r_o which, with r_o set equal to zero, connects the particle-burning time t_b with the initial particle radius $r_{o,i}$; it is:

$$t_b = \frac{\rho_C}{m_{ox,\infty}} \left[r_{o,i} \left\{ Kp \exp\left(\frac{-E}{RT_o}\right) \right\}^{-1} + \frac{\frac{1}{2} r_{o,i}^{\,2} (s + m_{ox,o})}{\Gamma_{ox}} \right] \qquad (20.3\text{-}12)$$

Inspection of this equation shows that the burning time is proportional to the square of the initial radius for physically-controlled burning. This accords with the behaviour which was encountered in Chapter 7 for liquid-fuel-droplet burning.

When the combustion process is controlled by the chemical "resistance", on the other hand, the burning time t_b is directly proportional to $r_{o,i}$.

The two extreme cases are thus:

$$T_o \text{ high:} \quad t_b = \frac{\rho_C r_{o,i}^2}{2\Gamma_{ox}} \cdot \frac{s + m_{ox,\infty}}{m_{ox,\infty}} \quad ; \tag{20.3-13}$$

$$T_o \text{ low:} \quad t_b = \rho_C r_{o,i} \Big/ \left\{ m_{ox,\infty} K p \exp\left(\frac{-E}{RT_o}\right) \right\} . \tag{20.3-14}$$

20.4 DISCUSSION

A) THE SURFACE TEMPERATURE, T_O, AND ITS INFLUENCE

A rise in particle temperature shifts the process towards the physically-controlled limit; a lowering of T_o tends to make kinetics control. But what controls the temperature T_o itself?

No detailed thermal analysis will be presented here; but the following qualitative remarks should be understandable; and a detailed account can be found in the specialist literature (e.g. Spalding, 1959, supplied as an appendix to this chapter).

- The particle temperature is controlled by the rate of exothermic combustion at the surface on the one hand, and by the loss of heat by conduction and radiation to the surroundings on the other.

- Because conduction, like diffusion, is inversely proportional to radius, whereas radiation is independent of it, radiation tends to become unimportant as the particle radius diminishes. As a consequence, the temperature of a particle which can "see" cold surroundings tends to increase as burning proceeds.

- The importance of radiation in the early stages of combustion permits coal particles to be ignited by injection, along with air, into a furnace space in which already burning particles are emitting intense radiant heat.

- If radiative heat loss is too intense, chemical reaction may be extinguished: a reduction of the surface temperature leads to a reduction in the reaction rate; this leads to a reduced rate of heat generation, and so to further reduction of temperature. The self-accentuating nature of the process makes heat-loss-induced extinction a sudden "critical-condition" phenomena, just like the excessive-flow-induced extinction which was discussed in Chapters 16 and 17.

- It is indeed a general feature of combustion systems that they can be extinguished either by excessive flow or by deficient flow. In the first case, the reaction is quenched by convection; in the second, it is quenched by heat losses. Discussion of these matters can be found in the papers referred to at the end of the chapter.

B) PRACTICAL CONSEQUENCES

Insertion of appropriate values of Γ_{ox}, ρ_C and $m_{ox,\infty}$ into equation (20.3-12), under the assumption that T_o is very large, leads to the conclusion:

$$t_b \approx 10^8 r_{o,i}^2 \quad , \quad (20.4\text{-}1)$$

where t_b is in seconds and $r_{o,i}$ in metres. Therefore, if $r_{o,i}$ has the typical value of 5×10^{-5}m, t_b is around 0.25 seconds.

Particles of this order of initial size are commonly produced by coal-pulverization equipment; and it follows that residence times of the order of one second or more should be provided by the furnace designer, so as to ensure that the particles are fully burned before leaving.

c) RESPECTS IN WHICH THE MATHEMATICAL MODEL DIFFERS FROM REALITY

It is wise, before leaving the discussion of coal-particle combustion, to recall the major ways in which reality differs from what has been supposed. Attention will be given first to features which have not been emphasised above; thereafter the more obvious short-comings of the model will be recalled.

(i) The influence of ash

Equation (20.3-11) implies that the particle radius diminishes as the carbon burns away. This is an over-simplification; for an ash residue is left, which may form a porous crust of diameter about equal to that of the original particle. This crust may present an additional resistance to the diffusion of oxygen; and it may also act to some extent as a barrier to heat loss by radiation from the burning carbon surface.

(ii) Swelling during devolatilisation

The evolution of volatile gases from a particle of coal is accompanied, as a rule, by an increase in it size. Bubbles form within the particle; and, by the time that devolatilisation is complete, the particle has acquired a porous structure. Its diameter may then be several times as great as that of the original coal.

This phenomenon reduces the mathematical model of the present Chapter to the status of a mere guide as to order-of-magnitude effects, and to qualitative trends: predictions of absolute accuracy are not to be looked for. However, it is seldom that knowledge of the particle-size distribution of the pulverized fuel is highly accurate; so the uncertainties produced by the swelling process are not uniquely serious.

(iii) Consumption of O_2

It has been tacitly assumed, during the analysis, that the oxygen concentration in the bulk of the gas surrounding the particles is uniform, and indeed equal to that of the air injected with the coal particles. Of course, this assumption is invalid; for, as the combustion of the coal proceeds, the oxygen is used up; and the designer prides himself on supplying only the smallest permissible proportion of excess air, so that the exhaust gases shall not carry away too much energy.

Therefore, a more exact analysis will take account of the reduction in $m_{ox,\infty}$ in the course of combustion. Often it will be sufficient to take $m_{ox,\infty}$ as possessing an "average" value, intermediate between that in the air supply and that of the combustion products leaving the furnace. Of course, the calculated burning time will be inversely proportional to this average value of $m_{ox,\infty}$.

(iv) Whereas the last-mentioned effect increases the burning time, the one to be discussed now decreases it: <u>this is combustion to the monoxide</u>, CO, instead of the dioxide, CO_2.

As was explained in section 20.2(c) above, which oxide is formed depends on the temperature of the carbon: the higher the temperature, the greater the proportion of the monoxide.

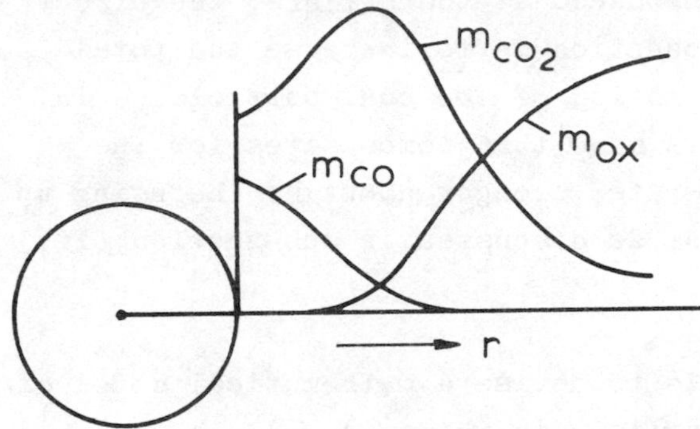

Fig. 20-4 Sketch of concentration profiles in the neighbourhood of a carbon particle at high temperature, for which CO is the first oxidation product.

When the surface temperature is very high, indeed, the conditions in the neighbourhood of a burning carbon particle may be as sketched in Fig. 20-4. No molecular oxygen actually penetrates to the surface; and the carbon may be said to be "burning" in carbon dioxide, CO_2, according to the chemical equation:

$$\begin{array}{llll} C & + \quad CO_2 & \longrightarrow & 2CO \\ 12kg & \quad 44kg & & 2 \times 28\ kg \end{array} \quad . \quad (20.4\text{-}2)$$

This situation is not unlike that of the burning droplet, discussed in Chapter 7 : and the CO acts as a kind of "fuel vapour", which is ultimately oxidized well away from the phase interface. The carbon-monoxide flame can indeed often be observed as a sheath of pale-blue gas, enveloping the particle.

Because the oxygen has less far to diffuse than when CO_2 is produced at the surface, the effect of the CO-production is to increase the total rate of consumption of the coal particle. In practice, this more than compensates for the reduction in rate, brought about by the using up of the oxygen, as discussed in sub-section (iii) above.

It is possible to devise a mathematical model of the carbon particle in which CO, CO_2, and indeed oxides of hydrogen, are all brought into account; and their various transport, thermodynamics and chemical-kinetic properties are realistically incorporated. However, such a model would lie outside the modest scope of the present work.

(v) Lastly, the effect should be mentioned of the relative velocity of the particles and the surrounding gas. As was mentioned in connexion with vaporising and burning droplets in earlier chapters, this effect is always to <u>increase</u> the burning rate and so to decrease the burning time. This is true whether the relative velocity is brought about by the momentum of injection, by the effect of gravity, or by the inability of the denser particles to follow the turbulent fluctuations of the lighter gases.

D) CLOSURE

In conclusion, it may be remarked that the mathematical model of coal-particle combustion, like many of the mathematical models discussed in this book, is valuable primarily as a guide to the designer's imagination; and it requires detailed augmentation before it can be used as part of a quantitative design procedure, in which order of magnitude accuracy is not enough.

However, an educated imagination is a necessity, whereas the detailed analysis can often be dispensed with. The reader who terminates his combustion studies at this point can therefore still feel confident that they have been of value to him.

20.5 REFERENCES

DOLEZAL R
"Large boiler furnaces".
Elsevier, 1967.

NACK H, KIANG K D, LIN K T, MURTHY K S, SMITHSON G R Jr & OXLEY J H
"Fluidised-bed combustion review".
in
KEARNS D L "Fluidisation Technology, Vol. II".
Hemisphere Publishing Corporation, Washington.

NUSSELT W
Z. VDI., Vol. 68, pp 124-128, 1924.

SPALDING D B & TALL B S
"Flame stabilisation in high velocity gas streams and the effect of heat losses at low pressure".
The Aeronautical Quarterly, Vol. V, pp 195-217, 1954.

SPALDING D B
"Some fundamentals of combustion".
Butterworth's, London, 1955.

SPALDING D B
"The stability of steady exothermic chemical reactions in simple non-adiabatic systems".
Chem. Engg. Sci., Vol. 11, pp 53-60, 1959.
(Supplied as an appendix to this chapter)

TU C M, DAVIS H & HOTTEL H C
"Combustion rate of carbon".
Ind. & Engg. Chem., Vol. 26, No. 7, p 749, 1934.

VULIS L A
"Thermal regimes of combustion".
McGraw Hill, New York, 1961.

EXERCISES TO FACILITATE ABSORPTION OF MATERIAL IN CHAPTER 20

MULTIPLE-CHOICE PROBLEMS

20.1 All the following statements are true except:

A Nearly all modern coal-fired power stations burn the coal in pulverised form.

B The greater part of the coal used in power stations is anthracite, which has a very low content of volatile matter.

C The coal is conveyed to the power station in the form of lumps or coarse grains, and is pulverised in grinding mills shortly before use.

D Power-station coal usually has a higher ash content and lower calorific value than coal employed in domestic installations.

E The pulverised coal is conveyed into the combustion space by a stream of air.

20.2 All the following statements are true except:

A The major part of the ash content of pulverised coal is collected as a dry powder. Although some industrial uses have been found for it, its disposal remains a troublesome problem.

B Some power-station boilers are arranged so that a large proportion of the ash collects in the furnace in the form of a liquid, which can be tapped off periodically.

C The ashes from coals originating from different fields differ significantly in chemical composition and relevant physical properties.

D Because of the build-up of ash on superheater tubes, power-station boilers usually have to be completely shut down for cleaning at least once per week.

E In the majority of power station boilers, the temperature in the furnace is kept sufficiently low for the ash particles to remain in the solid state.

20.3 The rate of diffusion of oxygen molecules across an imaginary surface in a gaseous mixture is proportional to the gradient normal to that surface of the concentration of oxygen, when concentration is expressed as:

A Mass of oxygen per unit volume of mixture.

B Number of moles of oxygen per unit volume of mixture.

C Mass flux of oxygen molecules divided by total mass flux of all molecules.

D Mass of oxygen per unit mass of mixture.

E None of the above.

20.4 All the following statements are true except:

A Fick's law of diffusion is a close analogue to Fourier's law of heat conduction.

B The diffusion coefficient of a component of a gaseous mixture usually increases as the temperature rises.

C The product of the diffusion coefficient and the gas-mixture density is usually of the same order of magnitude as the viscosity of the mixture.

D Light molecules, or atoms, usually have higher diffusion coefficients than heavy ones.

E The diffusion coefficient of a component of a gaseous mixture is substantially independent of both the mass fraction of that component and the pressure of the mixture.

20.5 All of the following statements are true of the model of a burning coal particle that is treated in the text except:

A The particle is a solid sphere of carbon.

B The ash is supposed to be uniformly interspersed with the carbon throughout the sphere.

C The Reynolds number of the particle motion is so low that only radial transfer of oxygen and other gas components needs to be considered.

D The density of the sphere remains uniform.

E The diffusion rates can be calculated from the steady-state equations, despite the fact that the particle diameter varies with time.

20.6 In order that the burning time of a model coal particle should be proportional to the square of its initial diameter, all the following conditions are necessary except:

A The temperature of the gas should be uniform at any instant.

B The temperature of the coal surface must be high enough for the oxygen to burn just as soon as it reaches the surface.

C The relative velocity of the particle and the bulk of the gas should be very low.

D The temperatures of the coal particle and of the bulk of the gas should remain constant.

E The oxygen concentration in the bulk of the gas should not vary.

The rate of oxidation per unit area of a small solid carbon particle, suspended in an atmosphere of air, is proportional:

20.7 when the surface temperature is low,

20.8 when the surface temperature is high,

to the diameter of the sphere raised to the power:

A -2.

B -1.

C 0.

D 1.

E 2.

20.9 At high temperatures of gas and solid, it is possible for carbon to react with oxygen to form carbon monoxide directly. By consideration of the extreme case in which carbon dioxide is entirely absent, one can determine that the particle will completely disappear in a time which, compared with the burning time for diffusion-controlled burning to CO_2, is approximately:

A Twice as great.

B $2^{\frac{1}{2}}$ times as great.

C Just the same.

D $2^{-\frac{1}{2}}$ times as great.

E Half as great.

For diffusion-controlled combustion to CO_2, the model considered in the text predicts that the burning time is approximately:

$\frac{2.67}{8} \frac{\rho c D_o{}^2}{\Gamma_{ox} m_{ox,\infty}}$, where D_o is the initial diameter. Of the features of the practical situation which differ from the mathematical model, the one tending to:

20.10 increase the constant,

20.11 decrease the constant,

20.12 leave the constant unchanged,

20.13 alter the constant in a direction that cannot be predicted,

is:

A The swelling of the particles, caused by the formation of internal gas pockets, at the same time as the volatile matter is released from the coal.

B The fact that the particles have a low temperature when they enter the furnace.

C The fact that at least some of the carbon burns directly to the monoxide rather than to the dioxide.

D The pressure of the gas varies somewhat because the furnace is so tall.

E None of the above.

20.14 The combustion efficiency of a boiler is likely to be diminished by all of the following except:

A An increase in the diameters of the pulverised-fuel particles.

B A lowering of the temperature of the particles entering the furnace.

C A proportionate increase in the flow rates of fuel and air entering the furnace.

D An increase in the volume of the furnace.

E An increase in the temperature of the walls of the furnace.

20.15 The reaction rate at a coal ∿ air interface of low temperature is very small
<u>because</u>
$\exp(E/RT_S)$ is then very large.

20.16 If the surface temperature of a carbon particle, suspended in air, were raised steadily; the burning rate per unit surface area would rise to a maximum and then fall
<u>because</u>
$\exp(-E/RT_S)$ also rises to a maximum and then falls.

20.17 Under conditions of diffusional control, the rate of burning of a particle decreases with rise of temperature of gas and particle

because

$\exp(E/RT_S)$ falls steadily.

20.18 If the mass of oxygen per unit volume is uniform throughout the space between two surfaces, no oxygen can diffuse across the space

because

Fick's law states that the diffusion rate is proportional to the mass per unit volume.

20.19 A cloud of inflammable dust, mixed with air, can burn almost as rapidly as a gaseous fuel

because

when the particle-size gas-to-surface diffusion takes place so rapidly as to introduce no significant delay.

ANSWERS TO MULTIPLE-CHOICE QUESTIONS

Answer	Problem number (20's omitted)
A	6, 13, 15, 19
B	1, 5, 7
C	8, 11
D	2, 3, 12, 14, 17
E	4, 9, 18, 16

APPENDIX TO CHAPTER 20

The stability of steady exothermic chemical reactions in simple non-adiabatic systems

D. B. SPALDING

Mechanical Engineering Department, Imperial College of Science and Technology, London

(*Received 28th January 1959*)

Abstract--The stirred homogeneous reactor, the solid catalytic surface and the solid fuel surface are shown to be governed by the same pair of algebraic equations, expressing energy and material conservation. A relation is derived between the temperature of the reaction region and two dimensionless controlling parameters, one representing the chemical loading, the other representing heat loss. The well-known phenomena of extinction, stable and unstable burning equilibria, and the existence of an upper permissible limit to the heat loss are then demonstrated. The paper does not present new physical insights, but aims at unifying, simplifying and clarifying this branch of reactor theory.

This reprint is supplied as an appendix to Chapter 20, rather than as an integral part of the text, partly so as to emphasise that it lies outside the scope of the elementary treatment of the book, partly so as to give interested readers an idea of the style and content of typical scientific publications in the field, and partly because of the intrinsic interest of the material. In the latter regard, attention is drawn to the fact that the paper emphasises the unity of the behaviour patterns of reacting systems which, at first sight, are different in character.

1. INTRODUCTION

1.1 *Purpose*

NUMEROUS investigations [1–11] have been made of the theory of systems involving steady exothermic chemical reaction with heat loss to the surroundings. The present paper concerns those dealt with in the first ten reference papers. Its intention is to point out the formal similarity between the systems, to simplify and unify their treatment, and to present particular solutions of the equations in terms of dimensionless variables. The paper is a tidying-up operation, and does not disclose new physical phenomena.

1.2 *Systems considered*

The first system is the ideal stirred reactor. This is a vessel through which a combustible gas mixture flows steadily; mixing is supposed so intense that the gas state (temperature and composition) is uniform at all points within the vessel. Unlike the early treatments of [12, 13, 14, 15], but in accordance with those of [7, 9, 9a], heat transfer is supposed to occur from the gas to the reactor walls.

The second system is the solid catalyst surface, past which reactive gas flows steadily. The surface temperature, and the composition of the gas immediately adjacent to it, are supposed to have taken up steady values controlled by the exothermic reaction at the surface and by heat loss from the surface ; the latter might be by radiation to cold vessel walls. This system has already been studied in [1], which also contains references to other early work in the field of reaction stability.

The third system is the solid-fuel surface, for example carbon, which is suspended in a stream of oxidizing gas, for example air. Once again, heat transfer other than by convection between fuel and gas is allowed for. The first analysis of this system, but without heat loss, was by WAGNER [16] ; analyses with heat loss have been made in [1, 2, 3, 4, 5, 6].

In the last two systems the reactive region is the solid surface. The gas composition and temperature adjacent to it are assumed uniform ; this is often the case in practice over sufficiently large regions for the analysis to be valid.

The three systems must be distinguished from others which exhibit similar behaviour in some respects but which are rendered harder to analyse by non-uniformity of the reaction regions. Such other systems are the laminar flame in a premixed gas, in which the temperature is non-uniform [10, 11], and the gaseous diffusion flame in which the fuel-oxidant ratio is non-uniform as well [17, 18, 19].

It will be supposed below that in each case the reaction rate depends on gas temperature and composition in accordance with relatively simple laws, i.e. increasing according to a power law with temperature rise and linearly with fuel or oxidant concentration. The extension to other reaction-rate functions, for example those of the Arrhenius type, presents no special difficulty. It is in any case always possible to find an exponent for the power law which fits the experimental data over a restricted temperature range.

1.3 *Mathematical features*

It will appear that the mathematical problem reduces to the simultaneous solution of two *algebraic* equations, which may be reduced, for the composition-dependence assumed, to a single quadratic equation. Solution is therefore easy. Trial-and-error procedures may be necessary with other reaction-rate functions.

The laminar propagating flame and the diffusion flame on the other hand involve the solution of *differential* equations ; it is for this reason that they are excluded from the present treatment.

In each case, the condition of the reaction zone will be shown to depend on the values of two dimensionless parameters. The first of these involves the ratio of the mass flow rate to the reaction zone divided by the maximum possible reaction rate for the system if operating adiabatically ; it will be called the *loading parameter*, L. The second involves the ratio of the maximum possible heat transfer rate from the reaction zone to the same reference reaction rate ; it will be called the *heat loss parameter*, Q.

The solution of the problem will be seen to have some striking characteristics. For fixed values of L and Q, in general three different reaction-zone conditions satisfy the equations. Only two of these represent physically stable conditions.

If L is too large or too small, two of the solutions become imaginary. If Q is too large, there is no real value of L for which three real solutions can be found.

2. The Equations

2.1 *The stirred reactor*

It will be supposed that the volumetric reaction rate, $\dot{q}'''$ in, say, cal/cm³sec, can be expressed by the formula :

$$\dot{q}''' = K\dot{q}'''_{\mathrm{m}}\, v\, \alpha\, \tau^n \tag{1}$$

where

$\dot{q}'''_{\mathrm{m}}$ is a constant of the fuel and oxidant streams and the reactor pressure ;

v is the reactor volume ;

α is the fraction of the initial fuel mass which is still unburned ;

τ is the reactedness (dimensionless temperature rise above unburned state)

$K = (n+1)^{n+1}/n^n$, a constant which ensures that the maximum possible value of $\dot{q}'''$ is $\dot{q}'''_{\mathrm{m}}$, and n is a constant of the reaction, chosen so as to fit approximately the experimental data.

Equation (1) implies that the reaction is first order with respect to fuel. The temperature dependence of the reaction rate is measured by the value of n.

It is further assumed that the heat loss from the gas to the reactor walls in unit time, $\dot{Q}$ is given by

$$\dot{Q} = \dot{Q}_{\mathrm{m}}\, \tau^{m} \tag{2}$$

where m is a constant depending on the mode of heat transfer. Once again, the exponent is shown so as approximately to fit the experimental data over the important part of the temperature range.

For constant gas specific heat, c, the steady-flow energy equation for the reactor can then be written as

$$\dot{m}\, c\, (T_{\mathrm{b}} - T_{\mathrm{u}})\, \tau = K\, \dot{q}_{\mathrm{m}}^{'''}\, v\, \alpha\, \tau^{n} - \dot{Q}_{\mathrm{m}}\, \tau^{m} \tag{3}$$

where

$\dot{m}$ = mass flow rate of entering mixture

$T_{\mathrm{b}} - T_{\mathrm{u}}$ = temperature rise of mixture in complete adiabatic steady-flow reaction.

The conservation-of-mass principle applied to the fuel or oxidant flowing through the reactor yields the equation :—

$$\dot{m}\, (1 - \alpha) = K\, \dot{q}_{\mathrm{m}}^{'''}\, v\, \alpha\, \tau^{n} / c\, (T_{\mathrm{b}} - T_{\mathrm{u}}) \tag{4}$$

wherein the left-hand side represents the difference between the inflow and the outflow, while the right-hand side represents the rate of disappearacne of the reactant due to chemical reaction.

Dimensionless form of the equation. We now introduce the dimensionless loading and heat-loss parameters by the definition :—

$$L \equiv \dot{m}\, c\, (T_{\mathrm{b}} - T_{\mathrm{u}}) / \dot{q}_{\mathrm{m}}^{'''}\, v \tag{5}$$

and $$Q \equiv \dot{Q}_{\mathrm{m}} / \dot{q}_{\mathrm{m}}^{'''}\, v \tag{6}$$

Introducing these definitions into (3) and (4), we obtain the pair of equations governing the behaviour of the stirred reactor. They are :—

$$L\tau = K\, \alpha\, \tau^{n} - Q\, \tau^{m} \tag{7}$$

and $$L\, (1 - \alpha) = K\, \alpha\, \tau^{n} \tag{8}$$

Here we leave the matter until the other systems have been dealt with. It will appear that they yield equations which are almost identical with (7) and (8).

2.2 *The solid catalyst surface*

It will be supposed that the reaction rate at the catalyst surface, $\dot{q}''$, in, say, cal/cm²sec, is given by

$$\dot{q}'' = K \dot{q}_{\mathrm{m}}''\, \alpha\, \tau^{n} \tag{9}$$

where

$\dot{q}_{\mathrm{m}}''$ is the maximum possible of reaction rate at the catalyst surface, with the given gas stream, when non-convective heat loss is prevented and Reynolds Analogy holds (these conditions lead to $\alpha = 1 - \tau$). $\dot{q}_{\mathrm{m}}''$ is a property of the catalyst, the reactants, and the stream temperature and pressure.

α is the concentration in the gas at the surface of the less plentiful reactant, divided by its concentration in the unreacted stream.

τ is the surface temperature minus the stream temperature divided by $T_{\mathrm{b}} - T_{\mathrm{u}}$, the temperature rise of the gas in adiabatic complete reaction.

Reaction in the gas phase is neglected.

Heat transfer from the catalyst to its non-gaseous surroundings, $\dot{Q}''$, is supposed to obey the equation

$$\dot{Q}'' = \dot{Q}_{\mathrm{m}}''\, \tau^{m} \tag{10}$$

where $\dot{Q}_{\mathrm{m}}''$ is the heat transfer rate when the catalyst surface is at the temperature of the adiabatic combustion products.

The catalyst surface is supposed to be engaged in steady heat and mass transfer with the combustible stream obeying linear laws, such that the convective heat transfer rate per unit area is

$$g\, c\, (T_{\mathrm{b}} - T_{\mathrm{u}})$$

and the reactant transfer rate per unit area is

$$\frac{g}{\sigma}\, m_{\mathrm{fu}}\, (1 - \alpha) \tag{11}$$

where

m_{fu} is the fuel concentration in the unreacted stream,

g is the mass of stream fluid reaching temperature equilibrium with unit area of catalyst surface in unit time,

g/σ is the mass of stream fluid reaching composition equilibrium with unit area of the catalyst surface in unit time, and

σ, according to the Chilton–Colburn analogy, is the Lewis No. (molecular diffusivity of fuel divided by thermal diffusivity of mixture) to the minus two-thirds power. When the Lewis No. equals unity, Reynolds Analogy holds for heat and mass transfer, though not necessarily for friction.

g is related to the more usual heat transfer coefficient h by

$$g = h/c \tag{12}$$

where c is the specific heat of the gas at constant pressure.

Conservation equations. The equations of conservation of energy and material can now be written for an element of the catalyst surface in the steady state. They are respectively:

$$gc\,(T_b - T_u)\,\tau = K\,\dot{q}''_m \,.\, \alpha\,\tau^n - \dot{Q}''_m\,\tau^m \tag{13}$$

and

$$\frac{g}{\sigma}\,m_{fu}\,(1-\alpha) = \frac{K\,\dot{q}''_m\,\alpha\,\tau^n}{c\,(T_b - T_u)/m_{fu}} \tag{14}$$

wherein it is recognized that $c\,(T_b - T_u)$ is equal to m_{fu} times the heat of reaction of the combustible.

Dimensionless form of equations. We now introduce the dimensionless loading and heat loss parameters in the form

$$L \equiv g\,c\,(T_b - T_u)/\dot{q}''_m \tag{15}$$

$$Q \equiv \dot{Q}''_m/\dot{q}''_m \tag{16}$$

which, substituted in (13) and (14), lead to

$$L\,\tau = K\,\alpha\,\tau^n - Q\,\tau^m \tag{17}$$

$$\frac{L\,(1-\alpha)}{\sigma} = K\,\alpha\,\tau^n \tag{18}$$

Comparison of (17) and (18) with (7) and (8) shows that the two pairs are identical except for the presence of σ in (18). This difference disappears if the Lewis Number is unity, i.e. when the Reynolds Analogy holds.

2.3 *The solid fuel surface*

We consider a solid fuel surface suspended in an oxidizing gas stream; it will be supposed, for simplicity, that only one oxide can be formed. The reaction will be be taken as first order in surface oxygen concentration, so that the reaction rate, $\dot{q}''$, in, say, cal/cm²sec is given by

$$\dot{q}'' = K\,\dot{q}''_m\,\alpha\,\tau^n \tag{19}$$

where

$\dot{q}''_m$ is the maximum possible of reaction rate at the fuel surface with the given gas stream, when non-convective heat transfer is absent and Reynolds Analogy holds (then $\alpha = 1 - \tau$). $\dot{q}''_m$ is a property of the fuel and of the gas stream composition, pressure and temperature;

α is m_{OS}/m_{OG}, the surface fractional mass oxygen concentration divided by that in the gas stream;

τ is defined as $(T_s - T_u)/(T_b - T_u)$ where T_s is the surface temperature, T_u is the temperature of the gas stream, and T_b is the temperature attained by a stoichiometric mixture of gas and fuel, the latter being supplied at T_b.

Equation (19) is formally identical with (9). Similarly, equation (10) can be taken as representing the non-convective heat loss from the fuel surface, as well as that from the catalyst, while the convective term is again represented by, $gc\,(T_b - T_u)\,\tau$.

The rate of transfer of oxygen to the surface, $\dot{m}''$ per unit area, is given mass transfer theory (e g. [20]) as

$$\dot{m}''_O = \frac{g}{\sigma}\,\frac{m_{OG} - m_{OS}}{1 + m_{OS}/r} \tag{20}$$

where

g, σ, m_{OG}, and m_{OS} are as above,

r is the stoichiometric ratio, mass of oxygen required for unit mass of fuel.

This transfer rate, in the steady state, equals the rate of oxygen consumption, S_0

$$\dot{m}''_O = m_{OG}\,\dot{q}''/c\,(T_b - T_u) \tag{21}$$

where c is the mean specific heat of the gas stream or products at constant pressure, and it is recognised that $c\,(T_b - T_u)/m_{OG}$ is the heat of combustion per unit mass of oxygen.

Conservation equations. Since (19) is identical with (9), equation (13) expresses the conservation-of-energy principle for the solid fuel surface also.

The conservation of mass is represented by combining equations (20) and (21).

After introduction of the same dimensionless loading and heat loss parameters from (15) and (16), the dimensionless energy- and mass-conservation equations becomes

$$L\,\tau = K\,\alpha\,\tau^n - Q\,\tau^m \qquad (22)$$

$$\frac{L}{\sigma}\cdot\frac{(1-\alpha)}{1+\alpha\,m_{OG}/r} = K\,\alpha\,\tau^n \qquad (23)$$

The former equation is identical with (7) and (17), which hold for the stirred reactor and for the catalyst surface. Equation (23) differs slightly both from (8) and (18), through the presence of the σ and $(1+\alpha\,m_{OG}/r)$ terms.

2.4 *Comparison of three systems*

We shall consider, for simplicity, the case in which the Lewis Number is unity. Then σ disappears from equations (18) and (23). Further, we note that m_{OG}/r is of the order of 0·1 for many practical systems; since, in addition, α is considerably less than unity under most conditions of interest, the term $\alpha\,m_{OG}/r$ will be neglected.

These simplifications ensure that equations (7) and (8) now hold for each of the three physical systems; their solution therefore is also valid for the three systems. From now on, therefore, the stirred reactor, the catalytic surface and the solid fuel surface will all be simultaneously under discussion.

3. Solution of the Equations

3.1 *Analytical solution*

In practice, the operator of a combustion plant controls the mass flow to the reaction region, while the heat loss relations are fixed by the geometry; the interplay of mass flow, heat flow and reaction then settles what temperature the reaction region will take up.

Correspondingly, we shall want to use equations (7) and (8) to tell us what τ is when L and Q have fixed values. It is however easier to develop a formula for L in terms of τ and Q. This follows by noting that, by eliminating $K\,\alpha\,\tau^n$ between (7) and (8), we find

$$\alpha = 1 - \tau - Q\,\tau^m/L \qquad (24)$$

After substitution of (24) in (7), there results a quadratic equation in L, with solution:

$$L = \frac{K}{2}\,\tau^{n-1}\,(1-\tau)\times \{1 - (Q/K\,(1-\tau)\,\tau^{n-m}) \pm \pm\sqrt{[1 - (2Q\,(1+\tau)/K\,(1-\tau)^2\,\tau^{n-m}) + + (Q^2/K^2\,(1-\tau)^2\,\tau^{2n-2m})]}\} \qquad (25)$$

3.2 *Graphical representation*

Equation (25) has been evaluated for various values of Q. The temperature exponents used were: $n = 8$, $m = 4$. The results are represented graphically in Fig. 1, with L as independent variable, τ as dependent variable, and Q as parameter.

It is evident that, for fixed Q, the $L - \tau$ relation is a closed loop; so for each value of τ there are in general two L's [as is deducible at once from the $\pm$ sign in (25)]; but also, for each L, there are in general two τ's (or rather three, since $\tau = 0$ is a solution for all L and Q).

The area occupied by the loop diminishes as Q increases. At $Q = 0{\cdot}155$ the loop encloses a tiny area around the point $L = 0{\cdot}163$, $\tau = 0{\cdot}64$; for higher values of Q there are no real pairs of values of L and τ which satisfy the equation at all.

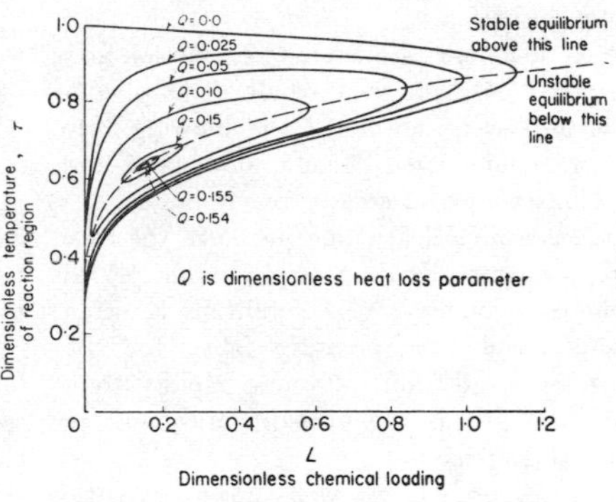

Fig. 1.

For a given Q the real range of L for which there is a real value of τ is bounded by the values for which the tangent to the loop is vertical. The corresponding points on the loops are joined by a broken line in Fig. 1.

The general shape of the curves of Fig. 1 is typical of those cases for which $n > m$, i.e. for which reaction rate depends more strongly on temperature than does heat loss, as is invariably the case in combustion practice.

4. Discussion of Solution

4.1 *Physical significance*

The loading parameter L, it will be recalled, is a dimensionless measure of the mass flow rate through the stirred reactor or of the mass transfer rate between the reacting surface and the gas stream. Inspection of Fig. 1 shows that, when L and the heat-loss parameter Q are fixed, the reaction region can have three possible temperatures, one of which is zero (i.e. equal to that of the gas stream).

Considering for a moment the highest of these three temperatures, (i.e. the upper half of a loop), we see that over a large part of the range an increase of L (blowing rate) causes the temperature to rise. This is a common phenomenon with burning carbon. Eventually however the temperature falls again, until, for an L value greater than that giving a vertical tangent, no real temperature can be found which satisfies the steady-burning equations: the flame has been extinguished.

For fixed heat-loss parameter Q, extinction also occurs when L is made very small. This phenomenon is also well-known: if the blowing rate is too low, heat losses become dominant and prevent continuance of steady reaction.

If Q exceeds a critical value (0·155 in the case considered) no reaction is possible at all. With fixed reactor geometry, this condition is often encountered when the pressure falls; for the reaction-rate usually falls off more rapidly than the heat loss rate as the pressure falls, and so their ratio Q decreases.

$Q = 0$ corresponds to the well-known adiabatic reactor studied by Wagner [16], van Heerden [12], Longwell [14] and others. The low-L extinction condition has now become $L = 0$, and so has ceased to be of practical importance.

4.2 *Stability*

It will now be demonstrated that the intermediate value of τ for fixed L and Q represents a condition which cannot be obtained in practice; for, although this condition satisfies the steady-state equations, the equilibrium is unstable.

Suppose that a reaction system is operating at such a condition, represented by U in Fig. 2. Now let the temperature of the system rise infinitesimally to U'. Will equilibrium be restored? The point U' lies on a steady-state curve which has a higher Q than that prevailing in the system. The system therefore cannot get rid of all the heat which is developed; its temperature correspondingly rises. So a small upward deviation from the steady-state point U causes the system to deviate still further upward.

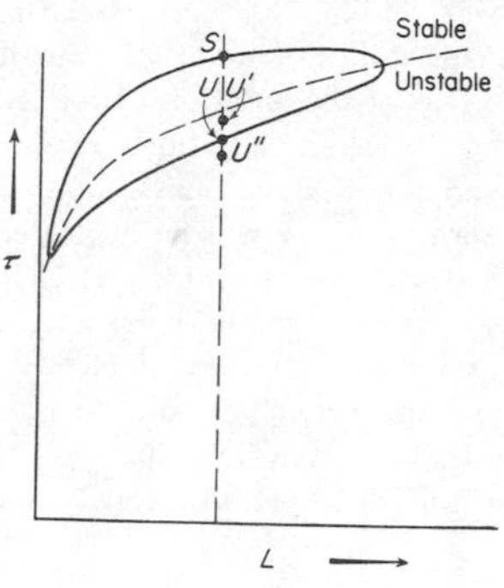

Fig. 2.

Similar arguments show that a downward deviation to U'' precipitates a further fall of temperature. The whole U branch of the loop is therefore unstable.

When points on the upper (S) branch of the loop are considered in this manner, it is easily seen that a small deviation from equilibrium so changes the heat development rate that the temperature tends to be restored to its equilibrium value for the L and Q in question. The S

branch therefore represents stable equilibrium states.

The trivial $\tau = 0$ solution is also stable.

4.3 *Remarks*

(*a*) *Influence of simplifying assumptions.* Equation (25) is only valid for all three systems when $\sigma = 1$ and $m_{OG}/r = 0$, and indeed when the reaction rate and heat loss depend exponentially on τ. In most practical cases these assumptions will not hold exactly. It should however be obvious that the insertion of more realistic functions into the equations will not fundamentally alter the character of the solution.

Extinction at the upper limit of L is caused by the reaction rate increasing more steeply than linearly with reactedness τ; extinction at the lower limit of L is caused by a steeper temperature dependence of reaction rate than that of heat loss. These conditions almost invariably prevail.

The actual run of the curves on Fig. 1, and in particular the upper limit of Q, naturally depend on the values of n and m or of other parameters (e.g. activation energy) describing the form of the reaction-rate and heat-loss functions. Each case has to be worked out as required. However the upper limit of Q is always likely to be less than unity.

Realistic reaction-rate functions differ from the exponential ones considered in not giving zero values at $\tau = 0$. This ensures that the $Q = 0$ curve, for example, does not quite touch the line $L = 0$ but sweeps round and becomes asymptotic to the line $\tau = 0$. Physically, this feature permits spontaneous ignition at low values of L.

Similar characteristics may appear where the temperature of the reservoir to which the reaction region loses heat is higher than that of the incoming stream. This is true of some furnace situations for example. Corresponding changes are easily made to the equations.

(*b*) " *Chemical resistance* " *to reaction.* In calculations of the rate of reaction at catalyst or solid fuel surfaces, it is common to suppose that mass transfer alone controls the rate, that is to say that the concentration of the gas-phase reactant is reduced to zero at the surface. We can use the present analysis to see whether this is true.

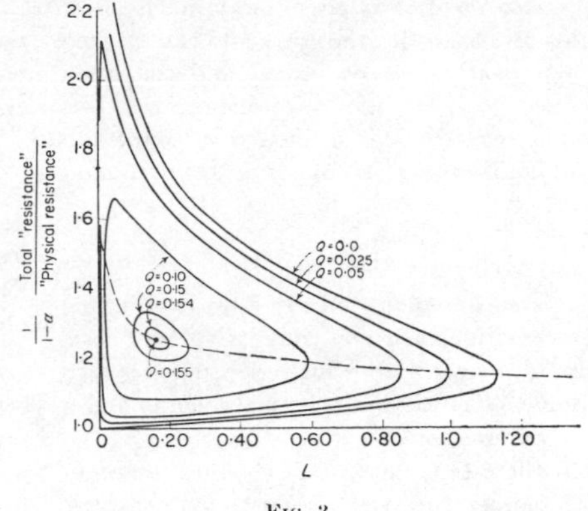

FIG. 3.

Figure 3 shows curves of $1/(1 - \alpha)$ versus L for various values of Q; the values $n = 8$, $m = 4$ have been used as for Fig. 1. Now it may easily be shown that $1/(1 - \alpha)$ is equal to the total " resistance " to reaction, divided by the " resistance " of the mass transfer process alone. " Resistance " is defined as stream reactant concentration divided by reaction rate at the surface.

Inspection of Fig. 2 shows that when heat loss is absent ($Q = 0$), $1/(1 - \alpha)$ has the value unity at low L and increases by 15 per cent over the whole range of stable reaction. This 15 per cent is the contribution of " chemical resistance."

When Q is finite, $1/(1 - \alpha)$ is always greater than unity, and can exceed 1·5 if L is very small. However, over most of the stable range of burning $1/(1 - \alpha)$ does not exceed 1·25. So a calculation of the burning rate which entirely ignores chemical influences will not usually over-estimate the rate by more than 25 per cent. A safe general rule, which will often give sufficient accuracy, would be always to reduce the rate calculated from the mass-transfer-control hypothesis by 10 per cent.

Of course different temperature dependences (n and m) of the chemical reaction and heat loss will modify the numbers in the above paragraphs. The larger n and m are; the smaller will be the amount by which $1/(1 - \alpha)$ exceeds unity. If the chemical reaction rate does not vanish at

$\tau = 0$, stable reaction is possible at much lower temperatures than in the present case; then $1/(1 - \alpha)$ greatly exceeds unity, and the main "resistance" to reaction is chemical. This matter has recently been discussed in connexion with ammonia synthesis by Bošnjaković and others [21, 22].

5. Conclusions

(a) Systems involving steady flow, exothermic chemical reaction, and non-convective heat loss, take up a temperature which depends on two dimensionless quantities, a chemical loading and a heat-loss parameter.

(b) If the rate of increase of chemical reaction rate with temperature itself rises with temperature, reaction can be extinguished by increase of the loading parameter.

(c) If the rate of increase of the chemical reaction with temperature exceeds that of the non-convective heat loss, reaction can also be extinguished by reducing the loading.

(d) If the heat loss parameter exceeds a critical value, which depends on the temperature dependences of chemical reaction and of heat loss, reaction is not possible at any value of the loading, unless significant chemical reaction takes place at the gas stream temperature.

(e) For reactions at solid surfaces, the rate of reaction calculated by neglecting the "chemical resistance" may give an over-estimate of the order of 25 per cent. The exact amount of the over-estimate depends on the reaction and heat-loss functions, and on the loading and heat-loss parameters.

References

[1] Frank-Kamenetsky D. A. *Diffusion and Heat Exchange in Chemical Kinetics.* Princeton University Press 1955. (Published in U.S.S.R., 1947).

[2] Smith F. W. Mass. Inst. Tech. Meteor Report No. 6, 1947.

[3] Spalding D. B. *J. Inst. Fuel* 1953 **26** 289.

[4] Silver R. S. *Fuel* 1953 **32** 138.

[5] Cannon K. J. and Denbigh K. G. *Chem. Engng. Sci.* 1957 **6** 155.

[6] Grigull U. *Chem.-Ing.-Tech.* 1958 **30** 40.

[7] Spalding D. B. and Tall B. S. *Aero. Quart.* 1954 **5** 195.

[8] de Zubay E. A. and Woodward E. C. Westinghouse Res. Lab. Sci. Pap. No. 1811, 1954; *Fifth Symp. Combustion* p. 329. Reinhold, New York 1955.

[9] Spalding D. B. *Third AGARD Combustion and Propulsion Colloq. Palermo, 1958.* p. 269. Pergamon Press 1959.

[9a] van Heerden C. *Chemical Reaction Engineering.* p. 133. Pergamon Press, London 1957.

[10] Spalding, D. B. *Proc. Roy. Soc.* A **240** 1957 83.

[11] Mayer E. *Combustion and Flame* 1957 **1** 438.

[12] Heerden van C. *Industr. Engng. Chem.* 1953 **45** 1243.

[13] Avery W. H. and Hart R. W. *Industr. Engng. Chem.* 1953 **45** 1634.

[14] Longwell J. P., Frost E. E. and Weiss M. A. *Industr. Engng. Chem.* 1953 **45** 1629.

[15] Bragg S. Unpublished work 1953. See also *Selected Combustion Problems* Vol. 2. Butterworth, London 1956.

[16] Wagner C. *Tech. Berlin Chem.* 1945 **18** 28.

[17] Zeldovich Y. B. *Zh. Tekh. Fiz.* 1949 **19** 1199.

[18] Spalding D. B. *Fuel* **33** 253.

[19] Agafanova F. A., Gurevich M. A. and Paleev I. I. Theory of burning of a liquid fuel drop. *J. Sov. Phys. Techn. Phys.* 1958 **2** 1689.

[20] Spalding D. B. *Some Fundamentals of Combustion.* Butterworth, London 1955.

[21] Bošnjaković F. *Chem.-Ing.-Tech.* 1957 **29** 187.

[22] Kopper H. H. *Chem.-Ing.-Tech.* 1958 **30** 40.

SUBJECT INDEX